Selected Writings from the Journal of the Mathematics Council of the Alberta Teachers' Association

Celebrating 50 Years (1962–2012) of *delta-K*

A volume in
The Montana Mathematics Enthusiast
Bharath Sriraman, *Series Editor*

The Montana Mathematics Enthusiast
Monograph Series in Mathematics Education

Bharath Sriraman, The University of Montana
Series Editor

Selected Writings from the Journal of the Mathematics Council of the Alberta Teachers' Association

Celebrating 50 Years (1962–2012) of *delta-K*

edited by

Egan J. Chernoff
University of Saskatchewan

Gladys Sterenberg
Mount Royal University

INFORMATION AGE PUBLISHING, INC.
Charlotte, NC • www.infoagepub.com

Library of Congress Cataloging-in-Publication Data

A CIP record for this book is available from the Library of Congress
http://www.loc.gov

ISBN: 978-1-62396-700-0 (Paperback)
          978-1-62396-701-7 (Hardcover)
          978-1-62396-702-4 (ebook)

To

*Kristen and Scout, for moving to the other (much colder, flatter) side of Alberta
so that I'm able to take on projects like this*

&

*Gary, who has supported my own math education and who pays me
the greatest compliment (humbly received) as a teacher of teachers*

&

*In memory of
Karen Virag, supervising editor of publications
at the Alberta Teachers' Association, who gifted us with her exceptional expertise
and gracious guidance throughout the editorial process.*

CONTENTS

1970s

1980s

1990s

2000s

THE MCATA CONSTITUTION

The MCATA Constitution, found on pages five and six of the January 1962 issue of the Mathematics Council of the Alberta Teachers' Association Newsletter, Volume Number One, Issue Number One, appears below.

MCATA CONSTITUTION

The MCATA constitution as amended at the inaugural conference appears below.

Name. The name of this organization shall be the Mathematics Council of The Alberta Teachers' Association.

Object. The object of this organization shall be: to promote and advance the teaching of mathematics throughout the province, especially in elementary and secondary schools.

Membership. The following are eligible for membership: (a) any member of The Alberta Teachers' Association or non-member covered by the Teachers' Retirement Fund, (b) any certified teacher in private schools, and (c) any member of the University of Alberta or of the Department of Education.

Fees. Membership fees may be established by resolution at the annual general meeting of this council.

Selected Writings from the Journal of the Mathematics Council of the Alberta Teachers' Association, pages xiii–xiv
Copyright © 2014 by Information Age Publishing

Finances. The executive committee shall have power to collect fees and to make expenditures. A financial statement shall be submitted to the annual general meeting.

Officers. The officers of this council shall consist of a president, a vice-president, a past president, and a secretary-treasurer, to be elected for a term of one year at the annual general meeting, and a member appointed by the Executive Council of The Alberta Teachers' Association.

Executive Committee. The executive committee shall consist of the officers and one MCATA member from each of the elementary school, the junior high school, the high school, the Faculty of Education, and the Department of Mathematics of the University of Alberta, appointed by the officers.

Committees. The executive may appoint from time to time such committees as are necessary to carry on the work of the council.

Liaison. Any communication regarding policy which this council wishes to make with any organization, government department, or other agency, within or without the province, shall be conducted through the Executive Council of The Alberta Teachers' Association or other regular channels of that Association.

Regional Councils. The executive committee of this council shall encourage the establishment of regional councils and shall have authority to determine regional boundaries and to establish regulations governing the organization of regional councils, consistent with this constitution.

Reports. The executive committee shall submit annually a written report of its activities to the Alberta Teachers' Association, prior to December 31.

Amendments. After three months' notice of motion to amend the constitution has been given to each member, this constitution may be amended by a two-thirds majority vote of the members present at any annual general meeting of this council, subject to ratification by the Executive Council of The Alberta Teachers' Association.

General Meetings. The Mathematics Council shall hold at least one general meeting a year. At least thirty days' notice shall be given for all general meetings.

FOREWORD

Colleagues,

We, Marj and Florence, are indeed honoured to write this welcome letter for this celebration of 50 years of publishing in *delta-K*. The work in this volume points to a story that can be told about mathematics education in the province of Alberta and in Canada. As we read through the articles we were reminded of the concept of change—and that mathematics, mathematics teaching, and mathematics classrooms have shifted. The articles selected by the editors highlight the shifts.

We want to first of all thank the editors, Gladys Sterenberg and Egan Chernoff, for their decision to pull together this publication. The act of selecting the articles to feature in this edition was onerous, and this special edition of *delta-K* will provide a valuable resource to mathematics educators.

As we looked over the articles and read the introductory and closing comments to each decade, we were reminded of our own journey as mathematics educators. We were beginning our school lives in the 1960s and we vividly remember when "new math" came to be. Florence tells the story of the new textbook she received in grade 7 and recalling the practicing required to "master" set notation. Florence also tells the story of her first grade teacher and the mathematics lesson that she remembers: counting sets of popsicle sticks, first into groups of 5 and then combining the groups of 5 into groups of 10.

Marj remembers frequently playing games that used mathematics (dominoes, cribbage, and Norwegian whist) as a family. Basic facts fluency was mastered without using rote memorization. Dominoes developed the skill of grouping numbers and looking for patterns, not unlike the 10-frames the elementary classes are using today.

We were secondary school students in the 1970s. We both remember the introduction of handheld four-function and scientific calculators. Florence remembers the slide rule and logarithm tables she used in high school mathematics, and then in December of her last year of high school receiving a scientific hand held calculator. Marj remembers using a slide rule in high school and using it on the government scholarship exams she wrote at the end of grade 12. One of Marj's most expensive graduation gifts she received was a handheld four-function calculator, one almost the size of a Kleenex box. How it changed the nature of the calculations! We also both started university during the 1970s and learned about what it was like to move to programmable calculators and computer programming. In the mid-1970s, the diploma examinations were discontinued in Alberta.

The 1980s were the beginning of our teaching careers; we were both secondary school mathematics teachers. During this time we remember calculator use in mathematics classes being the focus of conversation in staff rooms and we remember publications around problem solving and the use of calculators appearing. Our professional development conferences had many sessions around the use of calculators. In the 1980s, we saw the introduction of the graphing calculator. Florence remembers reading about the graphing calculator in the *Mathematics Teacher* and then, in 1986, purchasing her first one, a Casio. Marj's first exposure was at an National Council of Teachers of Mathematics (NCTM) regional conference in Edmonton in the mid-1980s. Marj's first graphing calculator was a TI-81 and cost almost the same as the four-function calculator she received as a high school graduation gift. The decade of the 1980s was also a time when computer programs such as VisiCalc (the first spreadsheet program for personal computers) and graphing programs started to be used in schools. The use of calculators and computers in mathematics classes did not become a part of the Alberta mathematics curriculum, though, until the late 1980s. During this decade, in 1984, we also saw the reintroduction of the diploma examinations and saw the introduction of the Provincial Achievement Tests in Alberta.

The 1990s was a time of vast changes in provincial curriculum; the technology was now available that allowed one to explore mathematical concepts through numerical analysis, algebraic analysis, and graphic analysis. The NCTM 1989 Curriculum and Evaluation Standards for School Mathematics impacted provincial curriculum documents, and there was more emphasis on students being able to communicate and explain mathematical ideas. Constructivism, as a theory of learning, was being discussed in professional development sessions, and there was the banding together of provinces and territories for the development of curriculum. A big change in teaching was the movement of the curriculum to one that was no longer "spiraled." What this meant was that the textbooks we used no longer covered the same material from grade to grade. For example, if the multiplication and division of fractions were in the grade 8 curriculum and textbook, then the grade 9 textbook would not use the grade 8 lessons to review the multiplication and division of fractions in the study of the multiplication and division of rational numbers. The textbook companies started to provide teacher resource manuals that were intended to help teachers transition with the change in the curriculum. Textbooks and curriculum documents also referred to mathematical processes like communication, visualization, reasoning, and problem-solving. It was in the 1990s that our professional practices led us into roles of leadership within our school division and the province. We became acutely aware of the differences in teaching practices and began to wonder how it was that each of us, individually and collectively, could contribute to the profession.

The 2000s were once again a time of changes in provincial curriculum, changes in the available technology, and changes in who was sitting in classrooms in Alberta. The NCTM 2000 *Principles and Standards for School Mathematics* cemented the focus in mathematics curriculum around the idea that *all* learners should have the opportunity to experience a high-quality mathematics program. Of ongoing debate within the province, and Western Canada, was the question of "what does a high quality mathematics program contain?" That is, what should learners learn in a high-quality mathematics program? For Florence, the 2000s were a decade when she became disillusioned with what was written in curriculum documents and realized that what really mattered, in her work, was that a high-quality mathematics teacher was available for *all* learners. It was the way in which teachers interpreted curriculum documents and made decisions about the experiences that they would offer learners that became the focus of her work in teacher education. Similarly, Marj also went from the classroom to working with teachers to help them improve their knowledge of the content and more importantly their pedagogy.

We invite you to reflect on your own practices and experiences as you read through this volume. As we read the articles, we learned about the experiences that researchers and authors were discussing in each of the decades and recalled our own experiences in schools. We were also reminded of the impact of *delta-K* and MCATA in our own teaching lives, right from the time we were studying with Drs. Tom Kieren and Sol Sigurdson in our teacher education programs at the University of Alberta.

To close, we would like to thank all of the contributors: Gladys Sterenberg, Egan Chernoff, Len Bonifacio, Olive Chapman, Tom Kieren, Werner Liedtke, Craig Loewen, Mark Mercer, David Pimm, Klaus Puhlmann, Elaine Simmt, and Daryl Smith; and all of the authors whose articles were selected to be a part of this volume. We are indeed fortunate to have the opportunity to honor the work of the journal and its contributions to our professional lives as mathematics educators.

And thank you for letting us reminisce about our relationship with mathematics education and our journey as colleagues and friends. We met in a mathematics class in our undergraduate program and then reconnected when we were teachers attending MCATA conferences. We were reminded as we read these articles, introductions, and commentaries of the changes that we experienced in mathematics curriculum, assessment, and teaching; and we were also reminded of the many great mathematics educators with whom we have had the chance to work in our careers. Alberta mathematics educators have been fortunate to have the leadership provided by organizations such as MCATA and publications such as *delta-K*.

Sincerely,

—Marj Farris, President
2010–present

—Florence Glanfield, Past President
1996–1998

PREFACE

Egan J. Chernoff and Gladys Sterenberg

EGAN

During the winter of 2011, while attending an executive meeting of the Saskatchewan Mathematics Teachers' Society (SMTS) in Saskatoon, it was brought to my attention, via a birthday cake, that the SMTS would be turning 50 in August 2011. At the end of the meeting, during the "other business" portion of the agenda, I mentioned that I was (now, having heard the news) interested in putting together a special issue of *vinculum: Journal of the Saskatchewan Mathematics Teachers' Society*, which would, somehow, celebrate 50 years of the SMTS and *vinculum* (in its many different renditions).

Driving home from the meeting, now liking the idea I floated to the executive more and more, I started to get ahead of myself. With access to 50 years of journal articles I could "easily" put together one special issue of *vinculum*. In fact, I could probably put together two special issues. By the time I got home, I decided to go with 10 articles from each of the past 5 decades. In other words, I wanted to represent/celebrate 50 years of the SMTS and its journal with 50 articles. However, my plans were contingent on the availability of back issues.

The next morning I phoned the Saskatchewan Teachers' Federation (STF), not sure what to expect. After getting through to the archives, I started describing (clumsily) the project, how the idea came about and, ultimately, what I was trying to accomplish. Their response, on the other

Selected Writings from the Journal of the Mathematics Council of the Alberta Teachers' Association, pages xix–xxiii
Copyright © 2014 by Information Age Publishing

end, was quite succinct: "That sounds wonderful. Yes, we have every issue from October 4, 1961, which is volume 1, number 1. Do you want to schedule your visitation now?" Stammering a bit (my idea had just turned into a reality), I mentioned that I would call back in the next few days and set up a time to visit.

Smash cut to Gale Russell and me walking down the steps to the STF archives. Simply put, the archives are meticulous. Further, the people we met "down there" were instrumental in making our daunting task, that is, choosing 50 articles to represent 50 years of the journal, much, much easier. For example, when we realized during the first day that we would only get through one decade (the 1960s) and, further, it was impossible to choose only ten articles at that time, MacKinley Darlington (and others), over the next few days, scanned the umpteen articles we had chosen, placed them in a Dropbox folder, sent us an email indicating that our articles were ready for viewing, and also asked when we would next be visiting the archives.

Gale and I decided that we would next pare down the number of viable articles for the sixties with a second cut, more specifically, from umpteen to (approximately) 30. Only then did we make our second appointment with the archives. In the end, the approach that we took for the 1960s became the approach that we adopted for each of the subsequent decades. Each visit to the archives resulted in umpteen articles for each decade. In total, our first cut resulted in approximately 500 articles. Subsequently deliberating over the scanned articles placed in our Dropbox folder, we again pared down each decade. In total, our second cut resulted in approximately 150 articles. Even though we had made our first and second cuts and our time at the archives was over, our work, essentially, had just begun. We now had to deliberate over approximately 150 articles and, with a third cut, choose only 50! (That is an exclamation point and not a factorial symbol.)

If you are wondering, yes, choosing 50 articles to appropriately represent/celebrate the SMTS and its journal took a very, very long time, resulted in many cuts (and "uncuts") and numerous quibbles (over particular articles for particular decades), which, more often than not, blew up into full-on arguments. In the end, however, we did it; we ended up with 50 articles to represent/celebrate 50 years of the SMTS. Thinking I was close to being done, I was unaware (at the time) that my journey with this project was just beginning.

As you can imagine, you learn a lot reading (in some instances, skimming) through 50 years of a journal. One thing I learned, while reading through the SMTS journal articles from the sixties, was that the British Columbia Association of Mathematics Teachers (BCAMT) and the Mathematics Council of the Alberta Teachers' Association (MCATA), and their respective journals, came into existence in similar time frames. For example, certain articles published in the SMTS journal in the sixties were originally published in the journals of the BCAMT and the MCATA. In addition, the SMTS journal would, in the information section, regularly reference the activities (conferences, journals, meetings and others) of the BCAMT and the MCATA. This got me thinking.

After sending a few emails back and forth between to the current editors of the *Vector: Journal of the British Columbia Association of Mathematics Teachers* (Peter Liljedahl and Sean Chorney) and *delta-k: Journal of the Mathematics Council of the Alberta Teachers' Association* (Gladys Sterenberg), I now had projects, similar to what Gale and I had done in Saskatchewan, underway in British Columbia and Alberta. (Of course, at this point, I decided to press my luck: I then sent emails to the current editors of the provincial mathematics teacher organizations and associations in all the other provinces in Canada and asked if they were interested in working on a similar project, to no avail). With three similar projects underway now, clearly, the scope of the project had increased. With the increase in scope, and with the blessing of Peter and Sean (in BC) and Gladys (in AB), I started presenting the idea of the books to different publishers.

In the proposal(s) that I sent to different publishers (yes, we were rejected a few times), I described that the book would be comprised of 50 articles from each of the five decades and, further, we would have modern-day Introductions to and Commentaries on the selections for each decade—all written by prominent members of the mathematics education community in British Columbia, Alberta, and Saskatchewan.

The three proposals:

- Selected writings from the *Journal of the Saskatchewan Mathematics Teachers' Society: Celebrating 50 years (1961–2011) of vinculum;*
- Selected writings from the *Journal of the Mathematics Council of the Alberta Teachers' Association: Celebrating 50 years (1962–2012) of delta-K;*
- and selected writings from the *Journal of the British Columbia Association of Mathematics Teachers: Celebrating 50 years (1962–2012) of Vector*

were picked up by Information Age Publishing and are now part of the *The Montana Mathematics Enthusiast: Monograph Series in Mathematics Education.* (Fittingly, Alberta is one of three Canadian provinces, with British Columbia and Saskatchewan being the others, that share a border with Montana). Excitedly, I emailed Gladys and told her it was time for us to get to work.

GLADYS

Egan's invitation was timely. As editor of *delta-K,* I had been thinking about possibilities for celebrating the contributions of mathematics educators in

Alberta during the past 50 years. In 1995, a special monograph highlighting the history of MCATA was published and the current executive was extremely interested in issuing a follow-up volume. This project surpassed expectations.

Drawing on my own collection of *delta-K* issues and those stored in the archives of the University of Alberta library and the Alberta Teachers' Association, I scanned copies of the research articles from the 1970s. Egan and I decided to independently rank the articles, taking into consideration the representation of repeated topics, curricular changes, prominent Albertan authors, research initiatives, and teaching ideas addressing the range of grades from kindergarten to twelve. Egan and I had many video conferences as we justified what we felt were the significant articles. We reached a consensus after much discussion, and this initial grappling informed the process for the other article selections across the remaining decades.

One important theme that emerged was the integration of research and classroom practice. It was clearly evident that post-secondary faculties of education and departments of mathematics were integral to the professional practice unfolding in the schools. We decided to include a special feature article for each decade in the Alberta book that demonstrated the prominent role of the researcher in the classroom.

Once articles had been chosen, we generated a list of mathematics educators and teachers who might help us make sense of the articles within contemporary and historical contexts. When I invited people to participate in the project as authors for the introductions and commentaries, I was overwhelmed by their graciousness and willingness to share their perspectives. After drafts of the authors' writings were submitted, we did an initial edit and then relied on the expertise of Karen Virag, the supervising editor of publications at the Alberta Teachers' Association, and her staff to complete the editing process. The manuscripts were returned to the authors for final review.

I feel very privileged to be part of this book publishing process. In many ways, it has helped me understand more deeply the contributions of teachers and educators to our understanding of teaching and learning mathematics. It has been an amazing project.

EGAN AND GLADYS

As you are about to read, the teaching and learning of mathematics in Alberta has a long and storied history. An integral part of the past 50 years (1962–2012) of this history has been *delta-K: Journal of the Mathematics Council of the Alberta Teachers' Association*. This volume, which presents ten memorable articles from each of the past five decades—that is, 50 articles from the past 50 years of the journal—provides an opportunity to share this rich history with a wide range of individuals interested in the teaching and learning of mathematics and mathematics education. Each decade begins with an introduction, providing a historical context, and concludes with a commentary from a prominent member of the Alberta mathematics education community. As a result, this edited volume provides a historical account as well as a contemporary view of many of the trends and issues in the teaching and learning of mathematics. This volume is meant to serve as a resource for a variety of individuals including: teachers of mathematics, mathematics teacher educators, mathematics education researchers, historians, and undergraduate and graduate students. Most importantly, this volume is a celebratory retrospective on the work of the Mathematics Council of the Alberta Teachers' Association.

INTRODUCTION

Gladys Sterenberg and Egan J. Chernoff

THE INAUGURATION OF THE MATHEMATICS COUNCIL OF ALBERTA TEACHERS' ASSOCIATION

The Mathematics Council of the Alberta Teachers' Association (MCATA) has a strong history of responding to the needs of teachers and providing opportunities for professional development in mathematics education. As a specialist council of the Alberta Teachers' Association, its inception was within the context of a dedicated leadership concerned with shaping the future and recognizing teaching as a profession. Formed in 1918, the Alberta Teachers' Alliance worked diligently to improve teaching conditions through salary schedules, teaching contracts, pensions, and the development of a professional code of ethics. In 1919, the Alberta Teachers' Alliance joined with teachers in other provinces to initiate the Canadian Teachers' Federation. In 1934, its name was changed to the Alberta Teachers' Association (ATA), and when the *Teaching Profession Act* was passed in 1935, it was recognized as a legal entity. Over the next ten years, the ATA worked to establish the university as the home of teacher education.

Within this context of leadership, the ATA established specialist councils to foster the professional development of teachers interested in common curriculum or specialty areas. The first meeting of the provisional executive of MCATA was held on October 29, 1960. The members were T. F. Reiger (chair), L. D. Nelson, S. E. L. Pollack, Max Wyman, Helen Morrison,

Selected Writings from the Journal of the Mathematics Council of the Alberta Teachers' Association, pages xxv–xxxiii
Copyright © 2014 by Information Age Publishing

and L. J. Scott (secretary). Over the next few months, this group met to write a constitution and plan the inaugural conference. The first conference of MCATA was held August 16–19, 1961 at the University of Alberta. Sixty teachers attended. The conference included a keynote address by J. H. Hlavaty, a member of the Board of Directors of the National Council of Teachers of Mathematics (NCTM), and presentations by N. M. Purvis, R. C. Jacka, W. F. Coulson, S. A Lindstedt, and L. W. Kunelius, members representing the Department of Education, Department of Mathematics at the University of Alberta, and the Faculty of Education at the University of Alberta. The conference also featured a demonstration lesson in elementary mathematics, a computer demonstration, group discussions on the future of MCATA, and a panel to evaluate the conference. The constitution was approved, officers were elected, five directors were appointed, the membership fee of $5 was set, and plans to issue the first newsletter were made.

The first newsletter was published in January, 1962 (volume 1, issue 1). The announcement that appeared on the first page indicated that members would be kept aware of current developments through news bulletins published four times a year. These bulletins included current news items and theoretical articles. As the content evolved to include more articles, MCATA decided to publish the newsletter as a professional journal. The newsletter was renamed *delta-K* in May 1971 (volume 10, number 3) and was published two or three times a year. As the membership grew, the need for more frequent contact emerged. In January, 1983, the newsletter was reintroduced and published as volume 1, number 1. In June 2006, *delta-K* became a peer-reviewed journal. Today, both publications remain vibrant ways to connect with the MCATA membership.

MCATA's service to members and commitment to publishing information and articles has presented a unique opportunity to celebrate its 50 years. In particular, articles in *delta-K* provide a glimpse into the activities of the council. Presented here is a brief summary of these activities.

ACTIVITIES OF MCATA

Appearing on MCATA letterhead, newsletters, and journals is the council logo. Originally it consisted of a Venn diagram with three striped circles representing students, teachers, and the mathematics curriculum with MCATA at the intersection. The logo has evolved over the years with the MCATA letters being superimposed on the Venn diagram logo. Currently the letters-only version is used as the logo. The name, MCATA, has remained throughout the history of the council. However, in 1971, the newsletter was renamed *delta-K,* a name chosen from those submitted by the membership. The meaning of the name was described in the first issue in May 1971 (volume 10, number 3):

> The chosen name represents *delta* (Δ), the fourth letter in the Greek alphabet used in mathematics to represent an increment or increase. *K* is for knowledge: knowledge of mathematics, knowledge of teaching mathematics and knowledge of new methods and developments in our discipline (p. 1).

The logos and names represent the focus of MCATA on the professional development of its members. Throughout the pages of the publications are examples of how the council organized annual conferences, offered regional workshops and seminars, produced publications, and maintained websites.

Annual conferences have remained the highlight of the council. Affiliations with other professional organizations such as the NCTM (from 1964) and Canadian Association of Mathematics Teachers (1967–1972) have added a national and international lens to such conferences, as evident in Table I.1.

In addition to annual conferences and in response to members' interest in curriculum changes, summer workshops were initiated in the mid-1960s; mini-conferences were prominent in the 1970s, 1980s and 1990s; and bi-annual symposiums were started in the 1990s and remain a part of MCATA activities offered today. Interests in providing service to local members resulted in the formation of regional councils in 1962. By 1993, these had disbanded as members became more involved in MCATA.

Since its formation, MCATA has provided innovative materials and resources to its members. Between 1964 and 1972, film circuits were popular. In 1963, a mathematics course at the University of Calgary developed by Sid Linstedt was reformatted as 16-mm films and a series of 50 films were distributed to 14 centres around the province for secondary mathematics

TABLE I.1 Conference Dates and Location

Conference Location	Year(s)
Edmonton	1961, 1962, 1969, 1973*, 1982, 1986*, 1988, 1991, 1994*, 1997, 2001, 2003*, 2005, 2007, 2009, 2011, 2013
Calgary	1963, 1966*, 1970, 1971, 1975, 1979*, 1983, 1987, 1990*, 1993, 1998*, 2004, 2010
Red Deer	1964, 1967, 1968, 1972, 1976, 1977, 1978, 1980, 1984, 1996, 2000
Vancouver	1965
Jasper	1974, 1999, 2006, 2008, 2012
Lethbridge	1981, 1985, 1989, 1995
Medicine Hat	1992
Canmore	2002

* NCTM meeting.

teachers who met once a week to study two or three of the films. In 1966, MCATA purchased ten films from the NCTM Elementary Film series, and these were distributed to centres throughout the province until 1974. Another film series from NCTM, *Mathematics for Tomorrow*, was purchased in 1966 and distributed to superintendents. In the 1970s, film circuits were replaced with materials circuits. Math Kits containing materials and teacher information were developed by Stu McCormick and were circulated throughout the province. In 1974, a Metric Kit was developed and MCATA arranged teams of teachers into the "Metric Missionaries" who toured Alberta to provide four-hour Saturday workshops. Eventually these were replaced by the mini-conferences, allowing more members to interact with one another at the local level.

Throughout its history, MCATA has worked closely with university departments of mathematics by supporting mathematics contests. Since the 1970s, MCATA has contributed to the Alberta High School Mathematics Prize Examination, the Calgary Mathematics Association Junior High Examination, and the Edmonton Junior High Mathematics Contest. Results are published each year in the MCATA newsletter, and examination questions and answers are published annually in *delta-K*. These contests honour mathematics students who excel.

Other ways of honouring excellence are evident in the awards initiated by MCATA. Currently, MCATA recognizes significant contributions of its members to mathematics education through the Mathematics Educator of the Year Award (Table I.2).

TABLE I.2 Recipient by Year of the Mathematics Educator of the Year Award

Year	Recipient	Year	Recipient	Year	Recipient
1984	Marshall Bye	1997	Donna Chanasyk	2004	Percy Zalasky
1985	Joan Worth	1997	Lynwen Hart	2005	Not awarded
1986	John Percevault	1997	Klaus Puhlmann	2006	Gerald Krabbe
1987	Bill Bober	1998	Not awarded	2007	Sandra Unrau
1988	Art Jorgensen	1999	Betty Morris	2008	Lori Ball
1989	Lois Marchand	1999	Gail Poshtar	2008	Kamie Klevgaard
1990	Joan Crawford	2000	Len Bonifacio	2009	Linda Arndt
1991	Cynthia Ballheim	2000	Shauna Boyce	2009	Brenda MacDonald
1992	Louise Frame	2000	Evan Fleetwood	2010	Ed Tobin
1993	Mary Anne Nissen	2000	Kathy McCabe	2011	Not awarded
1994	Bob Michie	2001	Roxanne Trouth	2012	Benita Greenwood
1995	Florence Glanfield	2002	Not awarded		
1996	Evelyn Sawicki	2003	Sharon Barry		

The Friends of MCATA Award is presented at the annual conference in recognition of service to MCATA. In 2002, MCATA created the Dr. Arthur Jorgensen Chair Award to encourage students enrolled in education programs in post-secondary institutions throughout Alberta to pursue and commit to mathematics education. The award consists of a one-year term on the MCATA Executive, with expenses paid to attend Executive meetings; a one-year membership in MCATA and NCTM; and an invitation to attend one MCATA conference with appropriate expenses paid.

MCATA also offers a $500 professional development grant for mathematics education initiatives that support current learning and teaching practices or current priorities as outlined by or through Alberta Education, local school districts, MCATA, NCTM, ATA, or other reputable education associations.

Publications have expanded from the newsletters and journal to include monographs and special issues. A series of monographs was published in the 1970s and 1980s that addressed issues and interests:

- *Manipulative Materials for Teaching and Learning Mathematics* (July 1973), edited by W. George Cathcart
- *Mathematics Teaching: The State of the Art: Proceedings of the Edmonton Meeting of the NCTM* (October 1973), edited by W. George Cathcart
- *Metrication: Activities, Relationships, and Humor* (June 1975), edited by K. Allen Neufeld
- *Timeless Activities for Mathematics K–12* (June 1976), edited by Bruce D. Harrison and Ed Carriger
- *Calculators in the Classroom* (November 1977), edited by K. Allen Neufeld
- *Reading in Mathematics* (December 1980), edited by John Percevault
- *Problem Solving in the Mathematics Classroom* (April 1982), edited by Side Rachlin and Judy McDonald
- *Microcomputer Development* (September 1982), edited by Ron Cammaert
- *56 Ideas: Make It, Take it* (December 1987), edited by William Bober and John Percevault
- *Communication in the Mathematics Classroom* (October 1992), edited by Daiyo Sawada

Special issues of *delta-K* include:

- *Teaching Mathematics in the Early childhood Classroom* (1987, volume 26, issue 3), a joint publication of MCATA and the Early Childhood Education Council, co-edited by John Percevault and Gordon Orlick

- *Technology and Mathematics* (1988, volume 27, number 1), edited by John Percevault
- *Mathematics for Gifted Students* (1989, volume 27, number 3), another joint publication with the Gifted and Talented Education Council and edited by Andy Liu
- *Early Childhood Mathematics* (2011, volume 48, number 2), co-edited by Lynn McGarvey and Gladys Sterenberg
- *Celebrating 50 years of delta-K* (2013, volume 50, number 2), edited by Gladys Sterenberg

Since its formation, MCATA has been involved in political action. Thinkers' Conferences were held to establish position statements, policies, and direction for MCATA. MCATA has initiated surveys on the needs and concerns of its membership to inform its action. These include the 1968 research report by Marshall P. Bye on the Project Calculator program; the 1984 report by Tom Schroeder and Louise Frame that examined the preparation and continuing education of mathematics teachers in Alberta; and the 2012 scholarly activities report of Alberta mathematics teachers by Erasmia Eliopoulos, Julie Long, Richelle Marynowski, Lynn McGarvey, and Gladys Sterenberg. A Blue Ribbon Panel was created in response to concerns about the mathematics 30 diploma exams and made 53 recommendations to Alberta Education and other stakeholders based on surveys of secondary students, teachers, and postsecondary institutions. In addition, countless letters and petitions have been distributed on behalf of the membership in response to changes in assessment, achievement testing, and new curriculum initiatives. The activities of MCATA have been rich and varied because of the commitment of its volunteers.

MCATA VOLUNTEERS

MCATA is directed by volunteer teachers and educators who contribute their time, talents, and enthusiasm to plan and implement programs and activities for the council. Significantly, this is a story about people who have a strong commitment to being great mathematics teachers and educators and to serving their professional community.

The inaugural provincial executive members of 1961–1962 were John Cherniwchan (president), T. F. Reiger (past president), Eugene Wasylyk (vice president), and Olive Jagoe (secretary-treasurer). MCATA presidents throughout the past 50 years are listed in Table I.3.

Because of strong leadership throughout its history, MCATA has remained dynamic. Membership fees began at $5 in 1961 and are currently set at $30 (the rate established in 1993). The number of members has ranged from

TABLE I.3 MCATA Presidents

Fred Tarlton (1962–1963)	K. Allen Neufeld (1975–1977)	George Ditto (1995–1996)
Tom Atkinson (1963–1964)	Robert Hold (1977–1979)	Florence Glanfield (1997–1998)
Len Pallesen (1964–1965)	Lyle Pagnucco (1979–1981)	Cynthia Ballheim (1998–2000)
Ted Rempel (1965–1966)	Gary Hill (1981–1983)	Sandra Unrau (2000–2002)
Marshall Bye (1966–1967)	Ron Cammaert (1983–1985)	Cynthia Ballheim (2003–2004)
Gus Bruns (1967–1968)	Robert Michie (1985–1987)	Len Bonifacio (2004–2005)
Murray Falk (1968–1969)	Louise Frame (1987–1989)	Janis Kristjansson (2005–2007)
Jim Kean (1969–1971)	Marie Hauk (1989–1991)	Sharon Gach (2007–2009)
Richard Daly (1971–1973)	Bob Hart (1991–1993)	Marj Ferris (2009–
W. George Cathcart (1973–1975)	Wendy Richards (1993–1995)	

fewer than 100 in the inaugural year to highs of almost 1100 in the 1960s and 1980s. The average of 500 is lower than the current enrolment of 640.

CELEBRATING 50 YEARS

The task of working on this project that celebrates 50 years of *delta-K* has been fascinating and enjoyable. For this book, Egan Chernoff and I have assembled a collection of articles that have appeared in *delta*-K throughout its history. We used the following criteria in choosing the articles. First, we collected and examined the tables of contents for all issues. We read each issue (yes, that was a lot of reading!) and looked for repeated themes. We noted curricular changes and related teaching and research ideas. Second, we looked for articles that represented a range of grades. Third, we focused on authors who were primarily from Alberta.

After short-listing all articles, we selected ten from each decade and invited prominent teachers and mathematics educators to provide introductions and commentaries. We are deeply appreciative of these authors for their investment in this project. Through their writings, we hope you will gain perspectives of the context of teaching and learning mathematics in Alberta. Each of these authors provides an in-depth investigation of emerging themes across the selected articles.

The articles have been left in their original format with the following exceptions. We have changed footnotes to endnotes to provide consistency, we have included the editorial notes and author descriptions from the original publications, and we have used gender inclusive language in most places. The formatting of earlier articles has been changed to align it with current protocols. Karen Virag, Supervising Editor of Publications at The Alberta Teachers' Association has been an integral part of the process of publishing this book. Together with Penny Harter, Kristina Lundberg, and Judith Plumb, she edited the commentaries and introductions and provided archival material and references. Permission is granted by The Alberta Teachers' Association to reprint all articles previously published in *delta-K*. Credit is also given to Joan Worth and Art Jorgensen (1995) who edited a special historical journal, *Thirty-Four Years and Counting*, that contributed greatly to the preparation of this introduction.

Throughout our study of the articles published in the past 50 years, we have been impressed with the dedication of teachers, mathematics educators, and mathematicians involved in MCATA. Articles written for *delta-K* have reflected the viewpoints of various people interested in mathematics education in Alberta.

As we celebrate 50 years of *delta-K*, we wish to convey our deep gratitude to previous editors who made strong contributions to our history (Table I.4).

TABLE I.4 Previous Editors

John M. Cherniwchan (1962)	Mary Beaton (1967–1970)	John Percevault (1986–1988)
W. F. Coulson (1963)	Murray Falk (1970–1973)	Linda I. Brandau (1988–1990)
J. E. Holditch (1963–1964)	Murray R. Falk (1970–1973)	A. Craig Loewen & John B. Percevault (1990–1992)
W. F. Coulson, J. E. Holditch, & Tom Atkinson (1965)	Ed Carriger (1973–1980)	A. Craig Loewen (1992–1994)
W. F. Coulson & Tom Atkinson (1965)	George Cathcart (1981–1982)	Arthur Jorgensen (1994–1996)
Tom Atkinson (1966)	Gordon Nicol (1983–1984)	Klaus Puhlmann (1997–2004)
Tom Atkinson & Sol Sigurdson (1966–1967)	Art Jorgensen & John Percevault (1985)	A. Craig Loewen (2004)

In his editorial of *Thirty-Four Years and Counting*, the late Art Jorgensen (1995) writes:

> As we look forward to the next 30 years, what will MCATA's future be? Will it grow and prosper and continue to be a voice to be heard, or will it wither and die? It is really up to the members. Today's executive members, like those of the past years, will do their best. Then they will move on and pass the torch to you. Personally, I have a great deal of confidence in our members, and believe that MCATA has a bright future. Let's make it so." (p. 4)

We believe he would be proud of the vibrancy of our community.

REFERENCE

Worth, J., & Jorgensen, A. (1995). *Thirty-four years and counting: The history of the Mathematics Council of The Alberta Teachers' Association.* Edmonton, AB: The Alberta Teachers' Association.

THE '60s—EVENTFUL AND MEMORABLE

Werner Liedtke

It's been said, "If you can remember the '60s, you weren't there." That is unlikely to apply to anyone involved in education or mathematics education during that decade. The ideas that were introduced and the changes that took place have been the source of many distant but unforgettable memories.

In the early 1960s, it was possible to begin a teaching career with a professional certificate from the University of Alberta, which was awarded after two years of study and the successful completion of practica at the elementary, junior high, and secondary levels. After attaining a professional certificate, I completed a bachelor of education degree in 1964, through evening and summer school classes. Here, I share memories from my experiences as an elementary teacher (1961–1967) and a graduate student (1967–1970). The latter included roles as faculty consultant, teaching assistant, and research assistant.

Any discussion of the 1960s should mention 1967, Canada's 100th birthday. A Centennial Train visited towns and cities. Students were given special flags, which they waved as they sang "Ca-na-da...now we are 20 million," the lyrics to Bobby Gimby's centennial song.

Selected Writings from the Journal of the Mathematics Council of the Alberta Teachers' Association, pages 1–6
Copyright © 2014 by Information Age Publishing

1

Some of the changes of the 1960s can be directly attributed to what took place in the 1950s. In 1957, the Russians sent the dog Laika into space. The *Edmonton Journal* informed its readers when it would be possible to view the Russian satellite Sputnik (launched October 1957) crossing the Alberta night skies. These events contributed to the belief that the Russians were far ahead and that changes in education were warranted that would enable us to catch up. Mathematics education was at the top of the list. For some, the need for change was reinforced when Russian astronaut Yuri Gagarin became the first man in space on April 12, 1961.

One major change that occurred while I was teaching Grade 5 students in Edmonton was the province-wide adoption of a new elementary mathematics program—*Seeing Through Arithmetic* (STA) (Van Engen & Hartung, 1958). The publisher also printed a little reference booklet for teachers that explained the structure of the arithmetic on which the program was based: *Foundations of Elementary School Arithmetic* (Van Engen, Hartung, & Stochl, 1965). STA included several new ideas and procedures, which resulted in some lively staff room discussions. The use of the repeated subtraction algorithm for long division had many colleagues puzzled. Even more puzzling was the distinction made between quotitive (measurement) division ($a \div b = [\ ]$) and partitive (partitioning) division ($a \div [\ ] = c$). STA outlined a specific marking procedure for word problems, with five possible marks for each problem: two marks for translating a word problem into an equation, two marks for calculating the correct answer, and one mark for a meaningful answer sentence.

The adoption of this mathematics program included support services offered by central office. A series of films about the structure of mathematics (i.e., numeration, properties of the operations) was shown. One film included a scene from a grade 1 classroom that made us smile. It showed a student answering a question posed by the teacher with, "Three plus two is equal to two plus three because addition is commutative." One teacher commented, "Is this for real? Will this allow us to catch up with the Russians?"

Consultation visits from resource teachers could also be requested. These teachers would arrive at the school with a trunk full of manipulative materials, which included Dienes Blocks—base 10 blocks designed by Hungarian mathematician Zoltan Dienes. For many of us, this was our first exposure to these blocks.

Around this time, Encyclopaedia Britannica (Canada) asked teachers from across Canada to submit ideas for activities that could be used in elementary mathematics classrooms. These activities appeared in a publication called *Mathex* (Sawyer & Nelson, 1966). An excerpt from the program's statement of aims reveals, in part, why the 1960s were eventful: "Mathematical education is at present in an unstable state. Everyone knows that it is due for many changes. These changes may be good or bad, wise or foolish;

the one thing we can be sure of is that things are not going to stay as they are" (Dobson, 1966, p. 2). *Mathex* had two main aims: to ensure that changes were indeed improvements and to make it possible for these changes to come with a minimum of discomfort for teachers.

During this time, open-area schools were being built in Edmonton. Teachers and their students were invited to tour these schools. During the visits, aspects of what were identified as active learning, discovery learning, and the laboratory approach were illustrated. The open area in these schools was well suited to the active learning approach and to many of the mathematics projects illustrated in the book *Freedom to Learn* (Biggs & MacLean, 1969). New equipment provided to all elementary schools in the 1960s included televisions, overhead projectors, and radios (along with schedules of special programming, such as the Alberta School Broadcasts program *Speech Explorers*).

I had the privilege of being the first student to complete the MEd and new PhD in mathematics education offered by the University of Alberta's Department of Elementary Education. My advisor, L. D. Nelson, would repeatedly show his enthusiasm for the early stages of learning when he reminded me, "Working with young children, that is where the action is!"—a very new perspective for someone with a secondary mathematics and science background. No doubt this enthusiasm was due to the many books and articles published in the 1960s about young children's thinking, particularly their thinking about aspects of mathematics. Below are selected examples of such books:

- Jean Piaget's *The Child's Conception of Number* (1952), *Logic and Psychology* (1957), and *The Child's Conception of Geometry* (coauthored with Bärbel Inhelder and Alina Szeminska, 1960), as well as his articles "Development and Learning" (1964) and "The Genetic Approach to the Psychology of Thought" (1967)
- John H. Flavell's (1963) *The Developmental Psychology of Jean Piaget*
- Millie Almy, Edward Chittenden, and Paula Miller's (1966) *Young Children's Thinking*
- David Paul Ausubel's (1967) *Learning Theory and Classroom Practice*
- Shmuel Avital and Sara Shettleworth's (1968) *Objectives for Mathematics Learning: Some Ideas for the Teacher*
- Jerome Bruner's (1960) *The Process of Education*
- Zoltan Dienes's (1960) *Building Up Mathematics*
- Edmund Sullivan's (1967) *Piaget and the School Curriculum: A Critical Appraisal*

My advisor's enthusiasm translated into the projects I worked on as a research assistant, including designing activities and problems for young children; testing the activities and problems in a preschool setting and with students in the early grades; filming children as they solved problems and interpreting their responses and moves; and analyzing, recording, and sharing observations about young children's thinking and problem-solving strategies.

As a graduate student, I was able to visit and observe students in elementary schools that offered different programs. These observations included group projects in an open-area school and students working on activities identified as math labs. An attempt to individualize curriculum and instruction led some schools to adopt programs created at the University of Pittsburgh. *Individualized Programmed Instruction,* or IPI, for elementary mathematics was based on detailed task analyses and hierarchies of specific cognitive objectives for each skill, procedure, and idea. Subsets of the objectives were translated into questions and tasks for students, which were recorded in small pamphlets. These pamphlets, along with several parallel tests for each pamphlet, were stored on shelves along a whole wall in the classroom. A test was administered upon completion of each pamphlet.

Most of the grade 2 students who were observed attempted to complete the tasks in different pamphlets. Upon completion, these attempts were marked. A score of 80% or more resulted in moving on to a pamphlet with tasks at the next level. A score lower than 80% meant that students had to repeat tasks at the same level until the desired score was reached. In this classroom were six adults who were marking students' booklets and steering the students to the next booklet of tasks—six managers, and not one who was teaching.

Part of the statement of aims from *Mathex* (Dobson, 1966) quoted earlier—that "mathematical education is . . . in an unstable state" and "it is due for many changes" (p. 2)—is illustrated by the ideas, comments, and questions from the authors of the articles written during this decade.

Kunelius predicted a gradual change for the geometry taught in high school and encouraged teachers to depart from the traditional approach.

McCall pointed out the shortcomings of the newer mathematics that had been introduced in American schools and that was coming to Canada. The author observed that the new mathematics appearing in Alberta was more for the mathematics philosopher, rather than mathematics for the scientific world, and suggested that a final test for students in grade 4 should be "how many kinds of problems each individual pupil can solve after having completed the course."

Krider suggested that the focus be on "transfer of learning" and cautioned about moving ahead slowly—while guarding against discarding good and valuable material merely for the sake of newness. The author asked, "Should changes stress pedagogy at the expense of content or should the preparation of future mathematicians be considered?" and concluded that "in education, if you're old-fashioned long enough, you'll be modern."

Coulson discussed approach versus content and suggested that in high schools, the emphasis should be on the former (i.e., "modern concepts such as group and field theory; topology; symbolic logic should be taught from a modern postulational and set-theory approach"). Coulson compared student discovery of relationships and structure with a programmed approach and concluded that in the latter there is little room for the role of a teacher.

Galeski addressed the dilemma that exists when new mathematics is introduced and the term *modern* is used as a descriptor. The author concluded that the meaning of *modern*—today, this year, or in this century—cannot be discussed in a logical and precise manner. According to the author, a discussion about "thinking mathematically" versus "knowledge of learning important facts" would be more appropriate.

The summary of research conducted by graduate students at the University of Alberta presented by Cathcart and Liedtke shows that the books and articles published about children's thinking and thinking about mathematics at this time had a great impact. New ways of gaining insight into children's understanding of aspects of number and geometry were probed, and implications for classroom teachers were discussed.

Falk's random notes would have been helpful for those who were being introduced to the overhead projector as a new teaching aid.

Larson used rate ratio examples to illustrate that new problem-solving strategies should not become the only problem-solving strategies.

Cleveland's main question—"How do students learn the language of mathematics?"—is relevant in any decade.

Werner Liedtke is a Professor Emeritus at the University of Victoria, BC. He is interested in strategies that foster early numeracy strategies related to: confidence/risk taking, sense making/number sense, and mathematical reasoning/flexible thinking in young children (especially his grandchildren!).

REFERENCES

Almy, M., Chittenden, E., & P. Miller. (1966). *Young children's thinking: Studies of some aspects of Piaget's theory.* New York, NY: Teachers College Press.

Ausubel, D. P. (1967). *Learning theory and classroom practice.* Toronto, ON: Ontario Institute for Studies in Education.

Avital, S. M., & Shettleworth, S. J. (1968). *Objectives for mathematics learning: Some ideas for the teacher.* Toronto, ON: Ontario Institute for Studies in Education.

Biggs, E. E., & MacLean, J. R. (1969). *Freedom to learn: An active learning approach to mathematics.* Don Mills, ON: Addison-Wesley.

Bruner, J. S. (1960). *The process of education.* Cambridge, MA: Harvard University Press.

Dienes, Z. P. (1960). *Building up mathematics.* London, UK: Hutchinson Educational.

Dobson, T. V. (1966). Introduction to Mathex. In W. W. Sawyer & L. D. Nelson (Eds.), *Mathex: Basics of mathematics experiences.* Toronto, ON: Encyclopaedia Britannica of Canada Ltd.

Flavell, J. H. (1963). *The developmental psychology of Jean Piaget* (with a foreword by J. Piaget). New York, NY: Van Nostrand Reinhold.

Piaget, J. (1952). *The child's conception of number.* London, UK: Routledge & Paul.

Piaget, J. (1957). *Logic and psychology.* New York, NY: Basic Books.

Piaget, J., Inhelder, B., & Szeminska, A. (1960). *The child's conception of geometry* (E. A. Lunzer, Trans.). New York, NY: Basic Books.

Piaget, J. (1964). Development and learning. In R. E. Ripple & V. N. Rockcastle (Eds.), *Piaget rediscovered* (pp. 7–20). Ithaca, NY: Cornell University Press.

Piaget, J. (1967). The genetic approach to the psychology of thought. In J. P. De Cecco (Ed.), *The psychology of language, thought, and instruction: Readings* (pp. 271–276). New York, NY: Holt, Rinehart & Winston.

Sawyer, W. W., & Nelson, L. D. (Eds.). (1966). *Mathex: Basics of mathematics experiences.* Toronto, ON: Encyclopaedia Britannica Canada.

Sullivan, E. V. (1967). *Piaget and the school curriculum: A critical appraisal.* Toronto, ON: Ontario Institute for Studies in Education.

Van Engen, H., & Hartung, M. L. (1958). *Seeing Through Arithmetic.* Chicago, IL: Scott Foresman.

Van Engen, H., Hartung, M. L., & Stochl, J. E. (1965). *Foundations of elementary school arithmetic.* Chicago, IL: Scott Foresman.

This chapter was originally published as:

Kunelius, W. (1962). On geometry. *Mathematics Council of the Alberta Teachers' Association Newsletter, 1*(1), 1–3.

CHAPTER 1

ON GEOMETRY

L. W. Kunelius

Editor's Note: At the Mathematics Council's inaugural conference in August 1961, Mr. Kunelius, at the end of his address, was asked the question, "What is the place of geometry in our high schools?" What follows is his answer.

The place of geometry in high school has been defended on several grounds:

1. It provides knowledge of important geometric relationships of which the educated adult has future need. In other words, it contributes to literacy.
2. It provides the best means we presently have for teaching elementary ideas concerning the nature of proof.
3. It has a peculiar or unique value for teaching habits of logical reasoning.
4. It provides an example of a logical mathematical system, the only example we have in our high school mathematics of a system founded upon undefined terms, definitions, and postulates.

Let us look at each of these. If the main purpose is to teach geometric facts—in other words, knowledge of basic geometric constructions and relationships—this can be accomplished by the methods of informal or intuitive geometry in junior high. The pupil at the end of Grade IX knows that the angles of a triangle are together equal to two right angles or 180°, he knows the conditions for congruence of triangles, similarity, and parallelism, though he cannot give formal proofs for them. Furthermore the number of geometric facts, which the average citizen should know and which are not now taught informally by the end of Grade IX, is few. Clearly the first reason is not enough to justify demonstrative geometry in the high school.

It provides the pupils with a means of acquiring a glimpse of the nature of a logical mathematical system. But Euclid's geometry as it is taught, or can be taught at the high school level is not very logical; it has several weaknesses. Much simpler and better examples of postulational systems can be developed in algebra.

It provides a means for teaching pupils something of the nature of proof and for practice in logical or deductive reasoning. It can do these provided it is taught with that end in view—if the role and nature of undefined terms, definitions, assumptions and of theorems derived from them are brought out; if hidden assumptions are care fully analyzed in both mathematical and non-mathematical situations; if the character of analysis, of indirect proof, of inductive proof, are understood. To do this we don't need to study several books of Euclid. Nor is any crime committed if, for example, the congruence theorems are left unproven and treated as assumptions.

I, therefore, predict that though the mastery of the more significant theorems (significant because they have frequent applications in life) will continue to be recognized, this will be overshadowed by other objectives for high school geometry, viz, (a) that pupils come to appreciate the postulational structure of geometry, (b) that pupils of come to appreciate that there are other geometries each based on its own set of postulates, (c) that algebraic or analytic methods will be introduced more widely, and (d) that pupils come to appreciate the nature of proof and the types of reasoning involved in proof.

These changes will come gradually. I believe that within the framework of the present text some changes can be made which would provide the teacher with more time to give greater emphasis to the structure of geometry and the nature of proof. I would hope that alert teachers who are thinking along the lines I have outlined above will take the initiative and the courage to depart from the traditional approach to the text in Mathematics 10. There are some commendable features in the text which presently receive but scant notice due to pressure of time and lack of appreciation of their significance. I would say: telescope the work on theorems.

CHAPTER 2

"POLYAN" MATHEMATICS

H. F. McCall

Editor's Note: Dr. McCall, principal of Seba Beach School, was awarded the Shell Merit Fellowship last year [1962]. The article below was intended to follow a discussion of the shortcomings of the mathematics that has already been introduced in many American schools and is being introduced to a degree in Canadian schools.

There is indication of the adoption of a "newer" mathematics, much more openly based upon induction, the reasoning of science, than traditional mathematics has ever been. The "newer" mathematics I have termed "Polyan," for it has been taught by the renowned Dr. G. Polya in many European and American universities for a good number of years. Besides the general mathematics of Professor Polya, there is considerable indication that a great deal of geometry will be placed in the primary grades. For those who would like to see the geometry books for primary grades, write for *Geometry for the Primary Grades, Books 1 and 11, and Teachers' Manuals* (Hawley and Suppes, Holden Day Inc., 728 Montgomery Street, San Francisco, California).

However, this is a mere detail. The really significant aspect of the "newer" mathematics is definitely Polyan, and, if you wish to acquaint yourself with something that is really interesting in mathematics, purchase *How To Solve It—A New Aspect of Mathematical Method* (Polya, G., Doubleday Anchor

Selected Writings from the Journal of the Mathematics Council of the Alberta Teachers' Association, pages 9–14
Copyright © 2014 by Information Age Publishing

Books, Doubleday and Co., Inc., Garden City, New York, $1.10) or *Induction and Analogy in Mathematics* (Polya, G., Princeton University Press, Princeton, New Jersey, $5.50).

The keynote to Polyan mathematics is the solving of problems. The rigorous, systematic, deductive science of mathematics is not scorned and discarded as useless, but a different experimental, inductive science of mathematics which should play just as important a part in the world is introduced.

The main purpose for mathematics should be solving of problems, not philosophic contemplation of the wonders of our number system or even the wonders of flawless deductive reasoning. In Dr. Polya's "Preface" to the first printing of *How To Solve It*, he says, "If [the teacher] challenges the curiosity of his students by setting them problems proportionate to their knowledge, and helps them to solve their problems with stimulating questions, he may give them a taste for, and some means of, independent thinking."

It seems advisable to give here a fairly extensive quotation from the "Preface" to *Induction and Analogy in Mathematics*:

> Strictly speaking, all our knowledge outside mathematics and demonstrative logic (which is, in fact, a branch of mathematics) consists of conjectures. There are, of course, conjectures and conjectures. There are highly respectable and reliable conjectures as those expressed in certain general laws of physical science. There are other conjectures, neither reliable nor respectable, some of which may make you angry when you read them in a newspaper. And in between there are all sorts of conjectures, hunches, and guesses.
>
> We secure our mathematical knowledge by demonstrative reasoning, but we support our conjectures by plausible reasoning. A mathematical proof is demonstrative reasoning, but the inductive evidence of the physicist, the circumstantial evidence of the lawyer, the documentary evidence of the historian, and the statistical evidence of the economist belong to plausible reasoning.
>
> The difference between the two kinds of reasoning is great and manifold. Demonstrative reasoning is safe, beyond controversy, and final. Plausible reasoning is hazardous, controversial, and provisional. Demonstrative reasoning penetrates the sciences just as far as mathematics does, but it is in itself (as mathematics is in itself) incapable of yielding essentially new knowledge about the world around us. Anything new that we learn about the world involves plausible reasoning, which is the only kind of reasoning for which we care in everyday affairs. Demonstrative reasoning has rigid standards, codified and clarified by logic (formal or demonstrative logic), which is the theory of demonstrative reasoning. The standards of plausible reasoning are fluid, and there is no theory of such reasoning that could be compared to demonstrative logic in clarity or would command comparable consensus. (p. v)

Another point concerning the two kinds of reasoning deserves our attention. Everyone knows that mathematics offers an excellent opportunity to learn demonstrative reasoning, but I contend also that there is no subject in the usual curricula of the schools that affords a comparable opportunity to learn plausible reasoning. I address myself to all interested students of mathematics of all grades and I say: "Certainly, let us learn proving, but also let us learn guessing." This sounds a little paradoxical and I must emphasize a few points to avoid possible misunderstandings.

Mathematics is regarded as a demonstrative science. Yet this is only one of its aspects. Finished mathematics presented in a finished form appears as purely demonstrative, consisting of proofs only. Yet mathematics in the making resembles any other human knowledge in the making. You have to guess a mathematical theorem before you prove it; you have to guess the idea of the proof before you carry through the details. You have to combine observations and follow analogies; you have to try and try again. The result of the mathematician's creative work is demonstrative reasoning, a proof; but the proof is discovered by plausible reasoning, by guessing. If the learning of mathematics reflects to any degree the invention of mathematics, it must have a place for guessing, for plausible inference.

There are two kinds of reasoning, as we said: demonstrative reasoning and plausible reasoning. Let me observe that they do not contradict each other; on the contrary, they complete each other. In strict reasoning the principal thing is to distinguish a proof from a guess, a valid demonstration from an invalid attempt. In plausible reasoning the principal thing is to distinguish a guess from a guess, a more reasonable guess from a less reasonable guess. If you direct your attention to both distinctions, both may become clearer.

A serious student of mathematics, intending to make it his life's work, must learn demonstrative reasoning; it is his profession and the distinctive mark of his science. Yet for real success he must also learn plausible reasoning; this is the kind of reasoning on which his creative work will depend. The general or amateur student should also get a taste of demonstrative reasoning: he may have little opportunity to use it directly, but he should acquire a standard with which he can compare alleged evidence of all sorts aimed at him in modern life. But in all his endeavors he will need plausible reasoning. At any rate, an ambitious student of mathematics, whatever his further interests may be, should try to learn both kinds of reasoning, demonstrative and plausible.

I do not believe that there is a foolproof method to learn in guessing. At any rate, if there is such a method, I do not know it, and quite certainly I do not pretend to offer it on the following pages. The efficient use of plausible reasoning is a practical skill and it is learned, as any other practical skill, by imitation and practice. I shall try to do my best for the reader who is anxious to learn plausible reasoning, but what I can offer are only examples for imitation and opportunity for practice.

The examples of plausible reasoning collected in this book may be put to another use: they may throw some light upon a much agitated philosophical problem: the problem of induction. The crucial question is: Are there rules for induction? Some philosophers say "Yes"; most scientists think "No." In order to be discussed profitably, the question should be put differently. It should be treated differently, too, with less reliance on traditional verbalisms, or on new-fangled formalisms, but in closer touch with the practice of scientists. Now, observe that inductive reasoning is a particular case of plausible reasoning. Observe also (what modern writers almost forgot, but some older writers, such as Euler and Laplace, clearly perceived) that the role of inductive evidence in mathematical investigation is similar to its role in physical research. Then you may notice the possibility of obtaining some information about inductive reasoning by observing and comparing examples of plausible reasoning in mathematical matters. And so the door opens to investigating induction inductively.

I shall here "solve" a problem using these less rigorous, inductive methods and proofs suggested by Dr. Polya.

> **Find:** the lengths of three mutually perpendicular edges x, y and z, of a box.
> **Given:** the volume, V, of the box.
> **Condition:** the surface S of the box is a minimum.

The first step in solving might be to change the problem to a simpler but analogous problem and solve it. Thus, let us find the lengths of the sides of a rectangle, being given the area (a) of the rectangle and the condition being that the perimeter (P) of the rectangle be a minimum.

Let the sides of the rectangle be X and Y units in length.

$$2X + 2Y = P$$

$$X + Y = P/2$$

At this critical juncture we well might make an educated guess, namely, that this rectangle will have to be a square if we are to obtain maximum area for minimum dimensions. (No fault should be found with this latter statement of slight aberration from the original postulates.)

Each side of a square with perimeter P is equal to P/4.
Each also would be one-half the sum of the two adjacent sides, i.e.,

$$\frac{X + Y}{2}.$$

The area of this square would be

$$\left(\frac{X+Y}{2}\right)^2$$

square units.

The area of the original rectangle would, in any case, be XY square units. Is XY as large as

$$\left(\frac{X+Y}{2}\right)^2?$$

We are, of course, presuming that it is not, unless X and Y are equal. The difference in area will, in any case, amount to

$$\left(\frac{X+Y}{2}\right)^2 - (XY) = \frac{X^2 + 2XY + Y^2}{4} - \frac{4XY}{4} = \frac{X^2 - 2XY + Y^2}{4} = \left(\frac{X-Y}{2}\right)^2$$

square units.

Now, there is only one way to make this difference amount to nothing, or, in other words, there is only one way to make the area of the rectangle as large as the area of the square. This would be to have X = Y. Then, of course, the rectangle is a square. The way to keep a large rectangle with minimum dimensions, then, is to have those dimensions equal—making the rectangle into a square.

By means of our intuitive recognition of patterns, we now may form a conclusion from this simpler problem: that what happens when we deal with two dimensions might happen in an analogous fashion when dealing with three dimensions. If this is true, then, to keep a constant volume for a box with the minimum surface area for the faces of this box, we would want each face to be a square, that is, the box would be a cube.

However, we should test this theory.

To do this, let us once more simplify matters by supposing that one dimension, Z, of the box is fixed, so that only the other two dimensions, X and Y, may vary. Now, if we are to maintain the large volume with minimum dimensions, and one of these dimensions is fixed, the problem becomes one of maintaining a large product of the other two dimensions while they are at minimum magnitude. But this is the equivalent of finding the minimum dimensions of a rectangle of constant area, which we already discovered. This means that X and Y must be equal, to make the faces of the box, to which Z is perpendicular, both squares.

But since X, Y, and Z are equal members of a democracy with no special privileges to either, all representing "a dimension" of the box, we would obtain exactly the same results by holding X constant in our imagination and having Y and Z vary—also by holding Y constant and letting X and Z vary. Therefore, all sides of the box are squares and the box is a cube.

The general ideas underlying this method I have tried to make apparent throughout the discussion. Use of analogy was made in this inductive process, simplifying cases, observing regularity of patterns, making tentative generalizations and testing the guesses—utilizing the concept of keeping our brains clear by despotically holding one variable constant when we are bothered by too many variables at a time.

I think I have said enough to give a general idea of this "mathematics for the scientific world" and to show its great difference from the "new mathematics" now appearing in Alberta, which is more like "mathematics for the mathematics philosopher."

It is my own personal hope that every mathematics teacher in Alberta can see that mathematics in any grade has value only to the extent to which it may be used to solve problems. The final test of the value of a mathematics course in, let us say, grade four, is in discovering how many kinds of problems each individual pupil can solve after having completed the course.

This chapter was originally published as:

Krider, E. A. (1963). Overview of change—Or a look at the forest before we
   can't see it for the trees. *Mathematics Council of the Alberta Teachers'
   Association Newsletter, 2*(2), 1–5.

CHAPTER 3

OVERVIEW OF CHANGE

Or a Look at the Forest Before We Can't See It for the Trees

E. A. Krider

Editor's Note: Mr. Krider is a former principal at Oyen. During the past year he has been a teaching assistant in mathematics education while working toward his master of education degree.

The development of the mathematics curriculum in North America has been closely associated with the changing views of transfer of learning. In the half century before 1900, the theory of mental discipline held sway and it was accepted that transfer took place more or less automatically. Mathematics was of the sequential type and generally all high school students were required to take it without regard to what practical use it might be put (Butler & Wren, 1960). At this time it was usual for subject matter specialists to determine the content of the mathematics curriculum.

In the first decades of the twentieth century, we see a reaction against over-emphasis on factual knowledge and also the theory of mental discipline being discredited. The emphasis on specific transfer as opposed to general transfer, the ascendency of pragmatic philosophy, the stimulus response psychology, and the increased proportion of the population in our secondary schools, all lead to more emphasis being put on skills and specific information in mathematics. In the twenties and thirties, the stress was on social adjustment and training for democracy—"preparing the well-informed citizen" (Harris, 1960).

As the first half of the century comes to a close we see the gap between the subject matter specialist and the educationalist at its widest and the scholars at the forefront of knowledge starting to demand a voice in designing school curricula. Another facet of the development of mathematics that deserves mentioning is the emphasis in the forties on classes for the less gifted and in the fifties on classes for the gifted (Harris, 1960).

Finally we come to the big turning point in the development of the mathematics curriculum in the mid-fifties. Here we see the results of the reaction to the extremes of progressive education, the stimulus–response psychology, the over-emphasis on skills and specific information, the over-emphasis on the social and the utilitarian aspect of education, the extreme negative views on transfer. The following quotation illustrates the changing view on transfer:

> Virtually all the evidence of the last two decades on the nature of learning and transfer has indicated that, while the original theory of formal discipline was poorly stated in terms of the training of the faculties, it is a fact that massive general transfer can be achieved by appropriate learning. (Bruner, 1960, p. 6)

The changes in ideas of transfer and the Gestalt psychology gave the reformers the psychological grounds for their movement. Bruner (1960) says:

> What may be emerging as a mark of our generation is a widespread renewal of concern for the quality and intellectual aims of education—but without the abandonment of the ideal that education should serve as a means of training well-balanced citizens for democracy. (p. 1)

With this movement we see the subject matter specialist moving back into the picture: "Curriculum programs such as SMSG and UICSM sprang from the dissatisfaction of the subject specialists with the preparation being given for their discipline in the schools" (Hughes, 1962, pp. 187–192).

Although this turning point seems to have taken place suddenly about 1954, the proponents of the need for radical change in emphasis were actively campaigning long before this. Professor Cecil B. Read of Wichita University lists quotations all taken from articles written between 1917 and 1932, registering the same complaints as voiced by the "revolutionists" of the fifties (Read, 1958, pp. 181–186). Why did these people suddenly become the authorities in the field of curriculum building? First, the gap between what was taught in schools and what was known in the field became acute because of the explosion of knowledge. Secondly, the shortage of scientists and mathematicians came to the public's attention with the first Sputnik. With the millions of dollars poured into the cause by the American government, the reformers were away. A number of professional groups attacking the problem of producing a new mathematics curriculum were set up. The three most influential groups are:

> The Commission of Mathematics of the College Entrance Examination Board (usually referred to as the Commission on Mathematics)
> The University of Illinois Committee of School Mathematics, headed by Professor Max Berberman (abbreviated UICSM)
> The School Mathematics Study Group headed by Professor Edward G. Gegle at Stanford (abbreviated SMSG)

These groups are made up of professional mathematicians, professional educators, psychologists, and usually practising teachers. It is hard to over-emphasize the impact that these three groups have made not only on mathematics curriculum but in the whole spectra of the school curriculum building (Bruner, 1960, p. 70). One cannot discuss recent mathematics curriculum change without referring to these groups.

In conclusion and at the risk of oversimplification, one might infer from this brief survey that the development of the mathematics curriculum in North America since the turn of the century has been a series of actions and reactions. If one is to extrapolate from this, we would expect a reaction to the modern approach to curriculum building as exemplified by Bruner, and the workers in the specific subject matter fields, to be discernible. In the case of mathematics this reaction is not only discernible, but is well established with a substantial following. C. Stanley Ogilvy, Hamilton College, Clinton, New York writes:

> After 20 years of propaganda in favor of the introduction of new mathematics, we can now discuss the beginning of a swing in the other direction. In almost every new issue of the *Mathematics Teacher* and the *American Mathematics Monthly* we find one or two articles cautioning us to move ahead slowly, to guard against discarding good and valuable material merely, to make room for something new for the take. (Ogilvy, 1960, p. 6)

And from a statement signed by 64 mathematicians in the United States and Canada:

> Mathematicians, reacting to the dominance of education by professional educators who may have stressed pedagogy at the expense of content, may now stress content at the expense of pedagogy and be equally ineffective. Math-

ematicians may unconsciously assume that all young people should like what present day mathematicians like or that the only students worth cultivating are those who might become professional mathematicians. (Ontario Mathematics Gazette, 1962, p. 5)

Could there be a little bit of truth in the statement that, in education, if you're old-fashioned long enough, you'll be modern!

REFERENCES

Bruner, J. S. (1960). *The process of education.* Cambridge, MA: Harvard University Press.

Butler, C. H., & Wren, F. L. (1960). *The teaching of secondary mathematics.* New York, NY: McGraw-Hill. Chapter I.

Harris, C. W. (1960). Mathematics. In *Encyclopedia of Educational Research.* New York, NY: MacMillan.

Hughes, P. (1962). Decisions and Curriculum Design. *Educational Theory, 12.*

Ontario Mathematics Gazette (1962, October). "On the Mathematics Curriculum of the High School". Bulletin 1, No. 2.

Ogilvy, S. C. (1960, November). Second thoughts on modernizing the curriculum. *The Mathematics Teacher.*

Read, C. B. (1958, March). What's Wrong with Mathematics. *School Science and Mathematics,* Vol. LVIII, No. 509.

This chapter was originally published as:

Chell, N., & Coulson, W. F. (1963). Can high school students learn some of the concepts of modern mathematics? *Mathematics Council of the Alberta Teachers' Association Newsletter, 2*(2), 12–17.

CHAPTER 4

CAN HIGH SCHOOL STUDENTS LEARN SOME OF THE CONCEPTS OF MODERN MATHEMATICS?

Nora Chell and W. F. Coulson

Editor's Note: Miss Chell recently received her bachelor of education degree from the University of Alberta with a major in mathematics. She was awarded the Clarence Sansom Memorial Gold Medal in Education by The Alberta Teachers' Association. Mr. Coulson is assistant professor of education at the University of Alberta, Edmonton. He is a member of the Senior High Mathematics Curriculum Subcommittee of the Department of Education.

The first part of this article deals with the question expressed in the title from the standpoint of what various writers have said in the literature. Later parts will deal with the same question but from the standpoint of the research which has been done.

To answer the question, one must first ask, "What is modern mathematics?" Some have defined modern mathematics as meaning modern in approach, while others have defined it as meaning modern content taught by modern approaches. Demsey (1962, p. 25) says, "As far as high schools are concerned the word modern applies more to approach than to content." By modern mathematics, Dr. Paul Beesack, assistant professor of mathematics at McMaster University (1960, p. 71) means, "axiomatic or postulational method which is characteristic of so much of the mathematical activity of this century." Rourke (1958, p. 225) states modern mathematics is "a means of broadening old ideas and introducing new ones to clarify, simplify, and unify our mathematical concepts." Zant (1961, p. 595) comments that modern mathematics does not imply a totally new program, but the "newer programs present essentially the same subject matter as before, but it is developed from a different point of view and use is made of a relatively small number of new concepts such as sets, algebraic systems, binary operations, axiomatic systems, etc." According to May (1958, p. 93), modern mathematics includes "such topics as logic, theory of set, Boolean algebra, and set theoretic approaches to relations functions, and other topics of mathematics."

For this paper, modern mathematics will be defined to mean modern concepts taught from a modern approach. Modern content will refer to group and field theory, topology, symbolic logic, and set theory while modern approaches will refer to the axiomatic approach or postulational approach and the set-theory approach. The literature will be reviewed first considering those articles which deal with the average student, then the slow student, third the above average and gifted student, and fourth with respect to mathematics clubs.

AVERAGE STUDENT

At St. Laurent High School, Montreal, experiments in modern mathematics were conducted with three groups of students: average, top, and students in mathematics clubs. According to Richardson (1960, p. 37), the average students in Grades X and XI benefitted from "the discussion of the postulational method and its use in mathematics."

Rourke (1960, p. 12), in his address to the Canadian Teachers' Federation Seminar in 1960, says:

> The boys at Kent School were using the set language in notation very well in Grade XII, so the next year we pushed it down to Grade XI, then down to Grade X and now, in Grade VIII we teach the idea of a relation as a set of pairs, because it fits very well there, with some of the work we are doing.

In his article, "Some Implications of Twentieth Century Mathematics for High Schools," Rourke (1958, p. 86) states:

> At the Grade X level, I made the definition of the function explicit and we talk the language of sets and set solutions. We don't get stuffy about it and we

allow ourselves some elliptical expressions. We may, on occasion call 2x + 3 a linear function, as indeed generations of mathematicians have done and likely will continue to do for some time. But we will keep in mind that in this elliptical talk, the function is really the set of ordered pairs (x,y) defined by the formula y = 2x + 3.

Zant (1961, p. 593), in describing an Oklahoma experimental program in modern mathematics used for regular classroom students, states:

> The program was considered very successful. Teachers reported much more interest and understanding on the part of the students than did those in classes taught from the traditional program. Students seemed to be as proficient in the standard skills of mathematics as those taking traditional courses, though no particular effort was made to make comparisons.

A summer school course attended by teachers and superior students is described by Montague (1959, pp. 22–23) in her article "A Demonstration Class in a National Science Foundation Summer Institute." The students formed a demonstration class, and the topics discussed included: The Development of Mathematics, Modern Concepts in Geometry, Modern Concepts in Algebra and Some Aspects of Applied Mathematics. Montague says:

> Follow-up correspondence with teacher participants has revealed that many of the concepts presented by the lecturers were used in their regular classes the following year. The teachers felt that, even though the demonstration group was a highly selective one, the ease with which they accepted new concepts indicated that the average students could also learn some of them.

From the above statements, it seems as though the average student can learn some of the concepts of modern mathematics.

SLOW STUDENT

Allen (1960, p. 14) states that the "set serves to make feasible a broadened introduction to algebraic symbolism and the sentence, be it equation or inequation." He also states that the introduction of sets, including the set concepts of union and intersection, motivated in particular the disinterested and slow learners (1960, p. 20). For the slow learner the material was new and different. (If sets motivated the students, is it not inevitable that they must have learned some of the concepts?)

From the results of the St. Laurent experiment, Richardson (1960, p. 38) states, "Sets of ordered pairs were used last year to present graphs to a group of students who were poor in algebra, and they showed good comprehension."

The opinions of two writers indicate that slow high school students can learn some of the concepts of modern mathematics.

ABOVE AVERAGE AND GIFTED STUDENTS

In British Columbia an experimental program of Mathematics 91 was established in 1959–60 "to determine whether or not it is possible to incorporate new subject matter and new approaches to traditional subject matter into the regular course" (Meredith, 1958, p. 210). The Chant Commission (Wilansky, 1961, p. 257) reviewed the experimental classes and found: "The results indicated that the classes could learn some additional concepts and also master the conventional material at a standard equivalent to that of other Mathematics 91 classes."

According to Richardson (1960, p. 38), advanced students were shown what is desirable in drawing up a system of postulates and discussed algebra from the viewpoint of the postulational method.

Rourke (1958, p. 85) states that the concept of the function (a set of ordered pairs (x, y) such that for each x there is exactly one y) as a kind of relation can be understood by many high school students, most certainly those capable of college study.

For the gifted child Allen (1960, p. 22) says, "...set terminology is felt to provide an unambiguous and succinct introduction to a number of concepts of beginning algebra."

Grossman (1961, pp. 75–81) reports that at the National Science Foundation Summer Institute for high school students at Columbia, special topics such as Boolean algebra, groups, rings, integral domains, fields, and the development of the number system from Peano's axioms were taught. Thirty-five students distinguished by their capacity and achievement participated in the 1961 course.

A similar report by Wilansky (1961, pp. 250–254) states that gifted students were taught topics of linearity, group theory, and multiplication from a modernistic view of Lehigh University during the summer of 1960.

According to Montague (1959, pp. 98–100), the demonstration class composed of gifted students at the National Science Foundation Summer Institute had no problems with the following topics: the development of mathematics, modern concepts in geometry, modern concepts in algebra, and some aspects of applied mathematics.

Nichols (1959, pp. 100–103) describes a school held at Florida State University in 1958 in which the talented students were divided into two groups. Both groups studied the fundamental concepts of set theory. In addition, one group was given "instruction in development of the number system beginning with the notion of set, leading to the definition of a natural number, integer, etc. The students also learned to perform arithmetic calculations using various number systems." The students learned the new concepts very rapidly.

Articles written about modern mathematics and the above-average or the gifted child indicate that these students most definitely can learn some of the concepts of modern mathematics.

Mathematics Clubs

Allen (1960, p. 20) suggests that students can learn the concept of the set and the set approach to algebraic symbolism as an extracurricular activity in a mathematics club. He also says:

> The extracurricular programs which have been introduced in many of our high schools have revealed a number of occasions in which the interest of a pupil whose classroom performance has been mediocre was stimulated by an unfamiliar mathematical topic, a problem in change, an exercise in space perception, or the diagrammatic representation of the relationships between sets.

Richardson (1960, p. 37) reports that the concepts of modern mathematics were taught in the mathematics club at St. Laurent High School.

From the statements of these two writers, one gains the impression that at least one way to introduce the ideas of modern mathematics is through a mathematics club. This will take no time away from the authorized course and still present the interested students with some challenging topics in a challenging subject.

CONCLUSION

In each of the articles quoted here, the authors indicated that high school students could learn some of the concepts of modern mathematics. Although traditionalists stated that they were definitely against modern mathematics, in no case was there a statement suggesting that high school students could not learn some of the concepts of modern mathematics.

The writers recognize that many more articles could have been mentioned. Limitations of space made the problem of selection a difficult one. In the next article the writers will take a short look at the research. Is it possible to find evidence that is not merely an expression of opinion that high school students can learn modern mathematics?

REFERENCES

Allen, H. D. (1960). Sets and the inequation. *The Teachers Magazine, XL*(200), 14–24.

Beesack, P. R. (1960). Modern mathematics, its evolution, logical structure, and subject matter. *New Thinking In School Mathematics*, Ottawa, pp. 71–82.

Demsey, F. C. (1962). Modern ideas for a basic treatment of algebra. *The Bulletin, 41*(1), 25–25.

Grossman, G. (1961). A report of a National Science Foundation Summer Institute in Mathematics for high school students at Columbia. *The Mathematics Teacher, LIV*(2), 75–81.

May, K. O. (1958). Finding out about "modern mathematics." *The Mathematics Teacher, LI*(3), 93–95.

Meredith, J. R. (1958). What's going on in mathematics. *The B.C. Teacher, XL*(5), 209–211.

Montague, H. F. (1959). A demonstration class in a National Science Foundation summer school. *The Bulletin of the National Association of Secondary School Principals, 43*(247), 98–100.

Nichols, E. D. (1959). A summer mathematics camp for talented high school students. *The Bulletin of the National Association of Secondary School Principals, 47*(247), 100–103.

Richardson, D. N. (1960). Experiments with modern mathematics. *The Educational Record, LXXVI*(1), 36–39.

Rourke, R. E. K. (1960). School mathematics in Russia and the United States. *New Thinking In School Mathematics*, Ottawa, pp. 1–14.

Rourke, R. E. K. (1958). Some implications of twentieth century mathematics for high schools. *The Mathematics Teacher, LI*(2), 74–86.

Wilansky, A. (1961). A research program for gifted secondary school students. *The Mathematics Teacher, LIV*(4), 250–254.

Zant, J. H. (1961). Improving the program in mathematics in Oklahoma schools. *The Mathematics Teacher, LIV*(8), 594–599.

The B.C. Teacher, "How the commissioners see it." *XL*(5), 207–208.

This chapter was originally published as:

Coulson, W. F. (1965). Discovery or programming. *Mathematics Council of the Alberta Teachers' Association Newsletter, 4*(1), 1–4.

CHAPTER 5

DISCOVERY OR PROGRAMMING

William F. Coulson

Is it possible that one sees two different trends in mathematics education as he surveys current publications from afar? On the one hand we have the experimental developments by such groups as the University of Illinois Committee on School Mathematics, School Mathematics Study Group, the Ball State Project, and the Madison Project. Upon closer inspection, one finds the discovery approach built into the material. Many present-day authors are trying to imitate their style and their approach to the material.

Each of the groups mentioned previously and most of the authors currently producing material for mathematics courses recognize and make use of pupil discovery of relationships and structure. Mathematics is recognized as a subject area that has a structure that is meaningful to the student. Thus, the students are able to discover relationships and to build one concept upon another. Mathematics does not become a series of isolated bits of fact and computations to be mastered until they become habit.

Taking the teacher and the teaching method into consideration, this approach is designed to give the student an opportunity to think for himself. The teacher must know a great deal about the material. He must know where

Selected Writings from the Journal of the Mathematics Council of the Alberta Teachers' Association, pages 25–27
Copyright © 2014 by Information Age Publishing

he has been and where he is going and must know how to guide the pupils to develop the structure for themselves. Questioning techniques take on greater importance. Rules must not be presented to the student to be applied blindly to a multitude of simple examples until the mechanics become habit and these rules can be recalled upon the receipt of the proper stimulus.

It has been argued that the statement of the generalization by the student is not necessary. Some even feel that this is a hindrance. The spoken or written language gets in the way of the mathematical ideas. When a student is able to apply the generalization in an unfamiliar situation, then he knows what it tells him.

Others argue that it is necessary for the student to express the generalization. Only then can one be certain that the concept is known and understood.

On the other hand, we have the type of curriculum exemplified by programmed materials. This development is not new, but in recent years it has gained impetus especially in mathematics.

One of the most significant points involved in programming is that the student is led down a very definite path. The material to be mastered is presented in very small steps to insure "understanding" and correctness. The student is able to proceed very slowly and along a path determined by the author of the program. At no point is he permitted to meditate upon a related topic. His attention is always directed toward the mastery of one specific concept.

If one sticks to the traditional definition of mathematics, one thinks of it more as a tool subject. Is this all that mathematics is in this day and age? Many very prominent people do not accept this view alone. Mathematics has become more than mere subject matter to be mastered because it is useful in some other field of endeavor. It is thought of more as a way of thinking, as an academic discipline to be studied for itself. A great many mathematicians study mathematics just for the sake of the mathematics involved and not because of its utilitarian value.

Looking for a moment at the mathematics curriculum or at mathematics education, can we note any relationship in the trends? How do they appear to be affecting the curricula in mathematics? What is the effect of each of them on the teaching of mathematics?

It would appear that the two ideas are not very closely related. They would seem to be worlds apart. The discovery approach adopted by the UICSM and the SMSG would seem to give the pupil credit for being able to think for himself, for being able to recognize meanings, for being able to direct his attention toward a series of related learning tasks.

The approach adopted by those who advocate programmed instruction would appear to deny the ability of the pupil to do independent thinking. A stimulus is presented to the student to which he must make one and only one response. Since this response is right 90 per cent or more of the time, he has little or no opportunity to analyze. His attention is directed toward a rather narrow, limited topic.

The effect on the curriculum in the one instance seems to be a freeing one. Pupils are given an opportunity to act as mature mathematicians. Observations are made. These are accepted or rejected by proof. New observations or relationships are introduced, not necessarily by the teacher or text. These, too, are accepted or rejected by proof from what has gone before. The pupil is an active participant in the development of mathematical concepts.

Programmed instruction tends to do the opposite, as far as the pupil is concerned. Pupils are not given an opportunity to make independent observations. They have little chance to analyze so that they might accept or reject a relationship. The opportunity to act as a mathematician is absent.

As far as the teaching of mathematics is concerned, one of these trends would permit the individual to develop as a skilled craftsman. The teacher would have a vast storehouse of knowledge which he would need to rely upon to keep the class moving in a correct fashion. For example, if a student wanted to solve quadratic equations, the teacher would know immediately whether or not this could be done with the knowledge possessed by the student. The teacher could then guide the student through the discovery of the various processes of solving this particular type of problem. Knowledge of his subject, then, is very important to the teacher who wants to follow the discovery approach used by the SMSG or the UICSM.

Programmed instruction would seem to leave very little for the teacher to do. When a student is unable to understand a specific point, the teacher could assist the student in mastering this concept. One other aspect of programming comes into play when a teacher builds a program of his own. During his labors, he becomes more intimately acquainted with the particular topic, with some of the problems involved in learning this topic, and with some of the problems involved in teaching this topic.

In this paper a brief look at two apparently divergent trends in mathematics education was attempted. Each teacher of mathematics must look more closely at each of these trends to see how they will or will not affect his teaching. It seems obvious that no teacher will remain untouched by these trends. Many people are advocating one or the other of these two approaches, people who are recognized as authorities in mathematics education. Perhaps it will be best for each teacher to conduct a little action research in his own classroom to help him decide. There can be no fence straddlers.

This chapter was originally published as:

Galeski, E. (1965). What is modern mathematics? *Mathematics Council of the Alberta Teachers' Association Newsletter, 4*(3), 4–6.

CHAPTER 6

WHAT IS MODERN MATHEMATICS?

Elizabeth Caleski

Editors' Note: Elizabeth Caleski was a student in the Faculty of Education. She completed the professional year leading to certification following a Bachelor of Science degree.

One of the phrases which we, as prospective teachers, hear over and over again is "modern mathematics." In attempting to learn something about this, one might take each word individually and try to form a satisfactory definition for each. From the standpoint of inexperience, one must turn to a search of the works of the more experienced and wise mathematicians of this era to find an answer to the question "What is Modern Mathematics?" First a record of some of the definitions of mathematics:

> The purely formal sciences, logic and mathematics, deal with those relations which are, or can be, independent of the particular content or the substance of objects. (Herman Hankel)

> Perhaps the least inadequate description of the general scope of modern Pure Mathematics—I will not call it a definition—would be to say that it deals with form in a very general sense of the term. (Hobson)

Mathematics in its widest signification is the development of all types of formal, necessary, deductive reasoning. (Whitehead)

Mathematics is the study of ideal constructions (often applicable to real problems), and the discovery thereby of relations between the parts of these constructions, before unknown. (Pierce)

Mathematics is preferably free in its development and is subject only to the obvious consideration that its concepts must be free from contradictions in themselves, as well as definitely and orderly related by means of definitions to the previously existing and established concepts. (Cantor)

Mathematicians assume the right to choose, within the limits of logical contradiction, what path they please in reaching their results. (Adams) (Eves & Newson, 1958, pp. 176–177)

Which brings us to the profound conclusion I think Bertrand Russell states beautifully, that "mathematics may be defined as the subject in which we never know what we are talking about, nor whether what we are saying is true" (Eves & Newson, 1958, p. 177)

Having arrived at this vague concept of what mathematics is, the topic implies the explanation of the term "modern," which is another impossibility. By "modern" do we mean today, this year, this decade, this century? When speaking in terms of math, it has been said that "modern" would include a period covering the last two or three hundred years. If this is the case, then the discussion of "modern math" would cover a vast array of topics that could probably fill dozens of textbooks.

Therefore, since it is unlikely that a definition of this topic can be discussed in a logical and precise manner, I would rather discuss what is entailed in the teaching of contemporary mathematics and which attitudes should be fostered. There is a certain feeling for mathematics that must be conveyed to the students of mathematics in order that they will be able to understand thoroughly what math is really all about and be able to apply their knowledge to the world of tomorrow. Professor Carl Raymond Hedrick vigorously supports the idea that the real values of mathematics reside in its processes. In his own words, he states that "if we allow our attention to centre on a special case, on a given fact, or on a particular skill, we shall be in danger of abandoning the process itself." (Fawcett, 1964, p. 454) It seems that today we have abandoned the emphasis on process—if there ever was one—and if "modern" curricular proposals do nothing more than redirect our attention to this most significant aspect of mathematics, they will have made a tremendous contribution to mathematics education. In a report from the Netherlands, summarized by Professor John Kemeny of the International Congress of Mathematics meeting in Stockholm, 1962, there is a recommendation that "stress should be laid on thinking mathematically and more value attached to this ability than to knowledge of a variety of less important facts" (Fawcett, 1964, p. 155). It has been said that "one of the great glories of mathematics is the possibility of moving its generalizations from the domain of probability into the domain of certainty within a given structure. It is this task to which we, as math teachers, have been called and at no time in the history of American education has the call been so inviting."

REFERENCES

Eves, H. W., & Newson, C. V. (1958). *An introduction to the foundations and fundamental concepts of mathematics.* New York, New York: Holt, Rinehart and Winston.

Fawcett, H. R. (1964, November). Reflections of a retiring teacher of mathematics. *The Math Teacher.*

CHAPTER 7

A COORDINATED REVIEW OF RECENT RESEARCH CONDUCTED IN THE DEPARTMENT OF ELEMENTARY EDUCATION UNIVERSITY OF ALBERTA RELEVANT TO MATHEMATICS EDUCATION

W. G. Cathcart and W. W. Liedtke

How do children acquire concepts? What limitation does development put
on the capabilities of logical thought? What are the characteristics of the
development of thought at any given age?

The answers to these questions would certainly aid in providing learning experiences and materials that are meaningful and beneficial for the
learner in that they could meet his needs at various stages of development.

Selected Writings from the Journal of the Mathematics Council of the Alberta Teachers' Association, pages 33–42
Copyright © 2014 by Information Age Publishing
All rights of reproduction in any form reserved.

Increasing interest in the study of children's thought processes and the manner in which they acquire certain concepts becomes evident when some of the recent research projects undertaken in the Department of Elementary Education in the field of elementary school mathematics are reviewed.

Perhaps the greatest quantity of work in the field of cognitive development has been conducted by the Swiss psychologist Jean Piaget, who has investigated the growth of children's concepts in a variety of areas. A simple ratio reflects the influence of Piaget's work on the studies done here. Since 1963, eight out of ten investigations into teaching and learning mathematics in the elementary school have been based on his work.

Piaget has concerned himself not only with the learner but also with the substance to be learned and its structure or logical organization. His theory and the results of his investigations provide a guide to some of the experiences necessary before certain mathematical concepts are formed or acquired. His method presents a way of evaluating an individual's stage of cognitive development.

Intellectual development, according to Piaget, is the result of four contributing forces: maturation, experience, socialization, and equilibration. These interacting factors contribute to the development of the intellect, and the results of his experiments indicate that a child becomes capable of 'adult thought' or reaches maturity at the age of 11 or 12. Characteristics of the development of thought before the age of about 12 are described by Piaget in terms of limitations or in terms of "things" children cannot do. These limitations give rise to three distinct developmental stages. They are sensori-motor (0–2 years), the pre-operational (2–7 years), and the concrete operational stages (7–11 years) (Baldwin, 1967, pp. 167–300).

The stages of most interest to people concerned with elementary education are the pre-operational and concrete-operational. The transition from the former to the latter stage is marked by the acquisition of various conservation concepts or the realization that properties of nature remain invariant under certain changes and transformations. The ages at which individuals acquire conservation concepts vary somewhat. However, from Piaget's experiments a certain order is apparent: quantity (continuous and discontinuous) and number (6–7 years), mass and length (7–8 years), area and weight (9-10 years), and finally volume (11–12 years) (Piaget, 1957, pp. 16–17).

In his book *The Child's Conception of Number*, Piaget (1952) describes the child's knowledge of, and stages of development in, such concepts as conservation of quantity, cardinal and ordinal correspondence, the properties of cardinal and ordinal number, the notion of series, the ability to classify, and the relation of classes to number. He tested what he called ordinary school children of primary school age and used a combination of observation, interview, and recording as his methods.

Brace (University of Alberta, 1963) replicated some parts of the work described above. His study was designed to determine to what extent preschool children have developed some of the basic concepts of number. He tested 124 preschool children on the following specific concepts:

1. rational counting
2. comparisons—group relations
3. conservation of number
4. cardinal number
5. ordinal number
6. place value as applied to the decimal system of notation

In addition, Brace attempted to clarify the nature of the relationships between the child's knowledge of basic number concepts and other variables such as the ability to count, sex, influence of older siblings, socioeconomic status or environment, and chronological age.

The most important finding of his investigation was the existence of a tremendous gap between the child's ability to count and his knowledge of the basic ideas of number. The relationship of counting to knowledge of cardinal number and the conservation of number was found to decrease with the age of the child, being less significant in older preschool children. That of counting to the child's concept of ordinal number increased with the age of the child. He found that chronological age and environment (socioeconomic status) contributed significantly to the knowledge of number concepts. Sex and the influence of older siblings were not related to the understanding of basic number concepts.

In his book *The Child's Conception of Geometry*, Piaget (Piaget, Inhelder, & Szeminska, 1960) made a number of suggestions about how concepts develop out of simple behavior patterns, and how a basic repertoire of geometric concepts is formed. He described how such notions as distance, length, change of position, ideas of measurement, and conservation of distance, length, area, and volume, lead to the construction of Euclidean space.

Pelletier (University of Alberta, 1966) subsequently used some of Piaget's tests to investigate grade 1 children's concepts of linear measurement. His subtests dealt with reconstructing relations of distance, conservation of length, measurement, and subdividing a straight line. Mathematics instruction of the 120 subjects tested was based on three different programs, and each program treated the topic of measurement differently. The three mathematics programs studied by these children were: *Numbers We Need* (Brownell and Weaver), *Seeing Through Arithmetic* (Hartung et al.), and a program based on Cuisenaire materials. In addition, Pelletier considered the simultaneous effects of sex, mental ability, and socioeconomic status.

The results of the study indicated no significant difference between the three groups tested. Many of the subjects were able to verbalize measurement terms, but they did not comprehend the underlying concepts. Such words as "distance," "length," and "measure" were not understood, and many subjects could not conserve length. Although no child had achieved complete mastery of the relevant concepts, children with higher mental ability were clearly superior in their understanding of linear measurement. Differences in sex and environment did not contribute significantly to the development of the concept of linear measurement.

Closely related to the construction of Euclidean space are the concepts of direction and scale. In his book *The Child's Conception of Space* (Piaget & Inhelder, 1963), Piaget deals with the child's knowledge of space and with intellectual development, that is, the development of intelligence as it operates in connection with spatial relations.

Towler (University of Alberta, 1965) selected four spatial concepts that he considered necessary for map reading and interpretation as well as for the graphing in mathematics. They were:

1. the concept of a reference system
2. the concept of distance
3. the concept of direction
4. the concept of scale

He examined the development of these concepts in elementary school children. In addition, he sought to clarify the relationships between the development of the concepts and sex, socioeconomic status, chronological age, intelligence, and grade level. He individually tested 120 children from grades 1 to 4.

The results of his study in general indicated a development of thought processes similar to those described by Piaget. However, the development of the concepts of a general reference system and scale occurred later in children studies by Towler than in those studied by Piaget. Towler found that intelligence, chronological age, and grade level were significantly correlated with the development of the spatial concepts under investigation, while sex and socioeconomic status were not.

In the elementary school mathematics program, attempts are usually made to develop and deepen concepts of time. A comprehensive report on the stages of development of time sense in young children has been made by Piaget (1955). The results of his experiments led him to the conclusion that the development of time concepts implies the progressive structuring of perceptual data by a sequence of logical operations such as seriation and duration.

Newman (University of Alberta, 1967) investigated the factors which influence the elementary school child's understanding of time duration and the stages by which such understanding is acquired. He tested 192 elementary school children from Grades I to VI on the following specific aspects of the concept of duration: understanding of verbal comparison of duration, understanding of graphic comparisons of duration, and ability to use measures of duration. Newman found that intelligence, age, grade, and socioeconomic status were related in a positive way to a child's understanding of time duration. He found that comparisons of duration were generally understood at the grade 3 level and measures of duration at the grade 4 level. The ability to use measures of duration was closely linked with the ability to understand verbal comparisons of duration. Grade 1 and 2 students understood verbal comparisons better than graphical comparisons of duration, but from grade 3 onward, both were understood equally well.

Conservation is often used as a criterion in the assessment of concept acquisition. Investigators who use the concept of conservation in this manner are often criticized for the method they employ. The "clinical method" which consists of observing, interviewing, and recording is attacked frequently. However, few critics offer an alternative approach.

One alternative was developed by Sawada (1966). He conducted a study of length conservation in 62 kindergarten and grade 1 children. The secondary purpose of his study was to ascertain whether or not an essentially nonverbal method of communicating the response criteria to the subjects would lower the age at which they gave evidence of possessing conservation of length. Sawada's nonverbal technique made use of calipers and a response apparatus to reward correct answers with a candy. The candy was placed behind one of three doors in a box. Over each door was a caliper with a model fit. That is, over one door a rod was inserted in the caliper so that it was too short, over the middle door the rod fitted exactly into the caliper, and over the last door the rod was too long for the caliper. The subject had another caliper that he could use to test the fit in the test situation. After a transformation was applied to the object, the subject was asked to open the door corresponding to the type of fit he thought there would be now without using his caliper after the transformation.

The results of this study showed that the threshold age at which 50 percent of the subjects conserved length was between five years and four months, and six years and two months. Thus, if the response criterion is essentially nonverbal, children are able to give conservation responses as much as two years earlier than they would if required to give verbal responses.

Sawada's major purpose was to explore the role of transformations (translations and rotations) in the child's conservation of length. He found that conservation of length cannot be solely explained in terms of the transformations applied to the objects exhibiting length. Neither could the factors extracted from the conservation test be defined solely in terms of the state properties of the objects to which the transformations were applied. Both state and transformation properties had to be used to interpret the factors extracted out of the test performance. Sawada also found that an illusion subtest was significantly more difficult than any of his other subtests. Age correlated significantly with performance on the test, but intelligence, as given by the "Detroit Beginning First Grade Intelligence Test" (1937), did not.

Should conservation be induced in young children? If Piaget is correct when he claims that conservation is necessary for any rational activity, then certainly the acquisition of conservation should be of prime concern to teachers. But why not let it take its natural course of development? One major problem is that there is an increasing pressure being placed upon teachers and curriculum planners to introduce more sophisticated mathematical concepts at younger and younger ages. The Cambridge Report (1963) is an indication of such a trend. If the mathematical concepts outlined in the report are to be understood by the children rather than merely memorized, then conservation in related areas is important. Thus, the inducing of conservation becomes a relevant problem.

Sawada's finding (referred to earlier) that the concept of conservation is present on the nonverbal level at an earlier age than indicated by Piaget might have an important bearing on the problem of teaching conservation to children. This strongly suggests that the preliminary stages of inducing the concept of conservation and mathematical concepts should be approached from a level that is as free from verbalization as possible.

Towler (1967), on the other hand, attempted to induce conservation by pointing out the important aspects of this concept in small groups. He used a sample of grade 1 students. Forty non-conservers and 40 partial conservers were assigned randomly to an experimental group and a control group. His major hypothesis was that learning of conservation could take place if the crucial aspects of conservation could be isolated and presented to the students in such a way that they could understand them. He hypothesized that these relevant variables were (a) an understanding that a quantity retains its identity during transformation, (b) an understanding of compensatory relationships, and (c) an understanding of the principle of reversibility.

On the basis of this hypothesis, Towler designed a training session for the experimental group, which he hoped would lead to the understandings mentioned above. To provide training in the identity relationship, Towler asked two types of questions. A drink was poured from a jug into a can, and the subjects were asked, "Does pouring change the kind of drink?" The second type of question asked "Does pouring change how much (amount) there is to drink?"

Another aspect of the training session was designed to confront the subjects with the compensatory relations of height and width. Containers of various dimensions were used so that the levels of the liquids varied. Students were asked to predict where the levels would be and to explain why the level was higher or lower in a given container than in the original standard. A final aspect of the training session was concerned with the ability of the child to relate what he had just discussed to pictorial representations of similar situations.

It was thought necessary to describe Towler's training session somewhat since he claims that it was highly successful in inducing conservation in the experimental group. A significant number of non-conservers and partial conservers in the experimental group acquired conservation and, with the exception of one subject, were able to retain conservation over a two- to three-week period. They also were able to transfer their learning to a new situation using different materials (discontinuous quantity). While the experimental group showed this improvement, there was no change in the control group's understanding of conservation.

There are some practical applications of Piaget's theory for mathematics education. Piaget says that thinking is an internalized action that has been developed as a result of physical actions. Therefore, it is important that children be given an opportunity to actively explore and manipulate physical objects in their environment. This implies that manipulative materials should be an important part of the mathematics program.

A recent study by Scherer (University of Alberta, 1968) examined the role of manipulative materials in problem-solving at the grade 3 level. Four classes were given concrete materials consisting of 50 counting sticks, 50 little tile squares, and 50 rectangles. Four other classes were chosen as a control group. Verbal problems presented consisted of the six types of additive and subtractive problems as outlined by the *Seeing Through Arithmetic* program. The experimental group used the concrete materials to represent the problem situations, while the control classes only discussed the problems. The treatment covered eight lessons at the rate of one lesson a day.

Scherer found that the use of manipulative materials was no more effective in developing ability to solve verbal problems than the use of discussion and printed materials. However, a number of methodological weaknesses may account for this lack of significant gain by the experimental group. The major weakness was the short duration of the experiment. Eight days may be an inadequate length of time in which to observe change. Furthermore, grade 3 students may have passed the age when concrete materials are most useful, at least for the type of problems they were working with. We still must maintain with Almy (1966) that Piaget's theory strongly suggests the importance of learning through activity. Certainly this is an area that requires more research.

From the research conducted in the Department of Elementary Education, a number of practical applications have arisen out of the concept of conservation. There is the need for teachers to be aware of the child's status with respect to conservation as it may be a clue to the cause of many difficulties. For example, teachers could be guilty of classifying a child as dull when in actual fact, he does not yet conserve the properties of the subjects that are being examined. This suggests that there may be a close relationship between achievement in mathematics and conservation.

Reimer (University of Alberta, 1968) gave a test of conservation of number, quantity, and length to a group of 81 grade 1 subjects. He found that both intelligence and conservation were closely related to achievement on the *Seeing Through Arithmetic* test; furthermore, achievement on this test was highly predictable for conservers—that is, conservers generally had high achievement scores. However, the achievement of non-conservers could not be predicted on the basis of their conservation scores.

Another implication of the principle of conservation is that it may be affected by experiences unique to certain children. Liedtke (University of Alberta, 1968) studied the conservation of distance and length in a sample of 50 bilingual and 50 monolingual Grade I students. A bilingual student was defined as a student who could converse as well in French as in English. Liedtke found that the concept of conservation of length was developed to a greater extent in the bilinguals, who also demonstrated a greater ability to make linear measurements. Perhaps there exist other unique experiences in our cultural setting that may affect the ability of young children to converse.

A practical application arising out of Pelletier's (1966) study of grade 1 children's concepts of linear measure is that the ability to verbalize measurement terms or identify measuring instruments is no assurance that the child understands the underlying concepts of linear measures. The implication is that before we teach a child to use a ruler to mark off inches, we need to be sure he understands the underlying principles such as transitivity, segmented length, and subdivisions.

A somewhat similar implication results from Brace's (1963) study of the preschool child's concept of number. He found that young children tend to confuse spatial relationships with number. Therefore, they must be given experiences with, and manipulation of, physical objects so that they can abstract the true concept of number apart from the physical properties of the objects. A variety of objects needs to be used in developing the concept of number so that number will not get confused with a particular shape, color, or size. This may point to one major weakness in teaching devices such as Cuisenaire rods, at least in the early grades.

Another study done in mathematics education in this department that does not fit the Piagetian framework deserves mention. Seward (University of Alberta, 1966) attempted to determine if exposure to the mathematical concept of ratio in grades 5, 6, and 7 affected students' success with tasks involving verbal analogies. Seward formed five groups on the basis of the amount of exposure to ratio instruction. The groups were:

1. ratio instruction in grades 5, 6, and 7
2. ratio instruction in grades 6 and 7
3. ratio instruction in grades 5 and 6
4. ratio instruction in grade 7 only
5. no ratio instruction

Each group was subdivided into three groups on the basis of intelligence, making a total of 15 groups. The only significant difference in achievement on a verbal analogies test was within the five groups of average intelligence. Using a Newman-Keuls comparison of ordered means, Seward found that the group who had ratio instruction in grades 5, 6, and 7, and the group who had it in grade 7 only did significantly better than the group who had no formal ratio instruction. A non-statistical examination of the data enabled Seward to conclude that the study of ratio enables students of average intelligence to perform almost as well as high-intelligence students on a verbal analogies test. In contrast, the average student who had no ratio experience did little better than the student of lower intelligence.

REFERENCES

Almy, M., with E. Chittenden & Miller (1966). *Young children's thinking.* New York, NY: Teachers College Press.

Baldwin, A. L. (1967). *Theories of child development.* New York, NY: John Wiley and Sons.

Brace, A. T. (1963). *The pre-school child's concept of number.* Unpublished master's thesis, University of Alberta, Edmonton.

Liedtke, W. W. (1968). *Linear measurement concepts of bilingual and monolingual children.* Unpublished master's thesis, University of Alberta, Edmonton.

Newman, W. O. (1967). *Children's understanding of time duration.* Unpublished master's thesis, University of Alberta, Edmonton.

Pelletier, J. D. (1966). *A study of grade one children's concept of linear measurement.* Unpublished master's thesis, University of Alberta, Edmonton.

Piaget, J. (1952). *The child's conception of number.* London, UK: Routledge and Kegan Paul.

Piaget, J. (1955). The development of time concepts in the child. In P. H. Hook & J. Zubin (Eds.), *Psychopathology of childhood* (pp. 34–44). New York, NY: Grune & Stratton.

Piaget, J. (1957). *Logic and psychology.* New York, NY: Basic Books.

Piaget, J., & Inhelder, B. (1963). *The child's conception of space.* London, UK: Routledge and Kegan Paul.

Piaget, J., Inhelder, B., & Szeminska, A. (1960). *The child's conception of geometry* (E. A. Lunzer, Trans.). New York, NY: Basic Books.

Reimer, A. (1968). *A study of first grade mathematics achievement and conservation.* Unpublished master's thesis, University of Alberta, Edmonton.

Sawada, D. (1966). *Transformations and concept attainment: A study of length conservation in children.* Unpublished master's thesis, University of Alberta, Edmonton.

Scherer, R. (1968). *Manipulative materials in the teaching of problem solving.* Unpublished master's thesis, University of Alberta, Edmonton.

Seward, R. K. (1966). *Relationship of mathematical ratios to verbal analogies.* Unpublished master's thesis, University of Alberta, Edmonton.

Towler, J. O. (1965). *Spatial concepts of elementary school children.* Unpublished master's thesis, University of Alberta, Edmonton.

Towler, J. O. (1967). *Training effects and concept development: A study of the conservation of continuous quantity in children.* Unpublished doctoral thesis, University of Alberta, Edmonton.

This chapter was originally published as:

Falk, M. R. (1969). Using the overhead projector: Some random notes.
Mathematics Council of the Alberta Teachers' Association Newsletter,
8(2), 11–13.

CHAPTER 8

USING THE OVERHEAD PROJECTOR

Some Random Notes

Murray R. Falk
President of MCATA

I am more fortunate than many of you in that I have had almost exclusive use of an overhead projector for the past three years. The following are a few random suggestions on the use of this device which I have found helpful and wish to pass on. Whenever possible, keep the projector and a screen in your room. You will find that you can then use it for a minute or two when the need arises, just as quickly put it away, and not waste precious time hunting for the machine. There are many occasions, particularly as you gain familiarity with the machine, on which you will not anticipate its use beforehand but you may find it best to handle a particular difficulty encountered by a student or the whole class.

I make all of my transparencies from India ink drawings on a 3M heat copier. I have found that the transparencies thus obtained are much clearer than those drawn in pencil. Therefore, the extra effort and care required are worthwhile. Furthermore, some of my transparencies are of such nature

Selected Writings from the Journal of the Mathematics Council of the Alberta Teachers' Association, pages 43–45
Copyright © 2014 by Information Age Publishing

that they can form a permanent part of students' notes. The 3M heat copier will also make fine ditto masters from the ink drawings.

Whenever color is required, it can be added by using gummed transparent material, also available from 3M. It is usually applied to the back of the transparency. Colored lines and handwriting are most easily made with one of the many kinds of fiberglass tipped pens on the market. Some experimentation is necessary to obtain those with a sufficiently fine tip and ink that adheres to the transparency or celluloid roll, yet may be removed easily with a damp sponge. Once you have acquired a suitable pen or set of pens, do not use them on paper, or the point will rapidly become too blunt.

If available, a primary typewriter or "Leroy" lettering set can be used to speed the production of professional-looking lettering on the masters. "Letraset" also survives the heat copying process and produces a fine result.

Some transparencies are best left unmounted. I refer especially to various types of graph grids (NXN, DXD, and others) which are most useful if taped to the projector under the transparent roll. Thus they may be used as often as required. Where the same graphs are to be used over again several times (in the same lesson which is being taught to different classes), I always put a small dot at the origin so that later I can get the graph in perfect register with the axes. With each type of axes, I also make ditto masters from the original drawing and give copies to the students on which they are too. This seems to be less confusing for them than commercial graph paper.

The figures that follow are sketches of the axes that I use most frequently.

Geometric constructions are effectively taught with the overhead projector. For this, ordinary transparent protractors, rulers, and triangles are suitable. A pair of compasses must be adapted somehow to hold the pens, which are somewhat larger than a pencil. Some adaptation in your technique must be made to hold the compasses so that your hand does not obscure what is being drawn. This is easily done by holding them at an angle of about 450 rather than vertically, as is the usual procedure. Therefore, the celluloid roll should be the heaviest possible in order to avoid tearing; also, the point of the compasses must be very sharp. I usually have the students make the same construction at their seats and watch them. By keeping my eye on those near me, I adjust the pace of the lesson so that they all can follow.

You can make or purchase a transparent slide rule. I have found this very helpful in teaching the use of the slide rule to a group of students who come early one morning each week for the instruction.

If you wish to review a mass of material—for example, the properties of the rational number system or of the ordered field of real numbers—you can often do this most efficiently with the overhead. In this case I have prepared a transparency with the desired material arranged in two columns and the corresponding items matched. One of the columns can then be exposed entirely and the other revealed item by item by sliding a piece of

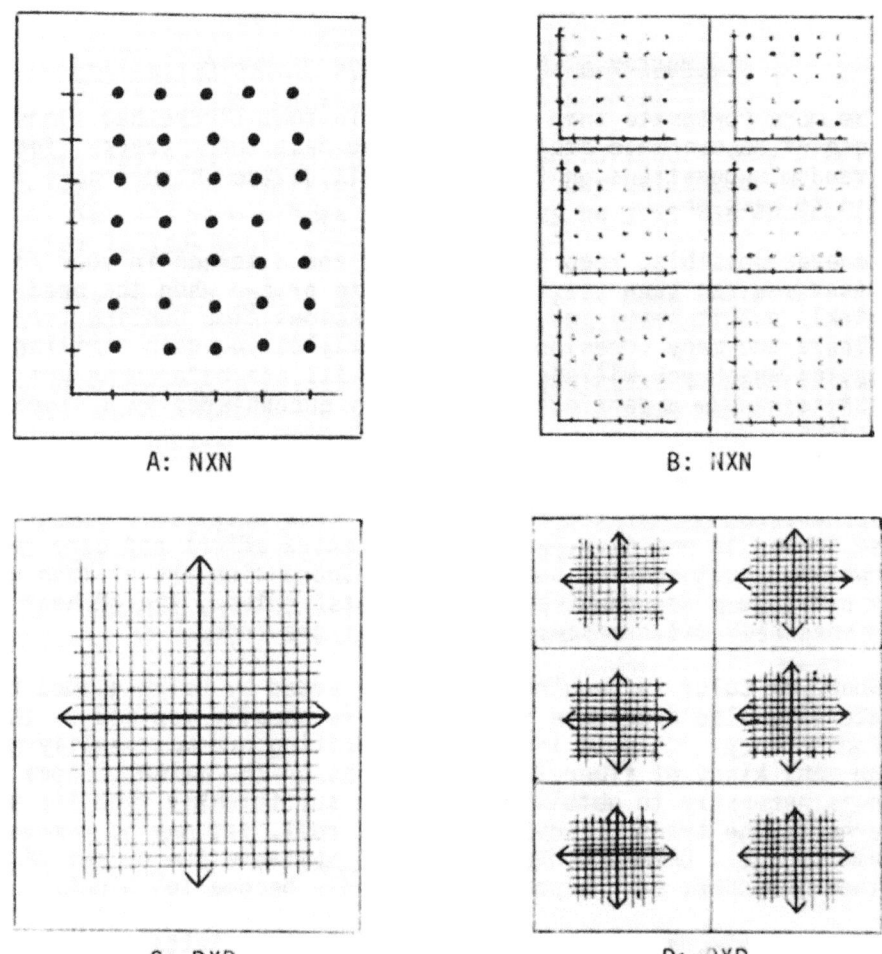

A: NXN

B: NXN

C: DXD

D: DXD

white paper down the stage. This has the added advantage that the teacher can see through the paper to anticipate the next item. A similar technique is useful in reviewing a list of formulas.

At times the occasion arises when you want to show a student's work to the whole class. If your school's heat copier is not too far from you, a transparency can be made of the student's work in a matter of about half a minute. However, in this case you must insist that students do their work in pencil. Ordinary ink or ball-point will not copy on a heat copier.

My suggestions have been of general nature. I hope that you will be able to adapt some of them to your own teaching methods.

CHAPTER 9

SOME NEW MATH IN OLD RUTS

H. L. Larson

Editor's Note: Mr. Larson is assistant superintendent of schools in Red Deer.

One of the pedagogical dangers we face in promoting new methods of teaching mathematics is that of falling into narrower and deeper ruts than our predecessors did. The temptation to immortalize a particular method that appears at first to be "the" means to an end is indeed very great. In so doing we may lose sight of many implications of the new method as well as shut out other refreshing ideas for handling the same problem.

A good example is the so-called rate-ratio method of handling percentage problems. One of the authorized texts for Alberta has accented this method to such an extent that many teachers and students alike have almost made a fetish of it. Mentioned below are some of the implications:

1. Whenever percentage problems appear, especially in the physical sciences, students begin to set up a proportional equation—almost without thinking. Unfortunately some problems do not lend themselves easily to this form of equation, and many students guess rather wildly the position of the "Unknown" symbol.
2. The computational bogey of "cross multiply" is still very much with us, despite certain refinements (Comparative or Trichotomy laws of

Selected Writings from the Journal of the Mathematics Council of the Alberta Teachers' Association, pages 47–49 order in Rational Numbers), which only a few teachers seem to be able to get across to their classes. We still find many students "cross-multiplying" in algorithms of binary operation as well as proportional equations.

3. It would appear that many students also are so well "trained" in solving proportional equations that when they meet other types for the first time, they have no method of attack at all—for example, the d = rt, V = lwh or I = prt types. Teachers of science are especially critical of students who have been steeped in this rather singular modus operandi.

To briefly conclude what could be a rather long article, may I suggest a comparison between the rate-ratio-proportional method and another method in handling one of the more difficult of the three types of percentage problems.

Problem: If I still have $50 left after having started with $200, what % did I save?

Rate-Ratio Method. We may reduce this to the form of $50 = $n% of $200.

Thence

$$\frac{50}{200} = \frac{n}{100}.$$

Or we may symbolize the problem directly by thinking "50 per 200 is equal to n per 100." The computation to close this sentence is as follows:

$$\frac{50}{200} = \frac{n}{100}$$
$$200n = 5000$$
$$n = \frac{5000}{200} = 25$$

However simple this may seem, we find many students getting the "answer" $n = 25\%$. This seems ample proof that the total operation has been mechanical. If the teacher insists that the student interpret the answer in the light of the original problem, one hears all kinds of agonizing sounds coming from students. In fact, some teachers do not seem to realize that the "answer" $n = 25\%$ is indeed very wrong.

Another method.

$$50 = n\% \text{ of } \$200$$

$$50 = n \cdot \frac{1}{100} \cdot 200 \qquad \% \text{ is a special fraction } \frac{1}{100}$$

$$50 = n \cdot \left(\frac{1}{100} \cdot 200\right) \qquad \text{Assoc of mult.}$$

$$\quad\ \text{P} \quad\ \ \text{F} \quad\ \text{F} \qquad\qquad\ \text{Multiply but don't reduce yet}$$

$$50 = n \cdot \frac{200}{100} \qquad\qquad \text{Using inverse form equation}$$

$$\quad\ \text{P} \quad\quad\ \text{F} \quad\quad \text{F}$$

$$50 \div \frac{200}{100} = n$$

$$50 \times \frac{100}{200} = n \qquad\qquad \text{Inverse of division}$$

$$25 = n \qquad\qquad\qquad\ \text{Simplification}$$

Most of the steps in the reasoning are shown here, but the student soon learns the shortcuts, based upon solid deductive logic. Note that the meaning of % is utilized in step 2; consequently, there is no repetition in the final result. Another advantage of this method is that only the basic field laws are used, and this is in keeping with the modern psychology of the "power of conceptual thinking."

However, the most important implication is that this method provides a modus operandi for a whole range of equations such as

$$V = lwh$$

$$150 = 10 \cdot w \cdot 3 \qquad \text{Given V, l and h. Find w}$$

$$150 = 10 \cdot 3 \cdot w \qquad \text{Comm. of Mult.}$$

$$\ \ \text{P} \quad\ \text{F} \quad \text{F}$$

$$150 = 30 \cdot w$$

$$\ \ \text{P} \quad\ \text{F} \quad \text{F}$$

$$150 \div 30 = w \qquad \text{Inverse form of equation}$$

$$\text{etc.}$$

May we conclude that while rate-ratio "method" of solving some problem situations is extremely useful, it can also be a narrow and even dangerous device if it becomes the only strategy.

This chapter was originally published as:

Cleveland, R. W. (1969). Reading in the field of mathematics. *Mathematics Council of the Alberta Teachers' Association Newsletter, 8*(2), 2–5.

CHAPTER 10

READING IN THE FIELD OF MATHEMATICS

R. W. Cleveland

Editor's Note: Ray Cleveland, an associate professor of curriculum and instruction at the University of Calgary, is well known as co-editor of *Seeing Through Mathematics.*

In one sense mathematics is a language. It has its own highly specialized symbols which can be arranged in order to communicate mathematical ideas. It is both an oral and a written language. It can be translated into English but at the same time introduces many new words into the English language. It is a science and at the same time the most indispensable tool of all other sciences. Its foundations rest on logic and philosophy. Its structure is just as aesthetic to the mathematician as the symphony is to the musician or the painting to the artist. It has an exciting history, starting with primitive man and continuing to the present. It is growing faster today than ever before. There are some aspects of language in mathematics that merit special attention. The following questions with their answers will provide this attention.

1. HOW DOES THE STUDENT LEARN THE LANGUAGE
OF MATHEMATICS?

The student has not added a new word to his mathematical vocabulary unless he understands the concept associated with the word. To introduce the word before the concept is developed causes frustration on the part of the student and inhibits learning. The task of the teacher is to arrange and implement student activities and experiences dealing with concrete objects and designed to have the student discover for himself a mathematical idea before the terminology is introduced. When a student discovers a new idea, he wants to tell others. Ensuing class discussion will give many children an opportunity to express their discoveries in their own words. Mathematical terminology then follows (not necessarily immediately) as a more precise and concise way to say what the students were trying to say. Note that at this stage the ideas are communicated orally. The written mathematical language follows for ideas already discovered and expressed orally, both in the natural language and the language of mathematics.

The above applies not only to individual words and special symbols but also to mathematical sentences expressed in English and in the special symbols of mathematics such as equations, inequalities, and statements of equivalence.

The next stage involves reading in textbooks or other sources about mathematical ideas for which the terminology has already been acquired. Eventually students will use these skills to represent mathematical ideas in written language in connection with problem solving, proofs, graphing, or even, perhaps at a later stage, creating their own mathematical systems. Some of these applications will involve translation from English to mathematical symbols and vice versa.

2. ARE CERTAIN ENGLISH WORDS USED
IN MATHEMATICS WITH MEANINGS THAT ARE TOTALLY
DIFFERENT FROM THE USUAL ENGLISH MEANINGS?

The answer is an emphatic "Yes!" This is a complicating factor in teaching the language of mathematics. It is helpful to contrast the usual English meaning with the mathematical meaning whenever possible. For example, in mathematics, the word irrational refers to a real number that cannot be expressed as the quotient of two integers, while the usual English meaning is related to unreasonable behavior. In mathematics, the word imaginary refers to a certain kind of complex number; the words group, ring, and field refer to certain types of mathematical structures. For the reasons implied above, a dictionary of the English language is not of much help in developing a mathematics vocabulary. In fact, definitions in mathematics undergo transformations from time to time as we gain new perspectives into mathematical ideas so that even a mathematics dictionary is not as reliable as one might hope.

3. IS THE STRUCTURAL ANALYSIS OF WORDS A HELP IN DEVELOPING A MATHEMATICS VOCABULARY?

There are many words in mathematics for which the answer to the above question is affirmative. To give a few examples: trigonometry can be analyzed from *trigon* (which means triangle) and *metry* (which pertains to measure). Of course, trigonometry today implies more than just triangle measure. It includes the study of periodic functions and many other ideas. Similarly, geometry was formed from words for earth and measure and thus originally meant earth measure.

4. ARE CERTAIN SYMBOLS OF MATHEMATICS USED AMBIGUOUSLY?

Very often different meanings are attached to the same symbol in mathematics, and it is also true that different symbols are often used to represent the same idea. In a sense this might be thought of as two-directional ambiguity. For example, "−" is used in at least three different ways: In "a − b" the dash indicates subtraction; in "−2" it shows that the number is less than zero (i.e., negative); and in "−x" it shows that you are referring to the additive inverse or opposite of whatever number x represents. For an example of two different symbols representing the same idea, "A" represents the complement of set A, and "A′" also represents this complement.

This two-way ambiguity is a real hazard in learning the language of mathematics. The situation is somewhat similar to the existence of synonyms in English. In fact, this analogy might be pointed out to students to alleviate the problem somewhat.

5. ARE THERE SPECIAL CONSIDERATIONS IN APPLYING MATHEMATICS TO PROBLEMS IN OTHER FIELDS SUCH AS SCIENCE AND BUSINESS?

This is very much so. To apply mathematics to other fields, it is necessary to (a) know enough about the other field to understand the problem, which would include knowing the language and also the technical aspects of the other field; (b) be able to select the information necessary to solve the problem and reject any superfluous information; (c) be able to translate the necessary information into a mathematical sentence such as an equation or inequality; (d) be able to use mathematical principles well enough to transform the mathematical sentence into a form simple enough to make the solution obvious; (e) be able to translate the solution back into the original problem situation to see if the mathematical solution is a practical solution.

6. ARE THERE ANY SPECIAL WAYS IN WHICH SEMANTICS AFFECTS THE LANGUAGE OF MATHEMATICS?

There are times in mathematics when it is important to distinguish between a symbol and the idea which is its referent. For example, one way to say what you mean by the sentence "a + b = c" is that "a + b" and "c" are two different names of the same number. It is also important to distinguish between a number system and a system of numeration. The first is a certain type of mathematical structure and is on the idea level, whereas the second is a system of symbols for representing numbers. For example, the system of Roman numerals and the system of Hindu-Arabic base ten numerals are systems of symbols for representing numbers, but the system of whole numbers and system of integers are examples of number systems.

7. IN WHAT WAYS DOES LOGIC AFFECT THE LANGUAGE OF MATHEMATICS?

The relationship of logic to mathematics is like that of parent to child, and indeed it is possible to derive all of mathematics from the axioms of logic.

In developing the language of secondary school mathematics, it is important that students have some knowledge of logic if they are going to get some feeling for mathematical systems and proof. For example, it is important that students know the logical meaning of connective such as "and," "or," "If . . . , then . . ." and "if and only if." They should also understand the idea of quantification in mathematical statements. For example, the statement "For all real numbers, x and y, it is true that $x + y = y + x$" is universally quantified because it tells us something about all pairs of numbers in the universe of real numbers. The statement "There exists a real number, x, such that $x + 2 = 5$" is existentially quantified, since it tells us that there exists at least one real number to which 2 may be added to form the sum 5. Certainly it is necessary to have some ideas of logic in order to use mathematical language in forming a proof.

8. WHAT NONVERBAL DEVICES ARE USED IN MATHEMATICS TO PRESENT OR CONVEY INFORMATION?

Graphs, charts, tables, sets of numerical data, manipulative devices such as slide rules or abaci, computers, diagrams, figures, and special symbols of many kinds are used.

9. WHAT LANGUAGE IMPLICATIONS ARE INVOLVED IN CONSTRUCTING AND PRESENTING A FORMAL MATHEMATICAL SYSTEM?

- Identifying certain terms as undefined.
- Using the undefined terms to form definitions.
- Using defined words to form other definitions.
- Accepting certain statements in the system without proof. These statements are called axioms.
- Using the definitions and axioms along with the rules of some system of logic to form new statements called theorems.
- Using theorems and the rules of logic to form other theorems.
- Investigating such properties of a set of axioms as independence, consistency, completeness and categoricalness.

10. WHAT ARE SOME REFERENCES ON READING IN THE FIELD OF MATHEMATICS?

Brune, Irvin H., *Language of Mathematics*, Twenty-first Yearbook, National Council of Teachers of Mathematics.
Johnson, Donovan A., et al., *Guidelines for Teaching Mathematics*, pp. 140–143. Belmont, CA: Wadsworth.

COMMENTARY

MATHEMATICS TEACHING AND LEARNING IN THE 1960s AS REFLECTED IN *delta-K*

Tom Kieren
University of Alberta

At the time I was writing this commentary, the August 2012 issue of *Scientific American* came out. Among black holes, ancient DNA, the neuroscience of joy, and gamma rays from clouds was an article entitled "Building a Better Science Teacher" (Wingert, 2012), which talks about mathematics teachers, as well.

While it is not my purpose here to review that article, I would like to note that it makes extensive reference to well-funded, large-scale projects, primarily sited in universities. These projects aim to provide better education for teachers in order to give them the rich background necessary for effective mathematics and science instruction, with special emphasis on subject matter education. President Obama has even pledged to train 100,000 highly effective mathematics and science teachers by 2020. Acronyms and catchy program names abound. Better science, technology, engineering, and mathematics (STEM) teachers and programs are needed. Projects like UTeach, UKanTeach, and Teach for America are aiming to educate

Selected Writings from the Journal of the Mathematics Council of the Alberta Teachers' Association, pages 57–73
Copyright © 2014 by Information Age Publishing

teachers for STEM at the high level anticipated, as well as researching the effects of such programs. Deborah Ball's work on the Learning Mathematics for Teaching project (University of Michigan) and Paul Cobb and his team's work at Vanderbilt University are highlighted in the article.

As the article suggests, "Not since the Russians launched Sputnik in 1957 have American policy makers, educators and businesses been so focused on improving math and science education" (p. 62).

NEW MATH AND THE 1960s

As the Mathematics Council of the Alberta Teachers' Association (MCATA) celebrates its 50th anniversary, it may seem like we're in the 1960s again. The 1960s were a time of change in political activism; in social goals and mores; in reactions to war; in music; and, indeed, in mathematics curriculum, learning, and teaching (not to mention Tom Lehrer's "New Math" song).

As was noted in several *delta-K* articles of the time, the sixties saw large-scale projects in redeveloping and rethinking school mathematics and its teaching (see the articles by Krider and by Coulson, for example). Examples include the School Mathematics Study Group (SMSG), based at Stanford University; the University of Illinois Committee on School Mathematics (UICSM); and the bellwether report of the U.S. National Commission on Mathematics in the late 1950s. These projects—led by teams of mathematicians, educators, and teachers (particularly the SMSG)—produced curricular materials that were widely tested and researched in the United States.

And, of course, Albertans were influenced by such work. Doyal Nelson, a long-time teacher and researcher in the Department of Elementary Education, University of Alberta, did his PhD at the University of Minnesota, and his thesis was part of the effort to research the effects of SMSG curricula in schools. Many other educators (in both universities and schools) came from or had done work in the United States, which meant that many in Alberta were able to both support and evaluate ideas from such projects.

Mathematics education in Alberta was definitely part of the new mathematics phenomenon of the 1960s. But this phenomenon did not simply reflect the various large-scale curriculum projects or commercial mathematics curriculum materials. In fact, new mathematics had many facets, some of which I will discuss here.

The first of these facets (and perhaps the most talked about) reflected ideas from larger centers and projects. These ideas included an emphasis on mathematical structures and on properties of operations and of number systems themselves (natural, integer, rational, real, complex), especially when applied to computations with numbers or with algebraic expressions. The idea was that students would not simply memorize algorithms but, rather, would understand and use valid properties to solve problems and set up and work through computations, as well as being able to explain, justify, or prove the validity of their actions—hallmarks of mathematical understanding. These projects also emphasized the careful use of mathematical language, as well as looking at numerical and algebraic operations and their characterization in new ways. The emphasis at the high school level was on the elementary functions—perhaps most vividly seen in the treatment of trigonometry from a circular functional rather than only a geometric point of view. While the emphasis in these new math curricula was on properties and structures, many of the newer approaches, especially in elementary functions, could be seen as facilitating later study of calculus and linear algebra and their applications in physics, engineering, economics, and statistics.

Of course, there were many projects in new math beyond those mentioned here. All of these projects focused on big ideas in mathematics and were likely seen as less "technique-al" by practicing teachers. The emphasis was on students seeing and using patterns or relationships when engaging in mathematical activities. Another emphasis was student discovery, guided by the teacher or done independently. The 15-year Madison Project, initially directed by Robert Davis out of Syracuse University, emphasized problem solving and the discovery of patterns of operating both algebraically and numerically. Henry Van Engen and Glenadine Gibb, at the University of Northern Iowa (and later at the University of Wisconsin and the University of Texas), developed and researched pedagogies focused on alternative algorithms for operating on numbers, such as the successive subtraction algorithm for division, and alternative approaches to proportions.

These ideas were picked up and used in Alberta, which was an early venue for the use of Seeing Through Arithmetic (STA) and Seeing Through Mathematics (STM)—programs that represented an early commercialization of the ideas of new math. One author in this volume, Ray Cleveland, was closely associated with STA and, especially, the STM project. Sid Lindstad, of the University of Calgary, did early work on such developments, both in Alberta and in collaboration with U.S. colleagues. The influence of Davis and the Madison Project found its way into teacher education activities at the University of Alberta through Sol Sigurdson's connections with Davis with respect to discovery learning. In the late 1960s, the doctoral dissertation of Sigurdson's student Sandy Dawson (1969) looked at the Madison Project's teaching and learning ideas as a fallibilistic approach to mathematical reasoning and thinking. (Fallibilism was a contemporary mathematical idea from the work of Imre Lakatos, a student of the famous philosopher of science Karl Popper.)

While the term *problem solving* has been used above, another new mathematics set of ideas became available to teachers through the work of mathematician George Polya, of Stanford University, in his remarkable books of the 1950s: *Induction and Analogy in Mathematics* (1954) and *How to Solve It: A New Aspect of Mathematical Method* (1957). In these works, Polya focused on what he saw as the living edge of mathematics—plausible reasoning—which complements the more formal, demonstrative part of mathematical reasoning (best seen in the form of deductive proofs). Included here is McCall's supportive article "Polyan Mathematics"—evidence of McCall's point that this aspect of new mathematics needed to occupy a larger place in the Alberta curriculum and examination programs.

This discussion might lead one to think that new math was a U.S. phenomenon and that its influences came into Alberta school mathematics mainly from the south. Even the *Scientific American* article seems to suggest that new math, both then and now, has been a U.S. reaction to world scientific or economic events. But the 1950s and 1960s were a period rich in ideas relevant to mathematics curriculum and teaching in many places in the world.

The Soviet Union (and countries where it had influence) enlisted the working support of prominent mathematicians and educators in developing school and teacher education materials in mathematics. The famous French Bourbaki society of eminent mathematicians, who were great advocates of formal and rigorous approaches to mathematical thinking, had significant influence on French school mathematics. One need only remember the work of Jean Dieudonné, who pushed for the replacement of classical demonstrative Euclidean geometry with more algebraic approaches to the study of geometry in schools. Polya was originally from Hungary, where there was a great emphasis on problem solving, especially extracurricular contests of problem solving (which took place in many countries at this time).

Many of the new mathematics ideas that gained currency and respectability in Alberta in the 1960s had some basis in the UK. For example, mathematics education pioneer Richard Skemp was strong in his support for students actively finding meaning in mathematics, especially connecting mathematical ideas to experiences, and he developed curricular materials around this approach.

At the same time, mathematics educator Zoltan Dienes (who was born in Hungary but moved to England as a teenager) was doing his creative work in developing materials and related games whereby children could learn mathematics through experiencing the mathematical properties of number, logic, or algebra. These included Dienes blocks (logic and sets), multibase arithmetic blocks (numeration systems), and various play-like settings for invoking algebraic thinking (particularly groups). He also created activities in which children and even secondary students could be observed to be doing mathematics across several levels of abstraction, as well as experiencing the same mathematical idea sets using different physical representations. Dienes was an originator of the idea that students think in structures as they play games or work with materials, thinking that was generalized by students through the use of informal and, later, more formal writing activities in mathematics. Dienes later lived and worked in Canada, and I met him when he worked with the Minnesota Mathematics and Science Teaching (MINNEMAST) Project, with Paul Rosenbloom at the University of Minnesota in the 1960s.

Finally, Caleb Gattegno, working in England, brought to the fore the uses of purpose-designed materials (e.g., Cuisenaire rods), which enabled children and secondary students to develop their own useful ideas and patterns for working with numbers in a way that generated general patterns.

Of course, most of us who were in Alberta in the 1960s remember the use of various new math materials in preservice and in-service teacher education but also in many classrooms. Cuisenaire approaches are mentioned in the Cathcart and Liedtke article in this book, and Doyal Nelson (with Floyd Robinson) produced an extensive research report on the use of this approach and materials, which was rather widespread in Canadian schools.

Bruce Harrison, now a professor emeritus at the University of Calgary, is perhaps the leading expert on Skemp and the implications of his work for mathematics knowing and curricula in schools. Harrison's long-term relationship with Skemp began with his rigorous research, for his doctoral thesis (Harrison, 1967), on Skempian ideas related to reflective thinking and mathematical knowing in students.

Dienes's materials, and to an extent his methodologies of structural thinking, could be seen in preservice and in-service classes for teachers as providing concrete approaches to the logic and properties of mathematics in working with children.

Thus, it is clear that new mathematics ideas, of a very different character from the large-scale U.S. projects, came to Alberta from many international sources and in many forms, and they were a rich source to be taken up and, in many ways, transformed for and through their use here.

The name Richard Skemp brings to mind another source of new ideas for the learning of mathematics in the 1960s. Skemp himself would have said that his work was based on ideas drawn from the epistemological psychology of Jean Piaget. The work of Piaget and various colleagues involved studies of children working in settings related explicitly or implicitly to children's conceptions of number, space, and geometry, and Piagetian theory emphasizes the importance of "groupement" structures from mathematics itself for thinking about and exploring the mathematical actions of children. Since the article by Cathcart and Liedtke deals with these ideas, through its review of the impressively large number of Piagetian studies done at the University of Alberta in the mid-1960s, there is no need to elaborate on these ideas here. The very presence of this collection reveals the impact of Piaget on mathematics education in Alberta. But I would be remiss if I did not give credit to Doyal Nelson for making Piagetian ideas salient for the many graduate students who worked with him, and also for prompting the use of Piagetian ideas related to mathematics knowing in courses for preservice and in-service teachers, particularly at the elementary level.

The reference to Piagetian ideas relating to mathematics knowing should remind us that the new math of the 1960s didn't suddenly arise in the late 1950s. MCATA has always maintained an association with a larger body, the U.S.-based National Council of Teachers of Mathematics (NCTM). As one of its resources for teachers, NCTM has a long history of producing year-books using ideas from experts on various mathematical topics. Many of the grounding ideas of school mathematics of the 1960s (at least for North America) can be found in the NCTM yearbooks *The Learning of Mathematics: Its Theory and Practice* (NCTM, 1953) and *Insights into Modern Mathematics* (NCTM, 1957). The 1953 yearbook introduces mathematics educators to the roles a Piagetian constructive view of cognition might play in mathematics knowing and in the provision for such knowing. It also points to the ideas of meaning, understanding, and the role of learning mathematics in a variety of ways and at a variety of intellectual, conceptual, and maturity levels. The 1957 yearbook provides sketches of the main new ideas (e.g., functions from the set of ordered pairs point of view) and approaches to mathematics that were to become the curricular currency of the 1960s.

Once again, as noted in Galeski's article, the ideas of so-called modern or new mathematics may have had their origins in the work of mathematicians as far back as the 19th century. Thus, the mathematical ideas and approaches of modern mathematics were not totally new (one need only look at Norman Miller and Robert Rourke's Alberta-adopted *Mathematics for Canadians* text in the late 1930s to see this), but in the 1960s, school-level students were brought into regular contact with such ideas and, through rich problem-solving activities and structured learning settings, were given opportunities to take on and take in these ideas in a variety of ways. At their best, these approaches to new math align with Jerome Bruner's contention, in his famous 1960 book *The Process of Education,* that school-aged children can profitably connect with any mathematical idea if it is presented in such a way that they can engage with it using the capabilities they have.

Another facet of new math (and education in general) that was just entering the scene was the use of computers in teaching. These uses were at that time experimental but diverse. Extensive computer-based drill-and-practice schemes were developed at Stanford that allowed students to work on their own through an ordered series of computational practice sets in which criteria-referenced decision algorithms both checked their work and provided appropriate next-step activities for them. This work simultaneously focused on researching the outcomes of such instruction. But work at the University of Pittsburgh, the University of Illinois, and Pennsylvania

State University (among others) aimed to provide more elaborate computer-guided instruction involving teaching new mathematics ideas primarily in urban centres.

While not focused on mathematics education, computer-assisted instruction development and research at a world-class level of innovation occurred at the University of Alberta, under the direction of Steve Hunka. Hunka also brought computer-assisted instruction ideas from the University of Illinois that were both mathematically and technically unique.

Of course, another child of the 1960s was the Logo programming language, developed by Seymour Papert and Wally Feurzeig. Papert related Logo's development and use to the ideas of Piaget, and it allowed schoolchildren to construct procedures in real time, many of which were related to and extended the mathematical ideas of geometry. This work was at least illustrative of constructivist approaches to new geometric ideas, which Papert himself brought, at a demonstration level, to Alberta at the invitation of Hunka.

Another programming language created explicitly for use by learners was BASIC, developed by John Kemeny at Harvard. An early use of BASIC in mathematics teaching and related curriculum development was the Computer Assisted Mathematics Project (CAMP) at the University of Minnesota, which allowed secondary school students to write programs related to a wide variety of mathematical topics (such as properties of systems, prime numbers, equation solving, linear systems, and sequences and series). I came to the University of Alberta in 1967 having worked on CAMP, and in my thesis I studied the use of computer programming as a tool in grade 11 mathematics, one of the first theses studying the role of such use of computer programming in enhancing mathematics learning.

It is evident that mathematics education in the 1960s was, like the era itself, swinging with new ideas. These ideas were related to providing curriculum and methodologies for teaching big mathematical ideas in new ways (e.g., discovery learning). There was a renewed but different focus on problem solving, from Polya's pointing to the complementary values of demonstrative rigour and plausible reasoning in mathematics. There was a psychological facet of new mathematics, illustrated by European influences, such as the role of concrete experience in learning mathematics in a meaningful way through reflective thinking (Skemp) and thinking in structures with mathematically inspired materials (Dienes). There was the new influence of the constructivist thinking of Piaget and its relationship to the development of mathematical thinking. And just entering the purview of mathematics educators were the various uses of the computer in teaching, which brought new possibilities for the mathematics curriculum as well as different ways of both learning and thinking about mathematical ideas.

ARTICLES FROM DELTA-K AND THE MATHEMATICAL
EDUCATION ETHOS OF THE 1960s

The articles in this volume speak well for themselves, so there is no need to review them in detail. Instead, I will try to sketch an answer to the following question: In what ways do these articles reflect the issues and facets of new mathematics in the 1960s?

Before turning to the articles themselves, I wish to assert that the articles selected for inclusion cover a wide variety of salient 1960s issues and speak to many of the facets discussed above. Thus, the Mathematics Council of the day can be seen as providing information to teachers that related to the concerns of the day. Now let me turn to a more specific analysis of how these articles address the mathematics education concerns of the 1960s.

I want to first consider McCall's article "Polyan Mathematics" and the first part of Coulson's article, "Discovery or Programming." McCall includes an extensive quotation from Polya in which he develops plausible reasoning strategies, including what he terms "guessing" and, especially, inductive thinking (not to be confused with mathematical induction), as a complement to or completion for traditional demonstrative, proof-oriented reasoning in mathematics. McCall then provides an elaborate example of finding the three dimensions of a rectangular solid of a given volume such that the surface area is a minimum. He starts by using the Polyan plausible reasoning strategy of solving a related two-dimensional problem: finding the dimensions of a rectangle of a given area with a minimal perimeter. Given his emphasis on plausible reasoning, I was surprised that he started with a general rectangle of sides x and y, giving $p = 2x + 2y$. Without discussing the nature of his plausible strategy, McCall then makes an educated guess that the solution is a square. He then does the algebraic solution to finding x and y, minimizing the difference between the area of the square of side $(x + y)/2$ and the area of the rectangle xy—in other words, his means of exploration of this minimization (when the sides are equal or $x = y$ is what would pass for a demonstrative proof). The reader is left with the question: Would there be less formal and more intuitive ways of reaching this solution more consonant with the idea of plausible reasoning, or could such a problem be solved by a grade 6 student without algebraic generalization? In his extension from three to two dimensions, McCall does provide an example of some nice plausible reasoning, but the means of his test of the plausible conclusion—of a cube as the solution—involves at least generalized if not formal reasoning. McCall has well illustrated the bi-play of plausible and demonstrative reasoning. He argues for the inclusion in schools of Polyan mathematics centered on problem solving—"mathematics for the scientific world"—and argues that the new math in Alberta schools at the time was "mathematics for a philosopher."

Coulson, on the other hand, values the contributions of programs such as SMSG and UICSM. He values mathematics as a subject worthy of study on its own terms, because its structure and relationships can be explored and discovered by students, rather than as a collection of facts, procedures, and techniques to be learned by rote. A big idea of his (which I think Mc-Call would agree with) is that students should be given credit for and opportunities to exhibit the ability to think mathematically for themselves. For this to happen, expert teaching is needed, both in terms of knowledge of mathematics and knowledge of how students might discover the patterns and structures of mathematics. (Does this not sound like the teacher knowledge aimed at in the *Scientific American* article of 2012?) Coulson sees the following as the observable fabric of student mathematical thinking: an observation in a situation, a proof of the validity of the observation, new observations based on previous ones, another proof. Surprisingly, this fabric is not that different in pattern from the one discussed by McCall, but it too raises questions about the nature, the role, and the level of plausible reasoning by students.

In the 1960s (and even today), these two articles forced teachers to think about how they might foster and observe plausible reasoning in action at any level. Both authors prompt us to ask just how teachers might provide the setting in which students can evaluate their own mathematical actions, perhaps through explaining to peers or other methods less formal than the word proof.

Coulson also contrasts what he sees as the pattern of a discovery method with programmed learning. He sees the former as opening the path for student thinking, while the latter necessarily limits students to moving through a series of small steps and, thus, confines their thinking. Programmed learning can be seen in the work of Robert Gagné in the 1960s, which posited a carefully ordered sequence of behaviourally stated objectives describing a task, an acceptable student behaviour, and a criterion-referenced method for observing such behaviour. A question pertinent to the time but not raised by Coulson is whether a program must follow such a rigid pattern. Writing in 1965, Coulson could not have anticipated the work on using computers in mathematics. The drill-and-practice programs of the mid-1960s might have been subject to his critique, but other more complex and open computer-based individualized instructional programs in mathematics would have challenged his comparisons. In other words, Coulson's contrast of discovery-oriented mathematics instruction with programmed instruction is limited by his rather narrow view of the latter and because he could not have anticipated the technology that would, over time, change the nature of both discovery in mathematics classrooms and individualized, computer-based instruction.

Murray Falk's article does focus on a technology relatively new at the time—the overhead projector. While the means of creating projections now seem crude, Falk shows just how they might be used, especially to allow students individually and collectively to think about mathematical ideas portrayed in a visual form. This technology enhanced the study of modern mathematics concepts such as sets and operations on them, and especially thinking of functions and relations as sets of ordered pairs and the structural nature of and applications of functions of various kinds. But what struck me while reading the article was the implicit idea—not new to the 1960s but, rather, eternally relevant to mathematics teaching—that preparation is key to effective instruction, even in taking advantage of sharing student ideas that arise serendipitously.

As noted above, McCall and Coulson present contrasting views of the impact of new mathematics in Alberta schools in the 1960s. While both offer valid and well-developed views, the nature of modern mathematics was at the time elusive. This is well illustrated by Galeski's article "What Is Modern Mathematics?" The article shows the difficulty of coming up with criteria that would be useful in choosing a mathematics curriculum for the 1960s. Galeski's exploration of the nature of mathematics yields discussion of the study of form in a general sense; of formal, necessary, deductive reasoning; and of Bertrand Russell and Ludwig Wittgenstein's characterization of mathematics as a form of language game. While I was taken with Kemeny's notion that "stress should be laid on thinking mathematically and more value attached to that ability than to knowledge [of particular facts]," even this left open the meaning of such thinking, as well as the nature and possible form of plausible and demonstrative reasoning and even of the nature of "modern mathematics."

Runclius's article is a brief response to a question raised by teachers at a talk he gave on the future of school geometry. In it, he presents one view of the nature of modern mathematics in more specific terms. He provides historical criteria for teaching geometry and then describes the features and emphases of a new geometry in high school. While the new geometry would still cover important facts drawn from congruency, properties of triangles, basic constructions, and so on, the key objectives would be appreciation of the postulational structure, appreciation of different geometries with other such structures, use of algebraic and analytic methods, and appreciation of the nature of demonstrative reasoning involved in proof. To me, this does not differ much from the "old geometry." Of course, in 1961, Runclius could hardly have imagined the geometries of chaos theory, or the art and mathematics of the Mandelbrot set, or geometric solutions to complex physical and statistical problems and models fostered through the use of computers. But, as Galeski might have remarked, Runclius's ideas of the modern in geometry were more 18th century than 20th. Since Donald Coxeter—a

Canadian who was considered "the man who saved geometry"—ideas such as various kinds of transformations, groups of reflections, and dimensionality and complex problems amenable to plausible reasoning at many levels of abstraction have been made available to students, providing them with a more living view of geometry. I am not trying to be critical of Runclius's off-the-cuff remarks but, rather, to point out that many times views of modern mathematics are spelled out in cautious terms.

This latter point of view relates to Krider's article. Krider points out what he sees as the cycles in school mathematics over the first 60 years of the 20th century. He views this on the basis of the psychological idea of transfer, seeing mathematics as a discipline being replaced by more specifically oriented pragmatic mathematics, and so on. This cycle was accompanied, in his view, by varying levels of conflict between mathematicians and educators in the 1950s and '60s, seeing a return of a contemporary positive view of general transfer arising from appropriate learning of the discipline (with attention also paid to citizenship), as well as Bruner's ideas on subject matter learning, which he sees as being accompanied by a wide split between mathematicians and educationalists on what should be taught in schools and how. Krider sees the curricular product of a mathematician-led project (such as SMSG) as dominating mathematics in schools, returning the emphasis to the subject matter in a way many perceive as radical. But in a call by a large group of North American mathematicians to return to a balance between content and pedagogy and a turn from mathematics that may only appeal to those who like what modern-day mathematicians like (demonstrative reason driven by more "pure" mathematics), Krider in 1963 saw the potential for turning away from what was then current.

The article by H. L. Larson in the later 1960s shows one aspect of what might be thought of as a middle-ground reaction to particular aspects of the modern mathematics curriculum. The article focuses on alternative approaches to setting up and solving proportional problems, particularly percentage problems. Larson suggests that the rate-ratio method in the new mathematics curricula favoured in Alberta was problematic for students and teachers alike, leading to the "computational bogey of cross multiply," where this technique was used blindly and was inappropriately generalized to other settings (e.g., adding fractional numbers). He proposes a return to what I remember from my school mathematics in the 1950s as an older percentage problem representational strategy in which variations of the equation $a = n\%$ of b would be matched with the various "cases" of percentage tasks. But Larson argues that the statement "50 is what percentage of 200?" leads to the equation $50 = n(1/100) \times 200$, which is nicely solvable using the properties of a field (new mathematics) and, further, is generalizable to volume problems like $V = l \times w \times h$ when three of the values are known based on "solid deductive logic" (again, new mathematics). Some reading

Larson in the 1960s might have said that he was favouring "old-tyme" computational cases, but in my reading, his thinking shows how explicit use of, for example, field laws allows students and teachers to look at computations in a non-rote, generalizable manner. And, of course, neither Larson nor the curricular proponents of the other general rate-ratio approach had the language to support or critique that which flowed from both mathematical and psychological work in the 1970s on multiplicative structures, which prompted students to think of $a{:}b = n{:}100$ in terms of using preservation of multiplicative structure relations either with respect to a to b and n to 100, or a to n and b to 100 in their solution, either of which leads to rather different, more intuitive mathematical operations.

Of course, Krider's view of the influence of new mathematics is quite different from that of Coulson. This raises the question of how well modern mathematics ideas (such as sets and number systems) were being learned and by whom. Chell and Coulson take on this question in their article "Can Students Learn Some of the Ideas of Modern Mathematics?" Of course, Larson, writing later in the 1960s, showed just how students might do such learning to their benefit at a junior high level. But Chell and Coulson looked to reports from teachers across Canada on their experiences. They started with a survey of ideas from mathematicians in Canada and the United States as to why such learning could be expected. The most concise of these arguments came from Robert Rourke, a well-known long-time teacher and curriculum developer from Kent School, in Ontario: Modern mathematics, he said, is "a means of broadening old ideas and introducing new ones to clarify, simplify, and unify our mathematical concepts [being taught in schools]."

Chell and Coulson then used a large number of Canadian sources to garner reactions as to whether using modern mathematics approaches— "modern concepts taught from a modern approach"—led to enhanced learning by students at various levels of ability. Rourke cited the Kent School experience, where set language and notation (e.g., relations as a set of ordered pairs and what follows from that logically and graphically) were at first introduced in grade 12 but were then pushed down to grades 11 and 10 and, later, to junior high because such ideas fit well with the concepts being taught there. Further, he talks about how this helped students better understand ideas such as linear functions, because the language prompted students to look at sets of pairs and solution sets. Of course, less formal language (such as the function named $2x + 3$) was used, but the set concepts allowed students to see that this expression, when used in a context, represented a set of pairs of numbers that could be explored and was not an isolated chunk of mathematical jargon. Similar positive statements about using modern mathematics were generated from a number of special summer courses and other trial projects, not surprisingly showing various positive effects in using modern mathematics with selected students at various levels. For example, in British Columbia, data from modern mathematics classes showed that students of various ability levels could learn many modern concepts while still mastering the conventional content as well as similar students in conventional classes did.

Chell and Coulson also reported on using modern mathematics with lower-performing students to good effect. For example, in Quebec, graphs (of sets of ordered pairs) were used with students who were "poor in algebra," with students showing "good comprehension." In general, Chell and Coulson found that students in various circumstances—from regular classes to special classes to math club experiences—were able to learn modern mathematics concepts under various regimes of modern instruction.

Of course, such observational reports would not satisfy the critics of new mathematics, or parents whose students were not working on computation like they used to. In my reading, I found an interesting sidebar on the effects of modern mathematics instruction with students in a brief report by Tom Atkinson in February 1965, the same issue in which Coulson's article appeared. He reported preliminary results in what seems to be a clever and rigorous study that compared 3,500 grade 7 students in Edmonton in 1961 who had not had modern mathematics instruction with 4,000 such students in 1964 who had had one or more years of instruction using Seeing Through Mathematics (STA), the Alberta modern mathematics elementary-level adoption, on a wide variety of computational tasks. The preliminary conclusion was that modern mathematics study under STA had not adversely affected students' mastery of basic computational skills.

Two other articles contain ideas that would affect teachers' approaches to teaching mathematics. The first, by Cleveland, is a brief but information-dense article on reading in mathematics, in which he presents his ideas as responses to nine questions related to mathematics as a language and to language use in mathematics in practical terms. The questions ranged from "How does a student learn the language of mathematics?" to questions about the effects of logic and semantics on such language to "What non-verbal devices (such as graphs, ordered sets of numerical data, tools such as slide rules and computers, diagrams) are used in mathematics?" His last big question—"What language implications are involved in constructing and presenting a formal mathematical system?"—clearly resonates with the other articles included here. Even in this brief article, there are many rich ideas, such as the following: "to introduce a [new] word before the concept is developed causes frustration on the part of the student and inhibits learning. The task of the teacher is to arrange student activities...dealing with concrete objects and designed to have the student discover a mathematical idea before the [formal] terminology is introduced." Students should then be given further opportunities to share their thinking with other students, addressing "discoveries in their own words," after which mathematical terminology can be introduced to give students more concise and precise ways to express themselves. Such advice resonates with ideas arising in the other 1960s articles in this collection. It might be said that the article provides a syllabus for a teacher workshop on reading and language use in mathematics learning, if not a full course.

The second article by Cathcart and Liedtke, which appeared in three parts in three issues of *delta-K* in the later 1960s, reports on a number of studies conducted by graduate students in the Department of Elementary Education, University of Alberta. These studies all involved Piaget's ideas on children's conceptual development related to mathematical ideas, particularly his stages of development (specifically, the transition from preoperational to concrete operational thinking in mathematics). This article bears reading because it represents an important early body of work on the relationship between Piagetian ideas and the mathematics knowing of children. This body of work reflects the influence of Doyal Nelson in bringing Piagetian ideas into the intellectual discussion around teaching mathematics, and it spurred continued work by graduate students both in elementary and secondary mathematics at the University of Alberta. It could also be seen as part of the genesis of a large project, led by Nelson and Sawada at the University of Alberta, that traced the thoughts and actions of children aged 3–8 as they worked with concrete-based mathematical problems. Further, reading these articles provides teachers with inputs as they think about classroom experiences in various aspects of elementary mathematics. Many of the ideas reported here resonate with Cleveland's thinking on the roles of concrete experience in children's mathematics.

I found the methodologies used in the studies very interesting and in many ways reflective of educational research in the 1960s. Although the ideas used and the reason behind the tasks developed for use in the studies were related to Piaget, for the most part these studies were not carried out as a Piagetian teaching experiment. Most of the testing involved large numbers of students using tasks individually and allowed for collecting data that could be tested statistically. These studies and the related tasks ranged across a wide variety of topics: various aspects of number with preschool students, linear measurement in grade 1, concepts of space and map reading in grades 1–6, time duration across the same grade levels, conservation of length in kindergarten and grade 1, training to conserve quantity under various transformations, relationship between linear measurement bilingualism, and many others. The findings varied. For example, Reimer found a high correlation between conservation of number and length and achievement on the STA achievement test in grade 1; however, in a brief study, Scherer found that the use of manipulative materials was no more effective than discussion in helping grade 2 students develop the ability to do word problems. One study reported in great detail was that of Sawada. He devised a clever way by which students could be tested for conservation of length by showing their responses using nonverbal means using callipers to select what they saw as the length of various objects under various transformations. Sawada found that in a group of 62 kindergarten and grade 1 students, the average age at which 50% gave an appropriate conservation response was between five years, four months and six years, two months—up to two years earlier than students who were asked to give a verbal response in the tasks. There were other interesting findings, such as the fact that both the state properties of the objects used and the transformations used were needed to explain correct conservation in this group. While age was a significant variable in performance, measured intelligence was not.

This rather extensive discussion illustrates the kind of insights into the roles that characteristics of concept formation and the nature of mental operations related to different mathematical concepts might play in the development of mathematical concepts in children. Like Cleveland's ideas on reading, a study of the ideas raised by these research works could provide part of the rich background necessary for teaching mathematics at the elementary level.

CONCLUSIONS

The articles on school mathematics published by MCATA in the 1960s make for interesting reading. They provide a broad view of mathematics education in Alberta in the 1960s. With the exception of computers and their potential use (which are mentioned only once), every facet of modern mathematics that I pointed to early in this commentary is considered in at least one article (and often in more than one). The selection of articles does justice to these facets in several ways, especially to the role of new mathematics in Alberta in the 1960s.

The reader will find contrasting views—even within one article. Thus, we can infer that there was no MCATA party line on new mathematics. I found the articles to be varied and individually quite rich. There are idea articles but also articles that provide concrete classroom-related examples. Both kinds of articles might feed the mathematics teaching thinking of the reader. The articles, while often providing support for a particular idea or practice, can rarely be read as being polemical in nature. These articles would have provided important ideas to MCATA members in its first decade and still make interesting reading today.

Of course, the ideas under discussion are from a different era and have been superseded by developments in curriculum, technology, teacher knowledge of mathematics, mathematical thinking, and methods (particularly in applications of mathematics and the use of computers in such applications), as well as knowledge of and thinking about mathematical knowing. Still, pieces of the writing here sounded very much like the writing on mathematics teaching in the 2012 *Scientific American* article. The sixties were indeed swinging with new ideas in mathematics education, and these articles show just how such ideas were taken up in Alberta and added to and transformed in rich ways.

Tom Kieren is a Professor Emeritus at the University of Alberta. He taught there from the late sixties to the late nineties teaching undergraduate pre-service teachers (primarily secondary but also elementary) and worked with a large number of graduate students. At the time of his retirement he had directed or served on the supervisory committee of over 70% of the doctoral students in mathematics education in Canada. He conducted research on the use of manipulative materials in mathematics learning; using computers in mathematics learning (including work on using Logo). He did extensive, well-known research on the fractional and rational number from the middle '70s into the '90s (working with Doyal Nelson, as well as Les Steffe at U of Georgia on some of this work). He worked with Susan Pirie (Oxford and UBC) studying mathematical understanding in students from ages 10 to university level as a recursive dynamical phenomenon. He worked with many U of A graduate students (and later colleagues) including Brent Davis, David Reid, Elaine Simmt, Lynn McGarvey, Florence Glanfield, and Jerome Proulx on an enactivist view of mathematics knowing and teaching. He was widely published in the field and continues to review 20–30 articles per year for various journals. Like his research that was conducted in Alberta schools with teaching colleagues such as Susan Ludwig, Beryl Tiffen, and Bob Frizzell, Tom has continued his work with children in schools in his granddaughters' mathematics classes in Edmonton and Calgary.

REFERENCES

Bruner, J. S. (1960). *The process of education.* Cambridge, MA: Harvard University Press.

Dawson, A. J. (1969). *The implications of the work of Polya, Popper, and Lakatos for a model of mathematics instruction.* Unpublished doctoral thesis, University of Alberta, Edmonton.

Harrison, R. B. (1967). *Reflective thinking and mathematics learning.* Unpublished doctoral thesis, University of Alberta, Edmonton.

National Council of Teachers of Mathematics (NCTM). (1953). *The learning of mathematics: Its theory and practice.* 21st yearbook. Washington, DC: NCTM.

National Council of Teachers of Mathematics (NCTM). (1957). *Insights into modern mathematics.* 23rd yearbook. Washington, DC: NCTM.

Polya, G. (1954). *Induction and analogy in mathematics.* Princeton, NJ: Princeton University Press.

Polya, G. (1957). *How to solve it: A new aspect of mathematical method* (2nd ed.). Garden City, NY: Doubleday.

Wingert, P. (2012). Building a better science teacher. *Scientific American, 307*(2), 60–67.

INTRODUCTION TO MATHEMATICS TEACHING AND LEARNING IN THE 1970s

Klaus Puhlmann

Reviews of and changes to the curriculum are absolutely essential if it is to remain relevant and effective in achieving the educational goals set by the province. Curricula are very much influenced and shaped by past developments. Such things as content, instructional methods, developments in other fields, development of the psychologies of teaching and learning, and even the political climate in a province or country at the time are considered when curriculum reviews occur. The mathematics curriculum of the 1970s was no exception to this rule.

It can be said that though particular attention was being paid to program revisions during the 1960s, the major concern in the 1970s was problems evidenced almost exclusively in test scores. Different schools or school systems responded to this concern in a variety of ways, but a clear-cut direction that looks toward the future was lacking during both decades. It should also be noted that the mathematics curriculum was of a relative low standard, and rote learning was the flavour of the day. This concern was addressed

Selected Writings from the Journal of the Mathematics Council of the Alberta Teachers' Association, pages 75–80

during the later years of the 1970s and will be dealt with toward the end of this introduction.

The 1970s mathematics curriculum was never subjected to a comprehensive review, but the changes that occurred were significant and reverberated through the entire range of grade levels. Though the New Math was introduced in the 1960s, it continued to be the central focus in the 1970s. Teachers struggled with the New Math, particularly at the elementary and junior high levels, because of their inadequate mathematics background and their inability to shift from the basic-skills movement and rote learning to teaching for conceptual understanding. Eventually, dissatisfaction with the New Math was expressed at many levels. Parents expressed doubt about whether the New Math really prepared students for the future. Teachers did not fully embrace the New Math either because the university math courses they took did not prepare them for it. The result was that teachers advocated for changes as well. Mathematics educators and mathematicians were divided on the effectiveness of the New Math curriculum. Some experts expressed dissatisfaction with the decline of rigour and discipline, the poorly written textbooks, and the inadequate preparation of students for university, while others felt that students achieved a much better understanding of mathematical concepts and the logic and structure of mathematics.

The mathematics curriculum of the 1970s added new content, issued new textbooks or pamphlets, and presented teachers with new challenges. Canada's adoption of the metric system of measurement was one challenge that teachers had to confront. Schools needed textbooks that were written using the metric system of measurement. However, many teachers needed assistance in bringing about this change, which was provided through in-services and the Mathematics Council of the Alberta Teachers' Association (MCATA) conferences and conventions. In addition, the National Council of Teachers of Mathematics (NCTM) and MCATA prepared a very helpful metric kit for teachers. However, many teachers, instead of focusing on teaching the metric system, focused on converting imperial to metric, making the metric system look cumbersome and disorganized. Only after teachers stopped converting from one system to the other did they begin to embrace the metric system.

Curriculum changes in the 1970s were largely approached in piecemeal fashion. The B-options at the junior high level are a case in point. In 1969, the Alberta Department of Education revised the curriculum and its program of electives and implemented the B-options in junior high schools. In addition to the core mathematics program that students were required to take and which was based on a clearly defined program of studies, the academic electives, like mathematics B-option, were completely unstructured courses with no course outline. As with all B-options, teachers were expected to develop a mathematics B-option program based on students' interests, the teacher's strengths, and the availability of resources. In addition, teachers were expected to be mindful that the content of the B-option would not substantially overlap with the core mathematics program. Again, the B-options were introduced in isolation, not as part of the larger picture. Schools that offered the mathematics B-option experienced a variety of challenges. Unfortunately, teachers encountered difficulties understanding the nature and philosophy of the B-option program and, indeed, the reasons for the implementation of such a program. The program, being completely unstructured, also raised questions about teachers' ability to develop curricula. The departure from the traditional approach, with respect to content and instructional activities, raised additional questions about students' perception of and adjustment to the mathematics program. The evaluation of the mathematics B-option in Alberta revealed that, by their very nature, B-options were subject to instances of high success and failure.

At the high school level during the 1970s, revisions of the program of studies were undertaken as well, often with little regard to the overall mathematics program. However, the Math 31 program remained largely unchanged. In Math 30, the statistics and probability units were expanded, and a revised unit on conic sections was introduced using a completely modified approach. Both changes were accompanied by textbook changes and, in the case of conics, a booklet dealing with conic sections was authorized. The 1970s also witnessed a reduction in Euclidian geometry and ultimately the removal of Euclidian geometry from the curriculum altogether. This was viewed by teachers as an unfortunate move, as many students excelled in geometry but struggled with algebra and trigonometry. Again, teachers were left to their own devices to implement these changes because there was no sound implementation plan for the changes.

The implementation of calculators in the classroom and computer-managed learning (CML) also had their start in the 1970s. The use of calculators in the classroom certainly initiated a debate that has continued to the present day. CML was just another teaching methodology for which teachers were ill prepared, and this lack of preparation meant that teachers would have to upgrade to meet new expectations. So, they attended summer sessions at university and organized and participated in workshops after school and on weekends as they struggled to implement CML in their classrooms. Software for conic sections was developed for teachers' use in the classroom, but only a few teachers were able to use it. The use of calculators, too, raised many questions by teachers, parents, and other stakeholders. Essentially confusion and concern arose over how, when, and where calculators should be used, and many questioned whether such use would further diminish students' acquisition of basic skills. Parents in particular were fearful that the use of calculators would impair students' acquisition of basic skills and impede their comprehension of mathematical concepts, and generally they opposed the use of calculators in the classroom. Despite this, teachers were asked to integrate calculators into math classes and to make them useful tools in the hands of students.

The mathematics curriculum of the 1970s truly lacked direction. The program of studies for mathematics at all grade levels was largely pure mathematics, theoretical in nature, rigorous, and narrowly confined to concepts that had to be taught. Consequently, the teaching methods were primarily teacher presentations and demonstrations, with students engaging in drill and practice and doing the textbook exercises. Teachers also looked for opportunities to participate in math contests and competitions, thus providing students with further opportunities to hone their math skills. Needless to say, during this time period students developed excellent algebra and computation skills, but whether they understood the concepts is another question. They certainly were quite capable of regurgitating their skills on departmental examinations (called diploma examinations today), which accounted for 100 percent of their final mark. Departmental examinations were discontinued in the mid-1970s, and now teachers are responsible for preparing their own final examinations and determining the students' final mark.

The 1970s saw the emergence of numerous instructional modes. Teachers were encouraged to use such things as manipulatives, games and puzzles, student projects, math lab experiments, class projects, field trips, and audiovisual resources in their classrooms. The intent was to motivate students and develop their interest in mathematics and help them become active participants in the learning process. Another intent, though, was to decrease students' boredom with the rigorous mathematics program. This required a considerable amount of adjustment by teachers—their past practice was quite different from what was now expected. Teachers took advantage of workshops, in-services, MCATA conferences, and seminars to learn about these teaching methodologies. Schools or school systems that had math leaders in their midst were the lucky ones because these people would organize in-services and workshops to keep teachers up to date. Many rural schools and systems lacked such leaders, and their teachers were literally on their own.

Throughout the 1960s and 1970s, there was considerable ferment in mathematics curriculum and instruction. It was not until the late 1970s that discussions began at the provincial level about the direction of the mathematics program. At the same time, the NCTM began to develop an overarching document called *An Agenda for Action* that contained eight recommendations for school mathematics of the 1980s. It was recommended that:

1. problem solving be the focus of school mathematics in the 1980s
2. basic skills in mathematics be defined to encompass more than computational facility

3. mathematics programs take full advantage of the power of calculators and computers at all grade levels
4. stringent standards of both effectiveness and efficiency be applied to the teaching of mathematics
5. the success of mathematics programs and student learning be evaluated by a wider range of measures than conventional testing
6. more mathematics study be required by all students and a flexible curriculum with a greater range of options be designed to accommodate the diverse needs of the student population
7. mathematics teachers demand of themselves and their colleagues a high level of professionalism
8. public support for mathematics instruction be raised to a level commensurate with the importance of mathematical understanding to individuals and society (these recommendations are from NCTM, *An Agenda for Action, Recommendations for School Mathematics of the 1980s*, 1980)

Specific actions were identified for each recommendation to ensure that the intended direction was being achieved. MCATA and the Department of Education adopted these recommendations as well as the subsequent *Standards for School Mathematics* documents and began an extensive review of the mathematics curriculum for Alberta that produced a new mathematics curriculum ready for implementation in 1982. Diploma examinations for grade 12 students in specified subjects were also implemented. The reinstatement of diploma examinations, now accounting for 50 percent of the students' final mark, was spurred on because public support for teacher-developed exams was waning because of the belief that students' final marks were being inflated and there was a surfeit of eligible students applying for scholarships. Universities also considered the introduction of entrance exams. When studies showed a very high positive correlation between teacher-assigned marks and diploma exam, diploma exams were here to stay. And as of this writing, in October 2012, they are still with us.

One final thing to mention about teaching and learning of mathematics in the 1970s is the dearth of trained mathematics teachers. At the elementary level there was an acute shortage of trained mathematics teachers. The overwhelming majority of teachers teaching mathematics had either no math background or, at best, a minimal number of math courses. At the junior and senior high level, the situation was only marginally better. Many teachers who taught mathematics at the various levels were clearly outside their field of expertise. This situation was particularly critical in rural schools and systems. It should not surprise anyone that these teachers called for support and guidance when changes in the mathematics curriculum were introduced. This notwithstanding, these teachers and their students achieved remarkable results.

Klaus Puhlmann, PhD, is a retired Superintendent of Schools who served Grande Yellowhead Public School Division in this capacity for 23 years. While his research interest is in interpreting mathematics curricula, he also has an insatiable interest in "all things mathematical." He taught at the high school and briefly at the university level.

This chapter was originally published as:

Nelson, D. O. (1970). Mathematical preparation of Alberta Math Teachers. *Mathematics Council of the Alberta Teachers' Association Newsletter,* 9(2), 13–15.

CHAPTER 11

MATHEMATICAL PREPARATION OF ALBERTA MATH TEACHERS

Donald O. Nelson

It may seem obvious that a teacher must know the subject matter which he or she is teaching, particularly in a subject such as modern 11 mathematics. The Mathematical Association of America considered the problem of the background in college mathematics which would be needed by teachers of mathematics. To do this, the Committee on the Undergraduate Program in Mathematics (CUPM) was formed. It recommended the following courses in university mathematics:

4 semester courses for elementary school teachers
7 semester courses for teachers of beginning algebra (grades 7 and 8)
11 semester courses for teachers in grades 9 through 12 mathematics.

The committee stated that these recommendations were minimal, and a comparison with recommendations of other committees and groups would show much higher standards being proposed.

Selected Writings from the Journal of the Mathematics Council of the Alberta Teachers' Association, pages 81–83
Copyright © 2014 by Information Age Publishing

Since the CUPM organization is a well-recognized mathematical body, it seemed interesting to compare Alberta's teachers of mathematics to this standard. The study became part of my MEd thesis at the University of Calgary. In May of 1968, questionnaires were mailed to a random selection of 209 schools in Alberta. Teachers of any mathematics were asked to complete the questionnaire, which dealt with their mathematical training and other selected topics. Results were tabulated from the 910 questionnaires returned.

Only 10 percent of Alberta elementary school mathematics teachers and 20 percent of Alberta junior and senior high mathematics teachers had sufficient mathematics background to satisfy the CUPM recommendations. Sixty-seven percent of Alberta elementary mathematics teachers reported no training in university mathematics. The CUPM felt that the elementary teacher in the self-contained classroom who must teach every subject should have taken four semester courses in mathematics plus whatever was required in the other subject areas. They expressed doubt as to whether or not this could be accomplished in only four years of a teacher preparation program. However, about 60 percent of Alberta teachers have less than four years of training (Wicks & Rieger, 1967).

At the junior high school level, 23 percent of the responding mathematics teachers reported no university mathematics courses. Fourteen percent of the senior high school mathematics teachers responded in a similar way. Only 62 percent of the junior high mathematics teachers and 51 percent of the senior high mathematics teachers stated that mathematics was their main field of interest. Since subject matter specializations is practised in these grades in many schools, it is assumed that anyone teaching mathematics would prefer the subject and would have some background beyond his own high school training. Has this large proportion of Alberta mathematics teachers been "mis"-assigned in their teaching positions? Is it not considered important that the teacher instructs in the subject for which he has been prepared? Or is it simply that there are not enough qualified mathematics teachers?

Ten percent of elementary teachers and 33 percent of secondary mathematics teachers expressed a desire to take further mathematics courses. It was found that the intention to take further mathematics courses was greater for the teachers who had previously taken one or more university mathematics courses than for those who had not. This may be an obvious observation, as most of those who have taken mathematics courses have taken them because they were interested in the subject and consequently would be interested in taking more courses. At the secondary school level, teachers' desire to take further mathematics courses was greater for younger teachers than for older teachers. This may be because the older teachers have completed their training to their satisfaction and, therefore, are not interested in taking further courses in any area.

If it is important that the modern teacher have a broad background in the subject he teaches, something must be done to persuade every mathematics teacher to further his own training to the point where he at least reaches a standard as set by the CUPM, which regards its own standards as minimal.

Inservice projects in mathematics may be an answer to the problem of insufficient background in mathematics, but it was found that approximately one-third of the respondents had not participated in such projects. Many inservice projects are of such short duration that they are relatively ineffective, but 26 percent and 37 percent of the elementary, and junior high and senior high mathematics teachers, respectively, reported over 15 hours of inservice activities in modern mathematics. However, the problem of reaching the teacher of mathematics who is not primarily interested in mathematics exists, as approximately 65 percent of the respondents who had no university background in mathematics had taken part in less than 15 hours of inservice modern mathematics activities, and thus the effect of these activities would be minimal.

Formal courses or inservice activities may not be the only answer. The teacher's initiative in self-study may give a good knowledge of the subject not measurable in the number of formal courses taken. The studying of topics in mathematics which appear in grades higher than the one taught by the teacher is a good project. Films on modern mathematics topics are available from our Mathematics Council, and could be valuable.

The training in mathematics of Alberta mathematics teachers may be lower than what is considered advisable in terms of formal courses; however, rather than bemoan the situation, we must be constructive. Have we as members of the Mathematics Council prepared ourselves at least to the level advocated by the CUPM? What can we do in promoting more interest and enthusiasm in activities which foster further mathematical knowledge in our own school and area? How can we draw people who are not primarily interested in mathematics into these activities? The answers to these questions are not clear, but they must come from our clear thinking on the issue. Each and every one of us will have to work on finding a solution to existing problems in this important matter.

REFERENCE

Wicks, J. E., & Rieger, T. F. (1967). *The Alberta Teaching Force, September, 1966.* (Research Monograph No. 12). Edmonton, AB: The Alberta Teachers' Association.

CHAPTER 12

THE CURRENT STATUS OF HIGH SCHOOL CALCULUS

Murray R. Falk

An article by Daly (1967) outlines the development of mathematics curricula in Alberta. He reports that in 1934, calculus was included as a part of the grade 12 course in analytical geometry. In 1951, he continues, "Calculus and analytic geometry were discontinued, but trigonometry was included as a separate three-credit course" (Daly, 1967, p. 13).

During the 17-year period following 1951, therefore, it was not possible for Alberta students to take a course in calculus at the grade 12 level. Two choices were available to them. They could take Mathematics 30 only, which was essentially an academic five-credit course in algebra, plane geometry and combinatorics, or they could take Mathematics 30 and Mathematics 31. The latter course was a course in trigonometry, carrying three credits. Most students who wanted to be admitted to a sciences program at a university in Alberta would elect both Mathematics 30 and Mathematics 31, as the former was a compulsory requirement, and the latter was "strongly recommended" by the universities.[1]

Although changes in textbook authorizations between 1951 and 1968 wrought a modernization of the approach to teaching Mathematics 30 and 31, the essential content of the courses remained substantially unchanged.

Selected Writings from the Journal of the Mathematics Council of the Alberta Teachers' Association, pages 85–92

During 1967 and 1968, reports circulated which indicated that the present high school mathematics courses had been examined, evaluated and were to be revised. By February, the Senior High School Curriculum Committee of the Alberta Department of Education had accepted *in principle* the recommendation that separate sequences of high school mathematics courses be adopted for students who were university-bound, non-university-bound, and business-oriented (*MCATA Newsletter*, 1968). For the approximately 35 percent of Alberta high school students who intended to go to university, the sequence consisted of Mathematics 10 (grade 10), Mathematics 20 (grade 11), and Mathematics 30 and 31 (grade 12).

The Curriculum Guide for Mathematics 31, which became prescriptive in September, 1968, outlined the revised Mathematics 31 course:

> Mathematics 31 is made up of two parts: (i) Trigonometry, and (ii) Calculus OR Vector AND Matrices (sic) (Province of Alberta, Department of Education, 1968, p. 2)

Each of the two parts was to consist of 80 periods (the periods 40 minutes in length). The choice as to which alternative to offer during the second half was left to the teacher or to the school, although some guidelines were given to assist in making the choice (Province of Alberta, Department of Education, 1968, p. 3). For the calculus alternative, the text *Calculus* by H. A. Elliott and others (Elliott, Fryer, Gardner, & Hill, 1966) was prescribed.

So far as the writer has been able to ascertain, only one Math 31 class in Calgary public schools was taking the non-calculus alternative during the 1968–1969 school year.

One assumes that the recent inclusion of calculus as a Grade XII level course in Alberta has been precipitated by similar action in the United States. The increase in offerings of high school calculus in that country dates back to 1959 when the recommendations of the Commission on Mathematics of the College Entrance Examination Board were published. The Commission held that calculus is a college-level subject, and high schools should prepare students to begin calculus when they enter college; nevertheless, it recommended that well-staffed high schools offer their ablest students a year of college-level calculus and analytic geometry (Commission on Mathematics, 1959, p. 1). Short courses in calculus at the end of grade 12 were not recommended (Commission on Mathematics, 1959, p. 15). Woodby (1965) concurred with this recommendation in 1965.

During the next few years, calculus was offered in more and more American high schools. Many of the students who completed it successfully were able to receive college credit through the Advanced Placement Program.

The Cambridge Conference on School Mathematics published its report in1963 (Cambridge Conference on School Mathematics, 1963). Two topical outlines were suggested for grades 7–12 mathematics. One included a logically complete course in calculus in grades 11 and 12, after the limit concept is understood. The second introduced calculus briefly in grade 9 and then returned to a complete and rigorous study in Grades XI and XII. Justification for both approaches was given (Cambridge Conference on School Mathematics, 1963, p. 49).[12]

There has been general agreement that calculus can be learned by Grade XII students (Woodby, 1965, p. 35). Many authorities quote Bruner (1960): "any subject can be taught to anybody at any age in some form that is honest." Moise (1962) puts it this way: "a student who can ever learn calculus can learn it in the 12th grade." But the controversy over whether or not calculus should be taught to Grade XII students has continued.

Ferguson (1960) believes that many more schools should offer calculus than do now, but that certain prior conditions must first of all be met. The school must have a curriculum in which elementary and intermediate algebra, plan and solid geometry, trigonometry, and coordinate geometry are completed by the end of grade 12. A college-level teacher must be available, students must be willing to spend eight to 10 hours per week on calculus homework, and must be adequately motivated. Ferguson cites figures from his own school to show that its program has been most successful.

Others would agree with Ferguson's assessment. Moise writes as follows: "Many schools teach in the 12th grade; and the experience of the last few years shows that this is quite workable if qualified teachers are available. But the *if* is important" (1962 p. 86).

Blank (1960) advocates a full year of essentially intuitive calculus in grade 12, giving the student insight into certain proofs without detailing them. Some of his reasons for favoring calculus are worth restating here:

1. Calculus is close to the spirit of secondary school mathematics in its manipulative and problem-solving aspects.
2. The concepts, methods and skills developed earlier in the high school mathematics program have applications to calculus.
3. Calculus provides the broadest possible base for students' later development.
4. Calculus plays a profound role in the intellectual and practical affairs of men.
5. Teachers feel more qualified in calculus than elsewhere, as present teacher education makes their having taken calculus more likely than any of the alternatives.

While the above endorsements of a grade 12 calculus course are not made without some reservations, the reader should be aware that there is another side to the coin.

Beninati (1966) favors a grade 12 course in probability and statistical inference or analytical geometry over calculus. He contends that a calculus program offered by inadequately prepared teachers may lead to subsequent failure in college calculus courses, due to the student's inadequate preparation and his meaningless manipulation of symbols.

Allendoerfer (1963) maintains that both the secondary schools and the colleges are heading for serious trouble in their teaching of calculus. While calculus is an essential part of a mathematical education, its importance in and relation to algebra, geometry and other subjects has been overemphasized. Calculus is being taught too early to students who should be studying other mathematical topics such as theory of polynomials, rational functions, algebraic functions, trigonometric functions, inverse trigonometric functions, exponential functions, logarithmic functions, and a full treatment of analytic geometry.

The only conditions under which Allendoerfer would endorse a grade 12 calculus course are stringent:

1. The topics (listed above) must have been covered prior to grade 12.
2. The teacher must be superior.
3. The course must emphasize an intuitive treatment of the limit concept and infinite series prior to calculus as such.
4. Integration, approached through sums of infinite series, must be taught before differentiation.

Brown (1963) reported the results of a survey of 49 mathematics department heads at colleges and universities in Ohio. Thirty-two questionnaires were returned. To the question, "For students who have already taken two years algebra, one year plane and solid geometry and a semester of trigonometry, what should be the emphasis for the remaining semester?" only one respondent (three percent) favored calculus. The topics favored over calculus were Analytic Geometry (53 percent), Probability and Statistics (16 percent), Function Concepts, Mathematical Structure and Analytic Geometry and Analytical Trigonometry and College Algebra (six percent).

When asked whether calculus should be included in the college-preparatory program, 69 percent responded "no," 25 percent answered "yes" (but in most cases there were conditions attached), and six percent were uncertain. Similar percentages were reported for answers to the question, "Should a short introduction to calculus be given during the senior year of high school?"

Buchanan (1965) reported a wider survey. Two hundred thirty-three questionnaires were distributed to college and university mathematics department heads throughout the United States. They were asked to rank seven topics for grade 12 in order of their preference, responding to the questions in accordance with the consensus of opinion of their staff members. Seventy-three percent of the questionnaires were returned. Calculus ranked sixth in order of preference, with only 18 percent choosing it first or second. A unit on limits was ranked above calculus on every questionnaire for which calculus was not ranked first.

A number of researchers have studied the effects of various high school courses on performance in college calculus courses. Chaney (1970) found that University of Kansas students who had studied limits of sequences, limits of functions and continuity in grade 12 performed significantly better in first-year university calculus than students who had only studied analytical geometry.

Tillotson (1963) studied 192 students in first-year calculus and analytic geometry at the University of Kansas. He found no significant difference in achievement between students who had taken an introductory course in calculus in grade 12 and those who had not.

McKillip (1966) used the multiple regression technique to predict students' scores in the first year of college calculus at University of Virginia. Scores predicted were derived on the presupposition that students had had no calculus in high school. Actual scores were subsequently compared with the predicted scores. Scores of students who had one to two semesters of calculus were not significantly different from predicted scores, whereas students who had taken two or more semesters of calculus in grade 12 attained scores that were significantly better than those predicted. McKillip's conclusion is that a grade 12 calculus course of less than two semesters must be justified on other grounds.

Maclay (1969) concluded that non-accelerated students should not be placed in a high school calculus course, even when their pre-training has been in a new mathematics program. Specifically, the SMSG High School Mathematics Program did not provide adequate background for Advanced Placement in calculus.

Wick (1965) found no significant differences in preparation for first-year college calculus as between SMSG and traditional high school programs. He noted a low but positive correlation between grade 12 mathematics average and success in first-year calculus. The conclusions were based on a study of 1962 students in six colleges and universities in Minnesota and Wisconsin.

Coon (1969) found that 60 SMSG students were superior to 122 conventional students in university calculus. The difference was significant at the .05 level for success and at the .01 level for achievement in a first-year course. However, he urges caution in interpreting the results of his study attributing the differences to "continued mathematical maturity based on different high-school preparation, not (necessarily) to learning in calculus."

Anderson and Whittemore (1967) found that students whose high school mathematics courses were 'modern' scored significantly better than students with a 'traditional' background, on a pre-university mathematics test (American College Testing Program: Mathematics (AC)). The correlation between high school grade in modern mathematics and grade point averages in university algebra and trigonometry and university calculus was found to be significant, while that between traditional high school mathematics score and the above GPAs was not found to be significant.

Prouse and Turner (1969) used the technique of multiple regression analysis to determine what combinations of 15 independent high school and university variables contributed to success in second-year university calculus. The high school variables which contributed most highly to success in Calculus 2 were Plane Geometry, Trigonometry, Algebra 2, rank in graduating class, Algebra 1 and ACT Mathematics standard score.

Woodby's conclusions and recommendations (1965, pp. 34–36) provide the best possible summary of the opinions and findings reviewed above.

CONCLUSIONS

1. The individual teacher is the most important factor in the development of a strong mathematics program.
2. There is a lack of agreement on the mathematics that should be taught in grade 12 courses for college-bound students.
3. Acceleration and enrichment have generally accompanied the development of a strong mathematics program.
4. Concern for calculus prevails.
5. There is little acceptance of a course in probability and statistics as the fourth- or fifth-year mathematics offering in the college preparatory program. There is even less acceptance of courses in linear algebra, matrices and computer mathematics.
6. Analytic Geometry as a separate course has achieved only slight popularity.
7. It has been demonstrated that college-level calculus courses can be successfully taught at the high school level provided the following conditions are met: (a) that there are enough capable students; (b) that the teacher is adequately prepared.

RECOMMENDATIONS

1. Experimentation with various grade 12 courses in mathematics should continue.
2. Colleges and universities should provide guidance in the development of grade 12 programs.

3. Courses in calculus and analytic geometry of the warm-up variety are not recommended.
4. Grade 12 courses in probability and statistics, linear algebra, or analytic geometry should be considered in preference to calculus for many high school students.

NOTE

1. See, for example, The University of Calgary Calendar 1969–70, p 51.

REFERENCES

Allendoerfer, C. B. (1963). The case against calculus. *The Mathematics Teacher, 56*(7), 482ff.

Anderson, J., & Whittemore, R. G. (1967). Predictive utility of certain criteria for advanced freshman mathematics courses. *The Mathematics Teacher, 60*(6), 619ff.

Beninati, A. (1966). "It's time to take a closer look at high school calculus. *The Mathematics Teacher, 59*(1), 29–30.

Blank, A. A. (1960, November). Remarks on the teaching of calculus in the secondary school. *The Mathematics Teacher, 53*(7), 537ff.

Brown, R. S. (1963). Survey of Ohio College opinions with reference to high school mathematics programs. *The Mathematics Teacher, 56*(4), 245ff.

Bruner, J. S. (1960, December). On learning mathematics. *The Mathematics Teacher, 53*(8), 610.

Buchanan, O. L., Jr. (1965). Opinions of college teachers of mathematics regarding content of the twelfth year course in mathematics. *The Mathematics Teacher, 5*(3), 223ff.

Cambridge Conference on School Mathematics. (1963). *Goals for school mathematics.* Boston, MA: Houghton Mifflin.

Chaney, G. L. (1970). The effect of formal study of the mathematical concept of limit in high school on achievement in a first course in university calculus. *Dissertation Abstracts*, Vol. 28, p 2884A.

Commission on Mathematics. (1959). *Program for College Preparatory Mathematics.* New York, NY: College Entrance Examination Board.

Coon, L. H. (1969). S. M. S. G. mathematics as a factor influencing success in freshman calculus. *Dissertation Abstracts*, Vol. 25, pp 4475.

Daly, R. R. (1967). Alberta School Mathematics, 1867–1967. *Annual.* The Mathematics Council of the Alberta Teachers' Association, 1967, pp 4–14.

Elliott, H. A., Fryer, K. D., Gardner, J. C., & Hill, N. J. (1966). *Calculus.* Toronto, ON: Holt, Rinehart and Winston of Canada.

Ferguson, W. E. (1960, October). Calculus in the high school. *The Mathematics Teacher, 53*(6), 451–453.

Maclay, C. W., Jr. (1969). The influence of two prerequisite programs on achievement in the high school advanced placement calculus course. *Dissertation Abstracts,* Vol. 29, pp 3917-8A.

MCATA Newsletter. (1968, February). *7*(2), p. 9.

McKillip, W. D. (1966). The effects of high school calculus on students' first-semester grades at the University of Virginia. *The Mathematics Teacher, 59*(5), 470ff.

Moise, E. (1962, Spring). The new mathematics programs. *The School Review, 70,* 88.

Province of Alberta, Department of Education. (1968, September). *Curriculum Guide for Mathematics 31.* Edmonton, AB: Department of Education, Province of Alberta.

Prouse, H., & Turner, V. D. (1969). Factors contributing to success in Calculus 2. *Journal of Educational Research, 62*(10), 439ff.

Tillotson, D. B. (1963). The relationship of an introductory study in high school to achievement in a university calculus course. *Dissertation Abstracts,* Vol. 24, p. 577.

Wick, M. E. (1965). A study of the factors associated with success in first-year college mathematics. *The Mathematics Teacher, 58*(7), 642ff.

Woodby, L. G. (1965). *Emerging twelfth-grade mathematics programs.* Washington, DC: U.S. Department of Health, Education and Welfare, Office of Education. p. 36.

This chapter was originally published as:

Saskatchewan Bulletin (1971). "New math" sparks lively debate. *delta-K: Journal of the Mathematics Council of the Alberta Teachers' Association, 10*(4), 1–2.

CHAPTER 13

"NEW MATH" SPARKS LIVELY DEBATE

Editor's Note: This article is reprinted from the *Saskatchewan Bulletin* June 1, 1971, a publication of the STF.

Four eminent Canadian scholars fought to a draw in a lively battle over the merits of "New Math" that highlighted the annual convention of the Saskatchewan Mathematics Teachers' Society (SMTS) in Regina on May 14–15.

The distinguished combatants were *Professor Ralph A. Stall* of the Department of Pure Mathematics, University of Waterloo; *Professor A. D. Booth,* Dean of Engineering, University of Saskatchewan, Saskatoon; *Professor Roger Servranckx,* Department of Mathematics, University of Saskatchewan, Saskatoon and *Professor James Beamer,* Department of Curriculum Studies, College of Education, Regina.

Dr. Staal, in his keynote address to the conference, conceded that "New Math" had produced valuable fall-out in the '60s, notably the development of the inquiry approach at the elementary level and some first class writing on teaching techniques. But he claimed these gains were outweighed by disadvantages in the "New Math" approach, including a decline in rigour and discipline, an overemphasis on notation, and the neglect of performance in "New Math's" overthrow of rote learning in favor of understanding.

Dr. Staal conceded that "New Math," 10 years after its conception, "is still alive and well but a little wiser," but the "ultimate verdict is in favor of its opponents." He concluded that "New Math" had suffered "transplant rejection" from its proper environment at the university level to lower educational altitudes.

The most unequivocal opponent of "New Math" on the SMTS panel proved to be Dr. Booth, who asserted that "new math will never be useful to society." As an engineer who had "to apply my math," Dr. Booth championed classical mathematics. Claiming that the present secondary school curriculum does not produce engineering students who can either "read, write, or do math," he declared himself as a "missionary who wants to remove Christianity in the form of new math from the curriculum." Dr. Booth said the amount of time spent on math studies in Canadian high schools was "hopelessly inadequate." Whereas Swiss and British students spend 1400 hours on math, Americans devote only about 1000 hours and Canadians 840 hours, just above the "banana republic level." He agreed with Dr. Staal that "New Math" had produced too much icing and not enough cake and a "nibbling" at math that resulted in "an absence of any necessity of real thinking." He criticized current math texts for posing math problems in such a way as to make "the answers immediately obvious." "The whole of mathematics is being taught at much too superficial a level," Dr. Booth concluded. "There is not enough discipline to achieve any respectable mathematics. New Math involves too many words, exactly what mathematics is not."

Dr. Booth's stubborn defense of Euclid later produced a brilliant blackboard battle with Dr. Roger Servranckx who defended "new math" procedures for bringing enhanced "clarity and simplicity to classical problems." However, he joined Dr. Booth's assault on the "textbook racket," claiming that new texts dealt with slogans but left content unchanged. The Belgian-trained mathematician claimed that "new math" was actually a myth, in that most math dates from 1850. "There is really no new math to remove from the curriculum." He agreed with Dr. Staal's contention that modern math procedures had involved themselves too much with notation, but that this was not the fault of the procedures, but the manner of using them.

The fourth panelist, Professor James Beamer, also recognized that "New Math" had not achieved its objectives but he disagreed with contentions that "it has failed to improve the school curriculum". He quoted recent studies that showed "significant gains in math ability" because of "New Math."

Professor Beamer said that the secondary school math curriculum must strive to serve more than university-bound students. He suggested a multi-level math curriculum which would include options that would serve the abilities of top students as well as a lab-oriented program for low achievers.

The stalemate produced by the panel discussion was summed up later by Harold Leibel, vice-principal of Regina Central Collegiate, who observed that "the experts can't agree if we have New Math, and if we have it, what it is." Mr. Leibel's remarks were made in a progress report on the Division IV math curriculum revision committee, as an example of the difficulties facing that committee.

This chapter was originally published as:

Promhouse, H. J. (1971). What is CAMT? *delta-K: Journal of the Mathematics Council of the Alberta Teachers' Association, 10*(4), 3–5.

CHAPTER 14

WHAT IS CAMT?

H. J. Promhouse

CAMT is an *association of associations.* Any individual who is a member of a provincial mathematics society is automatically a member of CAMT (Canadian Association of Mathematics Teachers), providing that the provincial society is itself a member of the natianal association.

HISTORY

The Canadian Teachers' Federation (CTF—all provincial associations are members) at its General Meeting (GM) in 1966 passed the following resolution:

> Be it resolved that the Canadian Teachers' Federation sponsor a Canadian council of teachers of mathematics.

As a result of this motion, a CTF mathematics teachers conference was held in Ottawa March 16–18, 1967. At this conference, the following resolution was passed:

Selected Writings from the Journal of the Mathematics Council of the Alberta Teachers' Association, pages 97–99
Copyright © 2014 by Information Age Publishing

Be it resolved that this mathematics teachers' conference establish a Canadian Association of Mathematics Teachers as provided by the Canadian Teachers' Federation GM resolution of 1966 and in accordance with the guidelines established by the CTF.

A planning committee of four was elected and given the following duties:

1. to prepare a constitution
2. to carry out all organization affairs
3. to arrange or approve a national conference in 1968
4. to issue at least one newsletter before June 30, 1968

By December, 1967, the planning committee had achieved all its objectives: the proposed constitution was completed in June, 1967; the first newsletter was distributed in December, 1967, and a national conference was organized and held in Ottawa on December 8–9, 1967.

FUNCTIONS

The functions of the CAMT are:

1. to provide a means of communication among provincial mathematics teachers' organizations such as the Saskatchewan Mathematics Teachers' Society (SMTS) and the Mathematics Council, ATA
2. to coordinate common activities
3. to provide effective liaison with the Canadian Mathematics Congress (CMC—this organization is concerned chiefly with internatlon al conferences on education
4. to provide effective liaison and contact with international organizations

To facilitate these functions, a national newsletter is issued, provincial journals are exchanged, correspondence is carried on with the councillors from each province, from CTF, CMC and NCTM are invited to executive and annual meetings.

CAMT has developed a close relationship with several other organizations, such as CTF, NCTM and CMC. At present CTF provides CAMT with secretarial aid and makes provision for a member on its executive staff to be an ex-officio member of the CAMT executive. (Dr. George Richert is presently the ex-officio member.) CTF works through our executive to get mathematics teachers to represent Canada at international conferences. In August 1969, two Canadian representatives attended the first International

Congress on Mathematical Education in Lyons, France. In 1972, a second international conference will be held in England.

NCTM is kind enough to pay the expenses of a representative (Dr. Joan Kirkpatrick) as a liaison officer at all CAMT general meetings and conferences.

The CMC has maintained a very close liaison with CAMT. Dr. Dulmage, chairman of the CMC educational committee, has taken a continued interest in our association from its initial organization to the present. He has been present at all CAMT meetings, including executive meetings, and has proven to be a tower of strength to the executive.

The CMC through Dr. Dulmage asked CAMT to nominate a member for each of the three committees set up by CMC. The names of the committees and CAMT's representatives on them are:

National Mathematics Contest Committee—Rev. J. Egsgard, Toronto
Summer Institutes for Teachers Committee—Rev. J. Egsgard, Toronto
Pre-Calculus Mathematics Committee—Doug Polvin, Montreal

It is also CAMT's purpose to provide a means of communication among mathematics teachers' organizations in the provinces and to coordinate any common activities. Heretofore, most communication has been north–south through the NCTM. Now, a means of east–west communication exists through our newsletter, which goes out to every member of a provincial mathematics organization. Our first conference, held at the time of out founding meeting in December, 1967, brought together for the first time representatives from the provincial departments of education and from provincial teacher organizations. The new mathematics curricula for the provinces were described and common problems were discussed. Our second conference was held in Toronto April 17–18, 1970 and representatives of the provinces described their use of Educational Television (ETV) in the mathematics classroom. This conference gave those committed to the use of ETV the opportunity to learn of new developments. Those undecided about ETV were able to gain knowledge which will be useful in making decisions about the future use of ETV as a method of instruction. The conference, attended by 150 mathematics educators from across Canada, was a decided success.

CAMT's only financial support now comes from its individual provincial mathematics organizations. We are presently exploring the possibilities of other means of obtaining financial assistance.

CAMT, though still in its infancy, has proved itself a necessary and worthwhile national organization. The support of the provincial organizations is required for it to continue to grow and provide much greater service and leadership in the field of mathematics to all Canadian teachers of mathematics.

CHAPTER 15

MORE TO IT THAN YOU THINK

Marion Loring

The New Math is supposed to teach concepts rather than just facts and operations, but I rather doubt if the *Seeing Through Arithmetic* series is doing this well enough in the primary grades. There are cases where the expression of concepts is too confusing.

Consider the sentence, "Five is greater than four." To master this, the child must learn five different concepts:

1. *Vocabulary.* The word "greater" is not in the average grade 1 child's vocabulary. Even the word "less" is not very often used, as these children tend to say, "He's got more than me," rather than, "I've got less than him." So the meanings of the words "greater" and "less" have to be taught.
2. *Word Recognition.* Grade 1 children at this stage are only just starting to read, so they will not know any of the words in the sentence by sight.
3. *Grammar.* The word "greater" applies to the first word in the sentence, which is the subject. We take this for granted, but it is not so easy for children. I have found some grade 2 children who didn't have this clear, as was shown by the type of mistakes they made. Whenever they had a sentence of the type, "____ is one greater than 7", they would

---

Selected Writings from the Journal of the Mathematics Council of the Alberta Teachers' Association, pages 101–103

write the correct numeral; but if the sentences had the missing number at the end, as in "7 is one greater than ____", they made mistakes. It appears that their thinking was something like, "Here is 7, and here is 'greater,' so I must make 7 greater," rather than thinking, "The first number must be greater." This is even more difficult for grade 1 students.

4. *Reading from left to right.* Children at this stage have not all grasped firmly the concept of expressing the order of spoken words by a left-to-right progression, but this is necessary if they are to put in writing the grammatical concept described above.

5. *Number.* Finally we come to the mathematical concept that the number five is greater than the number four.

When I finish analyzing in this fashion, I am filled with admiration for the children who do succeed in mastering all this. I have found it easier to use the symbols > and < first, because whichever way round you look at them, the point is always toward the lesser number, and therefore the "left to right" concept is not needed, so at first you just teach concepts (1) and (5). Then the children read what they have written, and so start to learn concepts (3) and (4). When they can read better, the words may be substituted for the symbols.

Another source of confusion is in the type of picture used to teach the concepts of "greater" and "less." This reinforces an error common among young children—that of thinking that the child who has four small candies has more than the child who has three large ones. Three bears are greater than two bears, but they are not greater than two elephants, which is what some of the pictures teach. To teach this is, in effect, to teach that three dimes are greater than two quarters. To teach the correct concepts, the pictures should be of objects that are the same, or at least of the same size.

The pictures illustrating subtraction are confusing also. I noticed that several grade 2 children, who could complete straight equations in addition and subtraction satisfactorily, made many mistakes when writing a mathematical sentence from a picture. An examination of the pictures showed why. First they are taught addition with pictures such as this **. They are taught to perceive two sets and to write first how many are in *** each set, then to complete the equation, $2 + 3 = 5$. When they see a picture representing subtraction, such as **, they perceive the same two sets, so they tend to write $3 - 2$. *** A picture representing subtraction should show the first number rather than the remainder. I found most effective way of clearing up this confusion was to put up this chart:

That this was helpful was shown by the fact that when I removed it the children asked me to put it up again.

But this is just a way of enabling children to do what the book requires them to do. I don't think subtraction can be efficiently taught through pictures. It needs manipulation of real objects and oral problems.

This confusion may be partly due to a failure to recognize that the operations of addition and subtraction are not the same as the equations expressing them. Two different concepts are involved. The operation is active; the equation is static. The operation involves change, motion, time; the equation is a universal statement of fact.

Because the numbers are the same, we are apt to think that the concepts are the same; so we need to look at this more closely. In the operation of addition there is one set of objects at the beginning; then another set moves toward it; then the two merge, forming a final resulting set which is greater than the first set. But an equation does not express change from a smaller number to a greater—it states that the two smaller numbers together are the same as one greater number. This can be expressed on a balance: a two-ounce weight and a one-ounce weight on one side will balance a three-ounce weight. But the operation of addition cannot be expressed on the two sides of a balance.

These are two different concepts, and I think the modern practice unsound, which insists on using the language of the equation and making children say, "Three plus two is equal to five," while they are observing a series of pictures to represent an operation in time. The old-fashioned teacher who permits the simpler sayings, "Three and two more make five," or, "Five take away two leaves three," does in fact teach the operational concept better. When children have performed the operation several times, they may realize the basic rule expressed in the equation.

It is difficult to present the concepts of change, motion, and time in pictures. Grade 1 children tend to look at all the pictures at once, rather than starting at the top left corner. This doesn't matter so much in addition, which can be expressed in one picture; but subtraction requires at least two. It might be better if grade 1 students had no textbook, but did all their work with real objects. This may seem too detailed a criticism, but if the proponents of New Math want us to teach concepts, let us think clearly what they are, and the most efficient way of teaching them. We must think in a more specialized way about grade 1. It is no use just planning a logical beginning to a series. The logical beginning may not be suited to grade 1. And it is there that we are laying the foundation; if that is not solid, what happens to the superstructure?

This chapter was originally published as:

Lindstedt, S. A. (1975). 1.4 kilograms of hamburger and a liter of milk please. *delta-K: Journal of the Mathematics Council of the Alberta Teachers' Association, 14*(3), 13–16.

CHAPTER 16

1.4 KILOGRAMS OF HAMBURGER AND A LITER OF MILK PLEASE

S. A. Lindstedt

Canada is adopting the Metric system of measurement. Mr. S. M. Gossage, chairman of the Metric Commission, has stated that he thinks we will be a "predominantly" metric country by 1980. In so doing we will join over 95 percent of the countries of the world; this will help our international trade and facilitate worldwide understanding in the fields of commerce, industry, and communications.

But, of course, international trade and communication is not the only reason for "going metric." There are other fundamental reasons for the adoption of this system of measurement.

A. The metric system is easier. Yes, it is. The units and subunits are all based on a decimal system and this means that conversion from one unit to another is just a matter of shifting the decimal point. For example, the length of a Canadian football field is 100.584 meters. The following chart shows how easy it is to change this measurement using other units of length.

Length of Football Field

1 0 0 5 8 4	millimeters	(mm)
1 0 0 5 8.4	centimeters	(cm)
1 0 0 5.8 4	decimeters	(dm)
1 0 0.5 8 4	meters	(m)
1 0.0 5 8 4	decameters	(dam)
1.0 0 5 8 4	hectometers	(hm)
0.1 0 0 5 8 4	kilometers	(km)

It is also easier to complete calculations. For example, if your granary is 20 ft. 6 in. long, 15 ft. 4 in. wide and 10 ft. 2 in. high, you need to
perform the following calculations to find out the number of bushels
it will hold: 20–1/2 × 15–1/3 × 10–1/6 × 6–1/4 × 1/8 bushels. That's a
pretty awkward computation—even an electric calculator would have
some difficulty with it.

The corresponding metric units to the same degree of precision
would be: length 6.25 meters, width 4.67 meters, height 3.10 meters.
To find the capacity of the granary you would complete this calculation: 6.25 × 4.67 × 3.10 × 1 kiloliters. Not difficult at all.

B. The metric system will simplify package sizes and make price comparisons much easier. For example, washing detergent is sold in a
great variety of sizes at various prices. In a recent survey I counted 28
different sizes on one shelf in a supermarket. Here are the sizes and
prices of eight which I selected:

5 lbs.	$2.43
75 oz.	$2.39
42 oz.	$1.35
40 oz.	$1.91
32 oz.	$0.89
28 oz.	$1.21
23 oz.	$1.09
16 oz.	$0.75

Quick now, which is the best buy?

Toothpaste, on the other hand, is now sold only in "metric" sizes. I
noted the following on display:

150 ml	$1.43
100 ml	$1.03
50 ml	$0.66

(ml is the symbol for milliliter)

You see, you have a much better chance to compare prices.

Well, what is this marvelous, elegant system of measurement? What are the basics? Because we have been taught and have used Imperial units such as the inch, quart, and pound, we may think that metric units are very numerous and very disorganized. Not so. There are *three* new basic units to learn for most of the everyday uses of measurement. They are:

1. The meter (symbol m), a unit of length. It is about half the height of an ordinary door.
2. The liter (symbol l), a unit of capacity. It is just a bit smaller than the Canadian quart—and, as it happens, just a bit larger than the American quart. (At least the use of the liter will eliminate that confusion.)
3. The gram (symbol g), a unit of mass (or "weight," as it is commonly called). It is a very small unit—less than the weight of a caper clip. For that reason the kilogram (symbol kg), which is 1 000 grams, will be in common use.

Now for each of the above three units, we derive larger and smaller units indicated by the following six prefixes:

For the bigger units

> kilo-, meaning "1,000 times"
> hecto-, meaning "100 times"
> deca-, meaning "10 times"

For the smaller units

> deci-, meaning "1/10 of"
> centi-, meaning "1/100 of"
> milli-, meaning "1/1000 of"

For different units of length we combine the above prefixes with the meter—

> a kilometer (symbol km) is 1000 meters
> a hectometer (symbol hm) is 100 meters
> a decameter (symbol dam) is 10 meters
> a meter (symbol m) is 1 meter
> a decimeter (symbol dm) is 1/10 of a meter
> a centimeter (symbol cm) is 1/100 of a meter
> a millimeter (symbol mm) is 1/1000 of a meter.

(Go back and review the example of the length of a football field.)

For different units of *capacity* we have a similar arrangement—we combine the same prefixes (and they keep their own meanings) with the—to get a kiloliter (kl), a hectoliter (hl), a decaliter (dal), for the bigger units, and a deciliter (dl), a centiliter (cl), and a milliliter (ml) for smaller units. Although these units do exist, we will probably not use all of them in everyday practice. We will use the big one—the kiloliter—for measuring the capacity of storage tanks, granaries, oil tankers, reservoirs, etc. We will use the very small one—the milliliter—for measuring the capacity of toothpaste tubes, medicine drops, shampoo bottles, etc., and we will use the liter itself for milk, paint, gasoline, oil, antifreeze, etc.

Similarly we combine the same prefixes with "*gram*" to get units of *mass*. The kilogram will be used in buying meat, vegetables, fruit, sugar, flour, fertilizer, lawn seed, cement. First class passengers on airlines will be allowed 30 kg of luggage; economy class must get along with 20 kg.

Even the kilogram (1000 grams) is a fairly small unit. Therefore a larger metric unit—the tonne (symbol t), sometimes called the "metric ton," will be used for larger quantities. The tonne is equal to 1 000 kilograms; it is about 10 percent bigger than the ordinary ton of 2000 pounds. It will be used to measure loads of wheat, gravel, sand, bricks. The milligram (mg) is a tiny, tiny unit of mass. It will be used to measure pharmaceutical quantities.

We will not become metric overnight, nor by a certain date. We will move into the system at various places at different times. Because the students in our schools of today will undoubtedly graduate into a metric world of tomorrow, we should in school programs. All weather forecasts will be using metric units of measurement during the year 1975—snowfall will be measured in centimeters, rainfall in millimeters, wind velocity in kilometers per hour and temperature in degrees Celsius. During the year 1976 we can expect the metrication of highway signs—distances in kilometers, speeds in kilometers per hour, the heights of mountains in meters. In 1977 all grain will be measured, for local sales, in metric tons. Even at the present time, we sell our wheat overseas in metric tons. Many household products will start to appear in "metric packages." As already mentioned, toothpaste tubes have been metricated. Heavy industries will take the first opportunity to replace worn-out or obsolete machines and tools with metric calibrated equipment. Many have already made the change—the Ford Pinto is a metric car manufactured in the States; International Harvester, I.B.M., Stelco Steel are going metric. General Motors has announced similar intentions. In sports we are already accustomed to the 100-meter dash, the 50-meter swim, the high dive from the 10m board; the new racetrack at Stampede Park in Calgary is one kilometer in length.

Some things will not change. We often use units of measure just as a manner of speaking rather than as an application of serious measurement. We sing the song "I love you a bushel and a peck" without really thinking of measuring out the love. But I hope we won't destroy the charm of these little expressions by insisting on the metric translation, "I love you 36.369 liters and an additional 90.922 deciliters."

This chapter was originally published as:

Van Brummelen, H. W. (1975). The teaching and learning of secondary
  school mathematics. *delta-K: Journal of the Mathematics Council of the
  Alberta Teachers' Association, 15*(1), 3–10.

CHAPTER 17

THE TEACHING AND LEARNING OF SECONDARY SCHOOL MATHEMATICS

H. W. Van Brummelen

INTRODUCTION

I am concerned about the state of mathematics teaching and learning at
the high school level in Canada today. The majority of students graduating
from our high schools have little if any idea of the role that mathematics
has played in the development of our culture; they have not been shown
the ways in which mathematics is used in our modern society; students in
college-bound programs, in particular, view mathematics as an isolated,
self-sufficient body of knowledge.

My concern is not of recent origin, although it has grown in degree over
the last ten years. When I had just started teaching nine or ten years ago, I
remember raising some questions about one of the first experimental programs being introduced in our district in Ontario.[1] I recognized then—as
I still do now—that the attempt to give students a thorough understanding
of a concept (rather than using a "cookbook" approach) is a valid one.
However, my questions about our goals in teaching mathematics ("Why do

we teach what we teach?" "How can we make mathematics relevant for the 90 percent of the students who will use very little mathematics after they graduate from high school?") were brushed aside as "too philosophical." This was the era of reform—and I, along with many others, was swept along by the tide of enthusiasm for the many improvements which the new programs did contain.

However, my misgivings increased over the years. It became clear that the development of new programs and courses was not guided by any comprehensive concept of the process of education. Rather than starting with the question why we should teach certain topics, the foremost question was how we should teach what rather uncritically was accepted as "modern mathematics." Instead of giving positive direction in teaching students to grapple with the problems dealing with the warp and woof of today's culture, today's mathematics curriculum is abstract and removed from the student's everyday experience. Rather than preparing him for life, the present mathematics courses by and large prepare him only for even more specialized courses in universities or at technical institutes. My views on the present state of mathematics in high school parallel those of Professor Morris Kline in many ways, although I do feel that in his latest book (1973) he has neglected the fact that many programs contained worthwhile improvements over the "traditional" texts they replaced.

*   *   *

What a person believes about the meaning and purpose of life determines his philosophy of education, his views on the nature of the child and of learning, and his goals for teaching mathematics (whether these be explicit or implicit). In this brief I will show that my basic faith commitment influences what mathematics I teach as well as how I teach it. At the same time, I recognize that persons with a different philosophy of life may reach other conclusions—such freedom exists in a democratic society. I believe that a school is accountable to the parents in formulating its philosophy of education and its curricular objectives. Parents should then be able to choose the type of school for their children whose philosophy of education agrees with their own sense of values.

For the sake of brevity, I have not substantiated every statement. For a more detailed exposition of my views, I refer you to my chapter on curriculum in *To Prod the Slumbering Giant* (Vriend et al., 1973) and to the mimeographed proceedings of a seminar on Mathematics in the Christian School (Van Brummelen, 1971).

GOALS OF CHRISTIAN EDUCATION

The curriculum of a school is the plan for learning that translates what one believes about man and his place in our society into a specific program of courses in the school. In our school, the curriculum aims at preparing the student for a Christian life of service in all areas of today's culture. Specifically, our school has the following four objectives:

1. The student's understanding of the unity and rich variety of the created world will be developed. He will systematically learn about matter, plants, animals, and man in society in order to gain knowledge of the structure, function, and interrelationships in creation.
2. The student will be directed to realize that he has a historic place in the world, to live obediently in the God–man covenantal relationship, and to fulfill the God-given task for man. The student will be guided to be a reforming influence in his society to help develop a Christian culture and life style according to the norms of God's Word.
3. The student will be directed to assume responsible discipleship of Jesus Christ in developing skills, knowledge, and insights to live the full Christian life as a church member, citizen, family member, neighbor, friend, worker, consumer.
4. The student will be led through a general, foundational program of studies. Sufficient competence will be developed so that the student can continue his education in vocational training, a community college, a Christian college, a technical institute, a school of nursing, or a university.

I mention these objectives because I am convinced that a definite philosophy of life, when worked out in a curriculum, will affect a specific course of studies in a substantial ways—and this applies to mathematics just as much as to any other subject. Conversely, if one pays lip service to a philosophy of life but does not work it out consistently in a curriculum, one may inadvertently end up teaching a course based on a philosophy at odds with one's own.

MATHEMATICS CURRICULUM IN TODAY'S SOCIETY

The dominant role of science and technology in our culture has led to material wealth but spiritual poverty. The paradox of modern society is that man has constructed a complex social machine to administer the technical machine he has built, but his whole "creation" stands over and above him and manipulates him. As a result, youth became alienated from our society during the 1960s. A reaction against our sterile culture set in—breakdown of authority, skepticism, protest, drugs, astrology, and reactions against scientism and individualism. The youth of 1974 is not as rebellious as that of five years ago, but it shows the continuing alienation in the appalling apathy among a large part of our high school generation.

The "modern" mathematics curricula at best have not counteracted, and at worst have abetted these phenomena. On the whole, the gospel (usually implicit) of modern mathematics texts is this: Mathematics should be done for the sake of mathematics; mathematicians do not concern themselves with real life or with the moral problems that society may face; you learn mathematical concepts mainly so that you can use these in developing more advanced mathematical concepts, but whether these are relevant in today's society or are important for historical reasons is immaterial. It is remarkable that the vast majority of even those students disliking mathematics "play the game" to please the teacher or to get their marks; however, when they come to physics class, most of them can't solve an equation because they have only "played the game" with the x's and y's in totally abstract situations. More serious is that most texts (and, therefore, most teachers) seldom show that mathematics is relevant, that mathematics does have a tremendous impact on our culture, that mathematicians must make moral choices in how mathematics is used (it is not without cause that computer centers were prime targets of violence several years ago).

In our school, we try to take the student and make his profession of Christianity a significant one. This means also the mathematics curriculum must lead the student to a deeper understanding of our modern society. He must be made aware of the historical roots of our civilization as well as of the present value systems, the aims and ideals, the ultimate loyalties and allegiances of Western culture. This may not be left to the English and social sciences departments; mathematics also shares in this responsibility.

THE RELATIONSHIP OF MATHEMATICS TO OTHER SUBJECTS IN THE CURRICULUM

One major weakness of the present high school curricula is that students are given almost no insight into the interrelationship of the various disciplines. A worthwhile curriculum is not a hodge-podge of ideas thrown together, but it must provide the student with a sense of unity and purpose, a sense of his responsibility to God, to his community, to society as a whole. It is not the task of the high school to create specialists. Some students will undoubtedly become specialists in their later life, but they must be taught to relate specific knowledge to the overall situation. The curriculum—also in mathematics—must teach them to ask: "How does my specialty affect other areas of knowledge and the whole of life? How can I use my specialty in contributing to the enrichment of human culture? How does my discipline contribute to a meaningful outlook on the future of society?"

The present mathematics curriculum is concerned mainly with techniques and the teaching of facts for the sake of cataloguing knowledge. The danger of such specialization can be seen from the way most texts handle a topic such as linear programming: Here is a technique to maximize profit; learn it and apply it to the following seven or eight problems. Unlike most exercises, these problems at least deal with situations the students can visualize occurring in everyday life. On the other hand, the implication of such a section is that the mathematician is concerned only with the mathematical technique and not at all with the other aspects of the situation. That a company should maximize its profit by hook or crook is a tacit assumption. The mathematician does not concern himself with the physical and biological aspects of the situation. It is not mentioned that maximizing profit might mean our resources are depleted unnecessarily or that it might upset the ecological balance. Nor does the mathematician concern himself with the psychological or social aspects (might he create unnecessary tensions between workers by putting this into effect?) or the legal aspect (might he not break the spirit, if not the letter, or certain of the government's laws?) or the ethical aspect (is it right to expect workers to have to work overtime or to be laid off at will? By maximizing profit, is the worker being reduced to a semi-robot who cannot do justice to his humanity?).

Our society needs people who know the techniques of mathematics—and these techniques must be taught. But such knowledge may never be taught as an end in itself—our society needs men and women who are aware of the consequences which the use of mathematical and scientific techniques may have, and who are able to make sound decisions on the basis of such knowledge.

The importance of mathematics lies in its applicability to other fields—not only to the physical sciences and technology, but also to biology, psychology, economics, and political science. A mathematical structure can often serve as a model for many seemingly unrelated problems. Therefore, we must include applications of mathematics to other fields in our curriculum. If this is done at the student's level, it will not only deepen his understanding of the mathematical concepts and techniques, but it will also help to make his studies in mathematics more relevant, leading to better motivation. While a subject such as economics cannot be reduced to mathematics alone, students must become aware of the usefulness of mathematical models in such subjects. Almost all major fields of human endeavor and innumerable situations in everyday life are likely to lead to significant applications of mathematics. We must find problems which are complicated enough to represent a situation honestly, but simple enough so that students have some chance to solve it. I admit that this is not an easy task—as the large number of artificial and insignificant problems in most texts indicate. Pollak (1970) discusses this in his *Applications of Mathematics* and states that applications are best chosen from classical analysis, linear algebra, probability and statistics, and computer science, if we keep in mind which fields will be of major importance in the future.

PSYCHOLOGICAL CONSIDERATIONS

Human development is not an automatic, natural process, but requires pedagogical influence and interaction and the exertion of formative power. Education always implies a deliberate attempt on the part of the educator to lead the student in a particular direction according to certain norms. At the same time, education requires a fundamental respect for those we seek to educate. A child does not develop into a person: he is a person from the start, though an immature one (DeGraaff, 1970).

Mathematics must be taught so that the student is shown how the subject helps him to take on a meaningful calling in life, and it must enable the student to be a full, responsible human being who is actively involved in the educational process: the student must participate, cooperate, and be given opportunities to initiate. Teaching mathematics well requires liveliness, inspiration, stimulation, care and genuine concern for a person's development. A student's opinions and reactions must be respected even if not always approved. To educate means to give direction to the development of a person's life.

However, such development is not a "linear" one (cf. "Relevance in Teaching Mathematics," 1973). The student must be able to explore, to follow up hunches, to start with a problem that's meaningful and try to solve it using his intuition, to go off on a tangent, to choose topics himself from time to time that interest him.

Unfortunately, few mathematics texts are written as if mathematics is exciting, as if it is a fascinating journey with beautiful, useful, and "relevant" results. If Euclid had been introduced to geometry in the same way as my grade 10 text (Wilcox, 1968) approaches it, he would likely never have been excited about mathematics. Pages and pages are spent on definitions and postulates and on seemingly endless similar exercises of proofs—and there are almost no results in the chapters that we "cover" in grade 10 which the students didn't already know before they started. This year, after discussing the historical background of geometry before the Greeks, including some of the incorrect results used by the Babylonians, I started deductive geometry with a theorem I'm not "supposed" to take until grade 11: a dissection-type proof of the Pythagorean theorem. We proved the theorem to the satisfaction of everyone involved, and then I pointed out some of the "gaps" in the proof. After that we worked backward to fill the gaps, taking five weeks to learn the same material that "officially" takes five months. So we now have more than three months to investigate other aspects of geometry.

Textbooks are not written for students; they are written for teachers (with the exception of such books as the SMP series[2]). And they are dull. The typical section has a couple of examples followed by a selection of similar exercises—and sometimes a written description that is usually too difficult to read for all but the best students. There is no attempt to put into practice what well-known educators have held to be psychologically sound, whether this be the use of Ausubel's "advance organizers," the intuitive approach of W. W. Sawyer, or the fact that Dienes has shown that the present sequence for teaching structure in mathematics is backward (Lamon, 1972). Perhaps not all of what these educators have to say is valid, but our textbook writers seem to think that none of it is. And while there are a great many creative teachers, the majority of teachers do not have the time or the ability to design their own curriculum and, therefore, usually depend on their texts.

We must structure the curriculum in such a way that the students are given the emotional freedom to respond to the teacher's guidance in their own unique way: to state their own views, to experiment, to investigate, to search and probe for answers. This means that general mathematics courses cannot and may not be just "more of the same." If students still cannot divide decimals properly when they come to grade 10, don't give them worksheets; give them challenging problems in an everyday context and let them use a calculator to work it out; maybe some of them will be motivated to see the usefulness of dividing decimals, and they may even want to learn the technique themselves. In the college-bound course, don't bind the students in a straitjacket of terminology, symbolism, and axiomatics. Give the students' intuition and imagination free rein; the good students will learn the correct symbolism and correct formal structure in due time, the poor students only get turned off by a too-early introduction to formalism.

Too often we treat students as if they were all mathematicians—with the exception that we don't even allow them the time needed by mathematicians to grasp ideas intuitively and work with them before attempting to become precise and rigorous. Calculus was used for many years before it was put on a rigorous logical basis. Logic and proof are useful tools in mathematics but do not form its essence. In our teaching, we must make clear that our everyday integral experience is the foundation of mathematics: everything in the world is subject to the cosmic law order. An intuitive approach to new topics with many different intuitive considerations is sound both from a philosophical and psychological point of view. We need problems that read: "Here is a situation—think about it—what can you say?" (Educational Services, 1963). Both teaching and learning in this way will be difficult at first, but it will also prove to be much more rewarding and meaningful than the stereotyped approach usually used now.

THE HIGH SCHOOL MATHEMATICS CURRICULUM

To be effective, a high school curriculum must be unified in purpose and direction. Our staff is working toward a model as described below. This model tries to stress the interdependence of all disciplines, for it is impossible to go very deeply in any one discipline without striking at the roots of another. All courses around the inner "core" are related to this core, and the objectives of each course tie in directly with the objectives of our school.

There are several "mandatory" subjects, including a course in mathematics. We hope to develop such a course during the next few years. Rather than stressing technical skills (although some basic ones would be taught!), the main goal would be to have the student come to an understanding of some of the basic structures of mathematics, as well as the place of mathematics within the structure of creation, how it developed through history, and how it is used and misused in today's society. Thus the mandatory course would emphasize the development and place of mathematics in Western culture as well as its relationships to the physical, biological, economic and aesthetic aspects of reality.

This course would be mandatory for all students—whether it be made part of our present courses or taught independently. With the present curricular structure in Alberta, it would probably become part of the present courses we offer, both at the general and matriculation levels. Eventually, however, we hope to teach this as one compulsory "module" in each of grades 10, 11, and 12 (such a module could be individualized to a large extent) with a number of optional modules available to the students each year—ranging from algebra to transformations and vectors to calculus to business mathematics. The optional modules would develop topics in much greater depth for students having a special interest in mathematics in general or in one or more of the topics covered by the modules.

With this type of structure, we hope that all students will learn to recognize the rightful place of mathematics as a science that describes and investigates in detail the numerical and geometric aspects of the universe around us. Students will gain comprehension to differing degrees, but it is our aim that all will learn to see mathematics as a functional tool to develop and unfold our society and our world. Specifically, we have set the following goals:

1. The student must gain an understanding of the concepts of number and space and their interrelationship so that he can abstract mathematical properties from concrete situations, so that he can theorize

about and develop such properties, and so that he can apply the results of such theorizing to new situations. For example, a mathematical model in physics is a mathematical abstraction that can be worked with in order to clarify and gain more insight into the physical situation.

2. The student must realize that mathematics serves as a functional tool in solving our everyday problems. The student must see that mathematics does not exist independently of other disciplines but contributes to the unity of all aspects of creation and helps him to analyze the quantitative and spatial situations in other disciplines.

3. The student must realize that mathematics is a developing science, and that throughout history it has influenced and, in turn, has been influenced by cultural forces.

If mathematics is taught to meet these objectives, the student will gain respect for the law structure of our universe and its dependability, and he will discover the order, patterns, and relationships that exist in creation. By having a better understanding of these ideas, the student will also appreciate more fully the aesthetic aspects of mathematics. The realization of our objectives may be helped by the development of the student's techniques and skills. The degree of facilitation should be individualized in relation to the student's abilities and direction.

CONCLUSION

It will be difficult to change the direction of mathematics education with such massive curricular reform behind us. Yet, considering (1) the shortcomings of present texts as pointed out in this brief (texts which the majority of teachers depend on) and (2) the dissatisfaction with the present "new mathematics" courses by parents, students and teachers (a change in the elementary curriculum in Edmonton public schools enabled the *Edmonton Journal* to play up the fact that "the new math is dead"), it is clear that fundamental reform in the teaching of mathematics is a necessity.

Mathematics may not be seen in abstraction from the rest of life. An intuitive rather than an axiomatic approach is called for until the student has a thorough grasp of concepts: mathematics starts with situations, not with theorems. Of course, the problem is not just one of writing courses with a different philosophy and using sound psychological methods based on an understanding of the nature of learning, but it also involves having such programs implemented at the classroom level.

How can this be achieved? First, new curricular materials must be developed. I hope that *Mathematics Canada* will be able to initiate and perhaps even sponsor workshops where groups of teachers and educators who are in agreement about their philosophy of education and also about the direction of mathematics education will write materials for the classroom, particularly for a "mandatory" course such as suggested in this brief. In order for such material to find its way into the classroom, a concerted effort will have to be made to disseminate such materials widely and help teachers with its implementation. To be useful, the materials should be written *for the student*, with extensive teachers' guides also available (such as are found in the SMP series, Books A to H). Persons writing materials should be thoroughly acquainted with recent significant work that has been done in the psychology of mathematics learning and make a conscious effort to structure course material accordingly.

Teachers and writers need more input from practising "applied" mathematicians with respect to meaningful applications that can be taught at the high school level. This will help both teachers and writers of curricular materials to show the interrelationship of mathematics with other branches of knowledge. At the same time, writers must be encouraged to show how the study of mathematics helps the school in achieving its objectives and how the subject material integrates with the rest of the curriculum.

Departments of education have given more "breathing room" in the past few years as far as choice of texts is concerned. However, very few teachers make use of the opportunity to depart from the "recommended text"—at least, this is the case in Alberta. Perhaps we have to move away from the all-inclusive texts to individual units, for which teachers' guides and such aids as transparencies are available.

* * *

It is true, of course, that it is far easier to criticize and find shortcomings than to write a good curriculum oneself. The problem of getting students to understand, appreciate, and be able to use mathematics will never be solved completely. However, our present mathematics curricula have the wrong philosophical and psychological basis, and we must move in new directions if students are to realize the proper place of mathematics within our universe. This cannot be done overnight, but a start must be made.

NOTES

1. This was the experimental version of *Mathematics 9* (Coleman et al), used in York county at that time.
2. *School Mathematics Study Books A to H* (Cambridge University Press: 1969 to 1971).

REFERENCES

DeGraaff, A. (1970). *The nature and aim of Christian education.* Toronto, ON: Wedge Publishing Foundation.

Educational Services. (1963). *Goals for School Mathematics: The Report of the Cambridge Conference on School Mathematics.* Boston: Houghton Mifflin, p. 11.

Kline, M. (1973). *Why Johnny can't add: The failure of the new math.* New York, NY: St. Martin's Press.

Lamon, W. E. (Ed.). (1972). *Learning and the nature of mathematics.* Chicago, IL: Science Research Associates. pp. 5lff.

Pollak, H. E. (1970). Applications of mathematics. In *Mathematics Education* (the Sixty-Ninth Yearbook of the National Society for the Study of Education, pp. 324–325). Chicago, IL: The University of Chicago Press.

Relevance in teaching mathematics. (1973). Excerpts from a letter from G. Edwards of March 15, 1973, unpublished).

Van Brummelen, H. (Ed.). (1971). *Mathematics in the Christian school* (mimeographed proceedings of a seminar on mathematics education). Toronto, ON: Wedge Publishing Foundation.

Vriend, J., et al. (1973). *To prod the "slumbering giant": Crisis, commitment, and Christian education.* Toronto, ON: Wedge Publishing Foundation.

Wilcox, M. (1968). *Geometry: A modern approach.* New York, NY: Addison Wesley.

This chapter was originally published as:

Biedron, B. (1976). Basics in junior high. *delta-K: Journal of the Mathematics Council of the Alberta Teachers' Association, 16*(1), 12–14.

CHAPTER 18

BASICS IN JUNIOR HIGH

Bernie Biedron

Editor's Note: This paper is reprinted from *The Mathematics Teacher*, Volume 4, Number 3, April, 1976.

Are basics the major emphasis in junior high school mathematics programs today? How well do your junior high students know their basics? How competent are your junior high students with their computations? These are only some of the questions which have rarely been seriously asked in the past ten years in the field of junior high mathematics. I believe it is time that we as junior high school mathematics teachers and for that matter, as mathematics teachers in general, seriously ask ourselves this question, for I believe that the basics are not the major emphasis in junior high school mathematics programs today and they definitely should be. Before proceeding any further, I will define the term *basics*. As far as I am concerned, "basics" refers to the basic elementary operations of addition, subtraction, multiplication, and division of whole numbers, fractions, and decimals.

The opinions expressed in this article are based not only upon scientific research but also upon personal experience. I have been teaching junior high school mathematics for the past seven years and I have also taught grades five and six mathematics for a period of four years. I really became concerned and preoccupied with this idea of basics approximately three years ago when I discovered, during the early portion of the school year, that several of my grade seven students had never mastered their three times multiplication table facts. Ever since, I have come to believe that, in general, junior high school mathematics students do not know their basics. Stop for one moment and think of any one particular class that you teach and see if you can honestly answer "no'" to each of the following questions. Do any of your students construct mini-multiplication tables and attach them into their mathematics notebooks in inconspicuous places? Do any of your students hesitate, for some time, when they are asked to answer very simple addition or subtraction questions? Do any of your students use fingers to facilitate computation? Now, I could go on and on, but if the answer is "yes" to any one of these questions, then your mathematics students do not know their basics. Students should learn to become so habituated to the basic elementary operations with numbers that they do not have to think about them. I believe that teachers should do all they can to make the basic elementary operations so habitual that students do not have to think about them any more than they think when they turn on the colour television set.

What is the problem? Why don't students know their basics? Two very good questions. Ever since the 1960s, when the new mathematics was first introduced, everyone closely involved with mathematics and especially teachers have been overemphasizing such concepts as set theory, groups and fields, combinations, permutations, bases, etc.—at the expense of the basics, and let's face it, it is much easier to teach bases than it is to teach the multiplication facts regardless of the method employed. Secondly, I believe that the mathematical game approach has gone too far. Mathematical games are an asset to any good junior high mathematics program, but they have their place and I say let's keep them there. Dr. Max Beberman, one of the founders of the new mathematics and quite active in it for some twenty years, claims that the students of the new mathematics cannot perform the elementary operations as well as those who were subject to traditional drill. In an article in the 1971 issue of *Mathematics Teaching*, Professor Beberman repudiated the entire new mathematics curriculum and was determined to pursue a totally new approach. Professor Edward G. Begle, director of the School Mathematics Study Group (SMSG) admitted that the SMSG curriculum has indeed minimized the acquisition of the basics.

Dr. Morris Kline, professor of mathematics at New York University, claims that the new mathematics has hurt the teaching of mathematics instead of improving it mainly due to the fact that students do not have a good understanding of the basics. Closer to home, professors of pure and applied mathematics at the University of Manitoba are seriously considering the implementation of a package of basics into the first year mathematics courses. All this means only one thing to me, and that is our students do not know their basics.

What is the solution? What can be done to ensure that basics are the major emphasis in junior high school mathematics programs? Let's begin right here in Manitoba. I believe that we urgently need a master curriculum with the basics and options for each particular grade level K–12 built in, therefore comprising of one complete package. The organization of the master curriculum should involve representatives from every facet of the mathematics education field. This would include, of course, representatives from the Department of Education, universities, community colleges, high schools, junior high schools, and so on. Every school division within the province could then take the master curriculum and adapt it to suit the various needs of that particular school division continuously emphasizing the basics and implementing some of the options as necessary this would provide for consistency in the teaching of the required basics and it would also provide for uniformity as far as students' standards are concerned. This would also mean that any one school division could virtually administer its own tests on the basic skills and thereby pinpoint the students' weaknesses. Furthermore, the province could also administer a standardized test in order to determine where the students are at as far as the basics are concerned.

I am convinced that a student cannot in any way come to possess a good understanding of the abstract concepts in mathematics—and I mean any field in mathematics—unless he or she has the basics at his or her fingertips.

I am also of the opinion that junior high school students should be required to learn the basics in one way or another, for there is no future ahead for them if they know only one third or one half of the basics necessary in the high school program. As far as I am concerned, basics were the major emphasis of junior high mathematics, and for that matter mathematics in general, prior to the 1960s when the new mathematics was introduced, and were intended to be the major emphasis according to those responsible for initiating the wave of the new mathematics, and always will be the major emphasis in mathematics regardless of the approach. Yes, basics should be the major emphasis in the junior high school mathematics program and also in any mathematics program at any level for that matter. Incidentally, basics are the major emphasis at our school.

This chapter was originally published as:

Carriger, E. (1977). Editor's report on calculator questionnaire. *delta-K: Journal of the Mathematics Council of the Alberta Teachers' Association, 16*(3), 24–25.

CHAPTER 19

THANKS FOR YOUR RESPONSE

Editor's Report on Calculator Questionnaire

Ed Carriger

Remember the questionnaire mailed to you with the September [1974] issue of *delta-K*? We asked you to respond to a number of questions with regard to the use of the hand-held calculator in mathematics classes, and we promised to publish a summary. Here it is: How do Alberta teachers feel about the use of mini-calculators in our schools?

The first impression gained from the answers is that we have a valuable instrument. Where it is more convenient, it can replace the slide rule and math tables in solving complex problems. It is used to increase the rate of solving complex problems where the principles to be learned are basically how to interpret data, particularly in areas of science and business.

Answers to specific questions reveal the following:

1. Many students have calculators available at the secondary level, and some have them at the elementary level.

Selected Writings from the Journal of the Mathematics Council of the Alberta Teachers' Association, pages 125–127

2. Most teachers have a calculator or have access to a calculator and use it.
3. Calculators belong in the secondary schools, and a few teachers in elementary grades would like to see them in elementary classes.
4. Calculators belong to all students, not just to the "good" students.
5. The use of calculators by students who cannot remember their basic skills needs to be further explored. Respondents to the question do not agree on this point.
6. The schools should not be made responsible for furnishing calculators.
7. The need for special courses in the use of calculators must be investigated more thoroughly. There is no agreement on this point.
8. Do parents favor or oppose the use of calculators? We don't know the answer to this question as yet. However, it is suspected that most parents are as unsure of this as they are of the "new" math, which they do not fully understand.
9. Special materials on the use of calculators are wanted and needed.
10. Most teachers would allow the use of calculators on tests designed to evaluate problem-solving ability and as an aid to improving problem-solving ability.
11. The use of a calculator leading to a breakdown in basic skills may be a liability. This needs extensive investigation.
12. Units of study on how to use the calculator need to be part of the program. We need to ask, "At what level?" From the responses it appears that such units probably should be introduced at the grade 6 or 7 level.
13. Special in-service training in the use of calculators should be given to teachers. (MCATA is working on this through the mini-conference and will help to supply expertise for professional development groups desiring workshops as part of professional days, institutes, and conventions.)
14. Calculators are helpful in inspiring students to do more math and to continue their studies in math.
15. There is no agreement on the extent to which machines will replace computation, and about one-third of the respondents have no definite position.
16. No consensus has been reached on the question as to whether the use of the calculator should be a "must" area. (It is speculated, though, that many students with limited basic skills will learn the simple operation of a low-cost mini-calculator regardless of what is being done in the classroom. Therefore, more teachers are seen moving toward the "must" position in time. Do you agree?)
17. The calculator is useful in other subject areas, particularly science, business, and industrial arts.

18. Senior high teachers want the calculator to become part of class-room tools now. Other teachers are not so concerned. Most teachers in the senior high schools use calculators when available to the students; junior high school teachers are divided, and elementary teachers are not using them.

Some of you may have answered differently, and you may feel differently. However, the above is the general feeling of those who responded to the questionnaire. The areas where we left doubt are those where the statistics left uncertainty about the majority opinion of the issue involved.

We will have to decide whether we want to teach the use of the mini-calculator as part of the mathematics program or whether we will leave it to the business education teachers to make it a part of the business machine and/or office practice program of studies, with many students becoming self-taught and getting only limited skill in its use.

Our school boards will have to decide whether the purchase of calculators should be the responsibility of each student or should be included in the budget as are texts and other school supplies. This will vary from board to board as it should, unless the ASTA decides to make a policy on this matter. Should we as professionals attempt to influence our boards? We should be prepared to make suggestions and/or recommendations when requested. Perhaps some of us will be in a situation in which a positive influence is desirable. At present, MCATA and the Alberta Teachers' Association have not made any policy statements or policy recommendations. Should MCATA make a statement with ATA approval? Let us know your opinion on this matter so that we can act on your behalf.

This chapter was originally published as:

Kieren, T. E. (1979). Constructive rational number tasks. *delta-K: Journal of the Mathematics Council of the Alberta Teachers' Association, 18*(3&4) & *19*(1), 30–33/12–17/10–12.

CHAPTER 20

CONSTRUCTIVE RATIONAL NUMBER TASKS

T. E. Kieren

FRACTION TASK 1: MEASUREMENT AND PARTITIONING

1. Take a piece of calculator tape and "work it" until it lays flat rather than curling up. Cut the ends so that they are perpendicular to its length.
2. Consider your piece of tape as a unit. Use your unit to measure the following objects:

table	_____ units
book	_____ units
your partner's height	_____ units
your waist	_____ units

Because your unit will not usually fit "evenly," you must subdivide your unit into 2, 3, 4, 6, 8, 12, 16... parts. You can do this by folding your tape appropriately. (For example, how can you fold "thirds"?) Write the names of the division lines on your tape.

Example:

Make the measurements using your divided tape.

3. What do you do if your divisions don't give you an even measure? Why can you always find numbers to represent your repeated partitions?

4. This activity is done to answer the following questions.
   a. Are fractional numbers always less than one?
   b. Counting is a useful mechanism in understanding whole numbers. What mechanism appears useful in understanding fractions?

FRACTION TASK 2A: MEASUREMENT, ORDER, AND EQUIVALENCE

1. Take a piece of calculator tape about 1 metre long and work it until it lays flat. Cut the ends perpendicular to the length and make them straight. Label the ends 0 and 1 right at the top of the tape.

2. Fold the tape lengthwise in two equal parts. Label as follows:

Because of space limitations you will want to use the formal forms 0/2, 1/2, 2/2, but *remember*—as children learn fractions, start with *word names first* and only later use ordered pairs of numbers.

3. Fold the tape lengthwise in three equal parts. Think before you act and *do it carefully*. Label the folds on the tape as follows:

4. Fold the tape into 6 equal parts, label the ends and the "sixths" folds appropriately. (Remember to add the label "2/6" to the "1/3" fold, etc.)

5. Fold the tape in 12 equal parts. Label the ends and the "twelfths" folds. (Remember to label the "2/3" fold with "8/12," etc.)

6. Fold the tape in 4 equal parts. Label as above. Fold the tape in 8 equal parts. Label as above.

7. Is 5/8 greater than 7/12? How can you tell? Make up a half-dozen ordering tasks using your tape.

8. List the fractions on the "1/2" fold.

   1/2, 2/4, ____, ____, ____, ____, ____.
   a. What can we say about these fractions?
   b. Why are there no "thirds" in this list?
   c. Give other sets of equivalent fractions from your tape.
9. How could you generate other fractions to go on the "7/12" fold?

FRACTION TASK 2B: MEANING OF ADDITION AND MEASUREMENT

1. Take your tape from task 2A. Hold the "1/3" fold directly on the "1/2" fold.
   a. Where does the "zero" end lie?

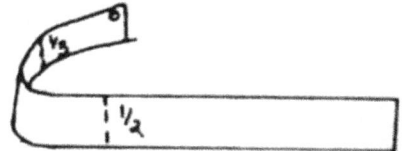

   b. Why?

   A mathematical sentence to describe this is:

   1/3 + 1/2 = 5/6

2. Repeat 1 above, but fold 1/3 on 7/12.
   a. Write the appropriate mathematical sentence.

      1/3 + _____ = _____
   b. Do other "additions" using your tape.
   c. Can you "add" fractions without like denominators?
3. What happens if you lay the "2/3" fold on the "5/6" fold?
   a. Can you figure out how much beyond 1 the tape extends?
   b. Complete this mathematical sentence:

      2/3 + 5/6 = 1 _____
4. Using the tape, do other additions whose sum is greater than 1. Write the related mathematical sentences.
5. Using your tape (and imagination) to solve the following:

   1/3 + _____ = 5/6   _____ + 1/12 = 11/12   5/8 + 1/2 = _____
6. Think up a way to show subtraction using your tape.

The first of a series of number tasks were published in the March 1979 issue of *delta-k*. The following exercises are a continuation of the series.

FRACTION TASK 3: OPERATORS AND MACHINES

1. Complete the table below:

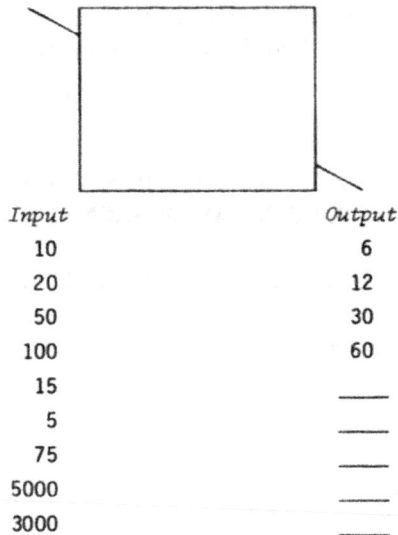

Input	Output
10	6
20	12
50	30
100	60
15	____
5	____
75	____
5000	____
3000	____

The name of this machine is a _____ for _____ machine. For every _____5_____ that go in, _____ come out.

2. For the above machine, complete this table:

Input	Output
____	9
____	300
____	600

How did you know these results?

You were using the notion of "inverse." A machine which would do the reverse of the 3 for 5 machine above would be a 5 for 3 machine.

3. Find a partner. Each of you make up a machine with a mixed list of eight inputs and outputs. Make sure you give three complete pairs. Exchange lists and see who can give the most correct answers. Here is a sample game machine:

Input	Output
15	10
9	6
60	40
6	___
___	2
24	___
18	___
___	20

This machine is a _____ for _____ machine. Its "inverse" machine would be a _____ for _____ machine.

4. Here is a mysterious machine's input and output list. Can you complete it?

Input	Output
10	10
3	3
727	727
___	46
29	___
11	___
7777	___
___	21

Can you name this machine? _____ for _____. A formal mathematical name for this machine is the identity machine. Its inverse machine would be a _____ for _____ machine.

5. Here are two machines.

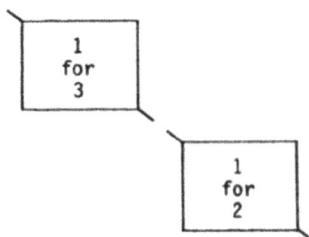

The *output* from the first machine is the *input* for the second machine. Can you complete the table below?

Machine 1			Machine 2	
I	0		I	0
30	10	----------	10	5
12	4	----------	4	2
24	8	----------	___	___
60	20	----------	___	10
90	___	----------	___	___
120	___	----------	___	___
___	2	----------	___	1
___	___	----------	___	4
___	___	----------	___	2000

Look at these machines carefully. What fraction could you use to automatically get the final result if 300 were put into the first machine?

We can write this result 1 for 3 followed by 1 for 2 is the same as 1 for 6. In more mathematical symbols, 1/3 _____ 1/2 = 1/6.

6. Use the machine idea to solve the following:

1 for 2 followed by 1 for 2 is _____ for _____.

3 for 4 followed by 1 for 2 is _____ for _____.

1 for 1 followed by 3 for 7 is _____ for _____.

3 for 5 followed by 5 for 3 is _____ for _____.

4 for 7 followed by _____ for _____ is 1 for _____.

7. What mathematical operations and what ideas are related to this approach to fractions?

FRACTION TASK 4: PART-WHOLE EQUIVALENCE

1. Using the set of 72 objects in front of you, complete the following list of all the ways you can divide 72 objects into subsets of the same size.

   (1) 36 sets of 2

   (2) _____ sets of 36

   (3) 24 sets of _____

   (4)

   (5)

   (6)

   *

   *

   *

   How many ways of partitioning the 72-object set did you get?

   Why is there such an abundance of ways?   (Remember Kennedy, pp. 268–273)

2. Looking at a list of partitions helps one see ways in which fractions can be expressed. For example, because there are four sets of 18 in 72, 18/72 can be expressed as 1/4.

   Complete the following lists of ways that the partitioning of 72 suggests for expressing various fractions.

   (a) 18/72, 1/4, _____, _____, _____

   (b) 24/72, 2/6, _____, _____, _____

   *(c) 10/72, _____, _____, _____, _____

   *(d) 4/72, _____, _____, _____, _____

   *(e) 17/72, _____, _____, _____, _____

   * Can you fill all the blanks? Why or why not?

   What can you say about the fractions in set (a) above? What can you say about all the sets of fractions above?

FRACTION TASK 5: MEASUREMENT AND ADDITION

1. Look back to Fraction Task 1 and use your tape from that task or make a new tape.

2. Measure the following objects as precisely as you can, and complete the following table.

Object	A Side 1	B Adjacent Side	C Both Sides in a Single Measure	D Sum of A + B
(a) book				
(b) table				
(c) room or part of room				

3. What appears to be the relationship between columns C and D in the table?

4. If the two sides of the room measure 6 1/2 and 4 3/8 tapes, we can relate these to the total length 10 7/8 with the following sentence:

$$6 \ 1/2 + 4 \ 3/8 \doteq 10 \ 7/8$$

Using your data, write three sentences which describe the relationship between the side lengths and the total.

Why do we use $\doteq$ instead of =?

Why, in theory or in elementary school texts, can we write:

$$6 \ 1/2 + 4 \ 3/8 = 10 \ 7/8?$$

5. Use two pieces of mayfair board which contain units marked off in eighths. Label the points starting at 1/8 with appropriate fractional and whole number names (for example, 3/4, 11/8, 3/2, etc.).

6. Use the two rulers to add 1/2 and 1/4. Result _____. Write a set of directions for Grade VI or VII children which would tell them how to add numbers using these rulers.

7. Use your ruler to add the following numbers:

    (a) 3/8 + 3/4 = _____

    (b) 5/8 + 3/2 = _____

    (c) 7/4 + 1/8 = _____

    (d) 17/8 + 1/2 = _____

8. Re-label your rulers using mixed numerals (for example, 1 1/2, 2 1/4, et cetera) or at least think of the partitions in those terms. Complete the following:

   (a) 3/8 + 3/4 = _____

   (b) 1 3/8 + 1 1/2 = _____

   (c) 1 1/4 + 7/8 = _____

   (d) 1 3/4 + 1 5/8 = _____

9. Use your ruler to answer the following questions:

   (a) 3/4 =   /8?

   (b) 5/2 =   /4?

   (c) 5/4 =   /8?

   (d) 5/4 + 7/8 = _____ This is the same question as   /8 + 7/8 = _____

   (e) 3/2 + 5/4 = _____ is the same question as   /4 + 5/4 = _____

10. The purpose of this task sheet has been to show two things:

    A. Fractions or rational numbers can be added! There is no question of common denominators!

    B. When you are making up a quick algorithm which uses symbols only, equivalence allows you to make use of the common denominator notion to do so.

The following number tasks are the last of a series which first appeared in the March 1979 issue of *delta-k* and again in the May 1979 issue.

FRACTION TASK 6: UNITS

1. Take a set of 10 different rods. Choose one rod to be your unit. Write fractional number names for all the other 9 rods in terms of your unit.

   Have your partner choose a longer rod as a unit and do the same task.

   Compare your green rods. Do they have the same name in both systems?

   Why or why not?

2. Could the following sentences ever be true? Explain.

   3/4 < 1/2

   5/4 = 1

   2/3 > 3/4

3. What kinds of learning problems are posed by the aspect of fractions as described above?

4. Find the piece of yarn at your table. If that piece is represented by 4/5, cut a piece of yarn from the ball which would represent 1. Describe how you did this.

5. Make up similar problems for children of age 10 or 11 to help them focus on the notion of <u>unit</u>.

6. On the table, find shapes labelled A, B, and C. Below draw the shapes of the figures represented by the given fractions if A, B, or C were considered as units.

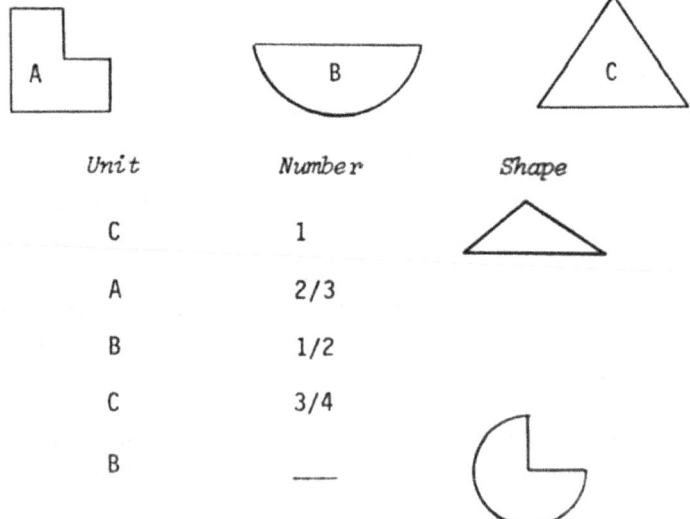

Unit	*Number*	*Shape*
C	1	△
A	2/3	
B	1/2	
C	3/4	
B	⎯	◖

7. Make up interesting and motivating exercises, similar to those in 6, but appropriate for Division II students. Use humor or fantasy.

FRACTION TASK 7: TEACHING

1. Make up fraction representation problems and fraction addition problems using
   a) Cuisenaire rods
   b) Graduated beakers or cans (what is the problem here?).

FRACTION TASK 8: RATIO NUMBERS*

1. From the box of rods, select a set, one of each color and order them. Associate a number from 1 to 9 with each rod.

2. Take a red and a light green rod. Describe the relationship between them in as many ways as you can.

```
        red   =  _____ light green          (a)
light green   =  _____ red                  (b)
 ____   red   =  _____ green                (c)
```

The ratio of red to light green is ____ to ____.   (d)

The ratio of light green to red is ____ to ____.   (e)

How are the numbers used to describe the relationships in (a) and (b) and (d) and (e) related?

3. Select two other pairs of rods and write mathematical sentences which describe the relationship between them.

4a. How are the red and white rods related? We can picture this relationship as follows:

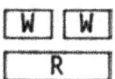

Find all the other rod pairs which have the same relationship.

List the set of ratio numbers used to describe these rod pairs:

{white to red, _____, _____,...}

{1/2, _____, _____}

How do you know physically that the rod pairs are in the relationship?

What can you say about the ratio numbers used to describe these rod pairs?

4b. Test two ratio numbers used to describe the relationships between pink and dark green rods.

Write an equation which describes this picture.

```
 d-g | d-g
 p | p | p
```

Find all the other rod pairs which share the relationship pictured above. What can you say about the ratio numbers which describe these rod pairs?

4c. Suppose there were a silver rod and a violet rod which were related in the following way:

$$3 S = 7 V$$

Write two sets of ratio numbers which would describe rod pairs having the same relationship.

{3/7, _____, _____, _____, _____}

{7/3, _____, _____, _____, _____}

How many ratio numbers would fall in such sets if one included all possible ones?

4d. How would you know physically that rod pairs would be represented by equivalent ratio numbers?

Extra for experts: Suppose we have a rod pair of colors x and y, such that ax = by

Give a ratio number which relates x and y and give five equivalent ratio numbers.

* The idea for this task sheet was taken from the work of Alan Bell and the South Nottingham Project in England.

COMMENTARY

MATHEMATICS EDUCATION IN THE 1970s

A Retroperspective

David Pimm

Although I was nowhere near Alberta during the 1970s, it was during that decade that I encountered for the first time many of the names cited in the articles collected here, such as Kline and Ausubel, Bruner and Pollak—touchstone figures to whom people cleave at various times and for a variety of reasons. In this piece, you will meet some of mine.

I came to the University of Alberta full-time in 2000—staying for the entire decade as a professor of mathematics education (following Tom Kieren's retirement)—but I had visited several times and briefly taught there in the preceding decade, during which time I was frequently somewhere in Canada. But for the first half of the 1970s, I was an undergraduate and then a graduate student in mathematics (mostly in the UK), and the second half I spent in graduate school in mathematics education (in the U.S.). In his commentary on the sixties, Tom Kieren refers to Richard Skemp and his connection with Alberta. Skemp taught me as an undergraduate at Warwick University in 1974, and it was to work with him as a researcher that I

Selected Writings from the Journal of the Mathematics Council of the Alberta Teachers' Association, pages 141–153

returned in the summer of 1979. (My first task was to help him organise the third International Group for the Psychology of Mathematics Education conference, an annual meeting whose 38th incarnation will take place in Vancouver in 2014.)

However, I had come across Skemp's name before that, scrawled on desks in my high school in the mid-1960s ("Skemp is mad," "Skemp is impossible"). I recall wondering what a Skemp was, for mine was the last class who studied "old maths" at my school, we unaware travellers on its trailing edge, with the replacement "new maths" champing hard on its heels (to mix metaphors atrociously). Between 1964 and 1970, Skemp had authored a series of "modern mathematics" secondary textbooks (one of the first, if not the first, in England), entitled *Understanding Mathematics*. Hence, a possibly unintended schoolboy irony lurked in those inked and carved rejoinders. This is not simply nostalgia on my part—the schism produced by the introduction of new mathematics in (at least) Western Europe and North America during the early to mid-1960s would resound throughout the following decade (as attested to by these 10 articles, to a greater or lesser extent). In fact, I believe that its issues and preoccupations can still clearly be heard (like some ghostly echo of a Biggish Bang) in our current clashes some 50 years on.

So, though I have been an active adult participant in the field of mathematics education for some 37 years, I was also present as a consumer for 15 years before that. (My elementary school was also one of those involved in the late 1950s in experimentation with Cuisenaire rods under the watchful eye of Dr. Caleb Gattegno, who was then working at the London Institute of Education before leaving for New York City.) I start with this brief biographical sketch in order to locate myself, as the phrase goes, to assist readers in reading my comments on this engaging and interesting ten-tet of articles.

While I have some specific comments in response to particular articles, there are also general themes and concerns that spoke to me from these pages. Themes related to curriculum (and those echoes from the 1960s new math reforms and curriculum projects); the appropriate or necessary background knowledge and preparation of mathematics teachers; the nascent use of technology in mathematics teaching; and tensions in the teaching and learning of mathematics (tensions either weighed and balanced or one-sidedly insisted upon), most specifically between "fluency and practice" and "understanding" (in its various guises). Those same themes, tensions, and concerns are still with us 40 years later, possibly and on occasion in a more refined or sophisticated form. Though with regard to the reverberating clangour of the more contemporary "math wars" (see, for example, S. Wilson, 2002), as well as the issue of the appropriate mathematical preparation of teachers (especially at the elementary level), arguably the opposite is true. And as ever, like it or not, university-based mathematicians seem always to be with us, too.

Here are a couple of my touchstone mathematics and mathematics education events from the mid-1970s: the computer-assisted proof of the four-colour theorem, and the posthumous publication of Imre Lakatos's (1976) *Proofs and Refutations*. Neither of these directly relates to the teaching of school mathematics,[1] but both engage with powerful forces related to the nature of mathematical proof and the changing and changeable face of the *doing* of mathematics, one which the Bourbaki-inspired new mathematics both supported and denied at one and the same time.

Below are my specific remarks on each article, followed by a broader look at the articles as a whole.

INDIVIDUAL COMMENTARY

Nelson's piece is concerned with the mathematical preparation of Alberta teachers and "the problem of insufficient background in mathematics," reporting on a comparison with standards offered by the Committee on the Undergraduate Program in Mathematics (CUPM), a U.S. body. The opening sentence places the article within the "modern" mathematics debate, whereby practising teachers may not themselves have studied the material they were now called upon to teach. What strikes me is how this accepted CUPM framing of "the problem" seems only to have a quantitative answer—namely, the number of courses that constitute a minimal acceptable level (as if they were interchangeable and more were evidently better). However, there is no discussion of the nature of appropriate courses, nor of how any given mathematics course might relate to the teaching of school mathematics, let alone how knowing *more* mathematics (again, a quantitative framing) is supposed to help a teacher teach mathematics. Somewhat recently, Brent Davis and Elaine Simmt (2006) (both currently at Alberta universities) published a thoughtful and challenging article relating to this issue, and in 2011 Susan Oesterle (at Simon Fraser University in B.C.) completed her interview-based doctoral study of 10 tertiary teachers of elementary Mathematics for Teachers courses, providing a complex look at the interlaced tensions with which teachers of such courses are concerned.

The appropriateness of calculus within high school mathematics education, the focus of Falk's longer article, is helpful as a reminder of the specific discussion in Alberta at that time, as well as setting it within a much broader North American framework. It also signals the influence of university-based mathematicians with regard to school matters (teaching calculus seen as a college prerogative). The 'just because it can be taught, should it?' controversy (see p. 87) reminded me of a late-1970s research study in Madison,

Wisconsin, where I was a doctoral student, in which grade 4 students were successfully taught operations with negative numbers, providing another instance of Bruner's much-quoted comment (see p. 87). But this teaching took much of the school year and raised the question of whether this was a good use of the students' time and mathematical attention.

Falk's piece ("Calculus Seen as an Essential Part of a Mathematical Education") also brought to mind how calculus, then as now, is the new emblem of an educated person (the way Euclidean geometry was in 19th-century England). The author's comments—concerned with how, despite there being an alternative for the second part of the Math 31 course (namely, visualisable linear algebra), it was almost never opted for—again attest to the perceived status of calculus. The question of specific curriculum status is a significant one, more recently seen in decisions about pure mathematics versus applied mathematics in this province (particularly in light of Falk's Conclusion #5). The lack of agreement about what should feature in the Grade 12 mathematics curriculum of the time makes for interesting reading, again in light of the dissension around both the recent Western and Northern Canadian Protocol (WNCP) and the Core Standards framework in the United States.[2] For a strong contemporary book on issues of teaching and one U.S. high school's mathematics curriculum (including the teaching of calculus), see Chazan, Callis, and Lehman (2007).

On a more mathematical note, Allendoerfer's requirement of "integration (via sums of series) before differentiation," which Falk cites, triggered two thoughts. First, working with integration as area rather than anti-differentiation allows for numerical approximations immediately to be made, as well as providing a clear meaning for the integral—the same way that defining π as the area of the unit circle does (as opposed to defining it as the arc length of its semi-perimeter). But it renders the fundamental theorem of calculus problematic, challenging, and unintuitive. My second thought was a memory of Dana Scott, the British mathematician, referring to analysis as a "pop-up" subject: just as you smooth down one difficulty in presentation here, another one pops up over there. There is, he claimed, no perfectly smooth presentation possible.

The third article is a reprint from Saskatchewan on a lively debate from 1971. I think the same debate, with similar voices from the same institutions, could be held today—perhaps under the aegis of the Fields Institute or the Pacific Institute for the Mathematical Sciences (PIMS). Professor Staal's acknowledgment of the benefits of the new math reminds me that there were two sources of novelty proposed: novel curriculum and novel pedagogy.[3] Historically, recollection of the latter got lost along the way in the rejection and vilification of the former. But the tension between rote learning and understanding is still with us. Some 15 years ago, I wrote a book (Pimm, 1995) in which I tried to show how the twin goals of fluency and understanding had a productive tension and interaction in mathematics education at all levels, one that was lost by opting for one over the other. Finally, the language of the debate as reported is also quite edifying: "transplant rejection," "missionary," "banana republic," "respectable mathematics." How we reveal ourselves when the blood is up!

The title of the fourth piece, "What Is CAMT?" (from July 1971), had me stumped. I had never come across mention of this organisation, presuming rather that the National Council of Teachers of Mathematics (NCTM) absorption of Canada under its *N* (there are other words one could use, I suppose: why is it not called ACTM or USACTM?) had been there right from that organisation's beginning in 1920. But, no, we learn from this piece that the Canadian Association of Mathematics Teachers (CAMT), under the auspices of the Canadian Teachers' Federation, was born in the mid-1960s and that two Canadian representatives attended the first International Congress on Mathematical Education (ICME) in 1969. The 32nd NCTM yearbook (NCTM, 1970), on the history of mathematics teaching in the United States and Canada, provides further information (pp. 430–431), referring in particular to the independence of individual provincial initiatives and reforms. Alberta was one of the earliest Canadian affiliates of the NCTM (p. 432), and MCATA co-hosted a summer meeting in Calgary in 1966 (p. 432). To date, I have been unable to ascertain at what stage Canada as a whole became affiliated with NCTM and whether that was at the same time as the demise of CAMT. It occurs to me that the subsequent national curriculum influence of NCTM in the United States (both in schools and in teacher education) through its various standards documents could perhaps not have been repeated here, due to the absence of CAMT as a national organisation.

"More to It Than You Think" is a spirited defence of the sophistication of the challenge of teaching Grade 1 students mathematics. Starting with a particular example, Loring takes on the new mathematics by analysing some of what a student needs to come to grips with in one mathematical sentence, only a small part of which is overtly mathematical. The implications of significant linguistic issues in mathematics teaching and learning is a topic I have spent my career exploring, and this short article points to many of them, including lexico-grammatical concerns (though she might have noted that *greater* signals itself as a comparative by its suffix, whereas *less*, rather than *lesser*, does not).[4] However, I was even more taken with her critique of mathematical diagrams (particularly the challenge of picturing subtraction[5]), contrasting the active operation with the static description (by means of an equation). In short, this piece is concerned with mathematics' elimination of time.

Historically, the 19th-century mathematicians Bolzano and Weierstrass were involved in what became known as "the arithmetisation of analysis" (doing away with the concept of variable moving points in calculus and defining continuity algebraically and statically). What we have here is an account of the arithmetisation of arithmetic. In summary, Loring's piece reminds us to pay attention to what is said, what is written, what is drawn, and what is meant—and, crucially, how all of these interrelate. These are new mathematics echoes that will be heard down the ages.

Van Brummelen's piece left me breathless: by its scope, by its charges, and by the continuing contemporaneity of its themes and concerns. While working on issues of new mathematics teacher induction a decade back, I spent a couple weeks in Shanghai with my then colleague Lynn Paine (see Paine, Fang, & Wilson, 2003). We learnt of the Chinese metaphor of the need for "teaching the whole fish" of mathematics. The problem with the mathematics curriculum is that it does not teach the tail of the fish (where mathematics has come from), nor its head (where mathematics is going, which is increasingly influenced, I feel, by certain sorts of electronic technology; see Rotman, 2008): it simply teaches the body (Van Brummelen's "isolated, self-sufficient body of knowledge"), in part a result of the willful amnesia of modernist mathematics itself (see Gray, 2008). Van Brummelen's question is one of *why* and not *how to* teach any particular mathematical idea.

The divorce between mathematics curricula (and their effects) and a commitment to the world and its problems is certainly alive today: authors[6] and conferences exploring working on social justice issues through mathematics;[7] the potential motivation of more human contexts, whether political or not;[8] and the recent links in the UK to requirements that *every* school subject (including mathematics) contribute to a student's moral education—to say nothing of the potential effects and costs of learning mathematics on certain learners.[9]

But the theological context of Van Brummelen's piece also reminded me of a paragraph from the Second World Conference on Islamic Education, offering a very different justification for the teaching and learning of mathematics:

> The objective is to make students implicitly able to formulate and understand abstractions and be steeped in the area of symbols. It is good training for the mind so that they may move from the concrete to the abstract, from sense experience to ideation and from matter-of-factness to symbolisation. It makes them prepare for a much better understanding of how the Universe, which appears to be concrete and matter of fact, is actually *ayatullah* signs of God—a symbol of reality. (as quoted in Pimm, 1995, p. 11)

This quotation feeds into the religio-philosophical flavour of Van Brummelen's piece. He claims that philosophy has influenced both how and what he teaches. And Thom (1973, p. 204) agrees: "In fact, whether one wishes it or not, all mathematical pedagogy, even if scarcely coherent, rests on a philosophy of mathematics." But at the end, I am reminded both of Van Brummelen's assertion that "mathematics starts with situations not with theorems" and of the urban youth's likely apocryphal riposte to a mathematics teacher going on about this or that problem: "Man, I wish I had your problems!"

The next article, a reprint from Manitoba this time, nicely had me up in arms: not just with its talk of "basics," not even just with its talk of "*the* basics" (the definite article presuming an agreed-upon universality), but with its isthmus-narrow specification of what is taken to be basic (arithmetic operations on various familiar sorts of numbers). There is slippage, perhaps in the heat of the moment, from talk of "the basics" to "basic skills," and from there to the somewhat oxymoronic phrase "good understanding of the basics." Once again, we are right back in the tension between fluency and understanding (despite Biedron's one-sided and, toward the end, rather confused presentation). Philosopher Alfred North Whitehead (1925) claimed that "civilisation advances by extending the number of important operations we can perform *without* thinking about them" (p. 59; *my emphasis*). But the key *teaching* issue is how to automate successfully, in order that such procedural fluency can then become un-thought, subordinated to other things.

And this concern runs right into the next piece, the results of a survey on calculator usage and teacher beliefs. As I am writing this piece (in September 2012), I have just started teaching a master's course for secondary mathematics teachers entitled "Learning Mathematics with Computers." I started by asserting that both *calculator* and *computer* are words with a long history and that, until the Second World War, their first reference was always to a human being, the one-who-calculates or the one-who-computes (as opposed to the thing-that-calculates or the thing-that-computes).[10] This has now changed, I suspect irrevocably. Devices have been fabricated to assist both mathematical practice and its teaching/learning for the past 5,000 years, with an interesting divide between those that are used throughout the culture (such as hand-held calculators) and those that exist only in school settings (such as Dienes apparatus or algebra tiles).[11] I had my current students read this survey article, and there followed a prolonged and stimulating discussion with regard to its relevance and virtual timeliness some 30-plus years later.

By contrast, the ninth article, "Metric by 1980," had an interesting historical feel. As someone school-educated in England pre-decimalisation of the currency (which occurred in 1971), a country where the metric system currently has but a bare beachhead in the shops and, frankly, in the schools, I found this interesting reading, though the convert's enthusiasm was evident. In a nice article from 1991, Janet Ainley asks, "Is there any mathematics in measurement?"—making the point that one reason for systems of different sizes for different quantities is that the quantities involved made human sense (i.e., with regard to human scale). The SI preference for naming units primarily in relations to thousands (grams and kilograms; metres and kilometres) can lead to issues (and not just George Orwell's complaint in his *Nineteen Eighty-Four* that a litre was too much beer while a half-litre was not enough), as well as so much salient, indeed significant, information being coded into unfamiliar prefixes. Susan Pirie's (1982, 1987) work on nurses and drug dose errors (what's a power of 10 between friends?) provides a powerful instance of this challenge. The system may indeed be systematic, but this does not prevent it from causing serious difficulties. And despite having lived in Canada for the past 15 years, when I go to the grocery store and ask for 100 grams of Stong's Own ham, I simply mean "some and not too much, please"!

The final set of articles are a collection of student tasks concerned with ratio and proportion (and addressed, in places, directly to them) produced by Tom Kieren. This was the primary mathematical area that absorbed his research interests for more than 30 years. Interestingly, nothing is offered other than a set of tasks: no rationale, no sense of either where they came from or what they might be good for. At the time, I believe, there was a shortage of innovative, educational tasks of this sort, systematically foraging around a single area, building to more than the sum of their parts, quite different from standard textbook fare. (It is quite a different world we live in nowadays, drowning in material as we are.) Tom's work in this area has been both seminal and generative. It is certainly well-known and highly regarded far from Alberta, as well as within it.

SOME CLOSING COMMENTS

There is no possible way that 10 brief articles can capture a decade. Nevertheless, I would like to end this commentary by looking across the pieces a little more generally and teasing out three broad themes that emerged during the 1970s (with a little generosity with respect to temporal boundaries). Those themes are (a) technology; (b) associations, organisations, and journals; and (c) the centrality of curricular issues.

With the exception of the calculator piece, there is not a great deal of discussion about technology: educational television is mentioned, and even the calculators are described as "desk-top" more than "hand-held." No mention is made of the computer—unsurprisingly, since it was only in 1977 that the Apple II emerged, with its seemingly limitless 4KB of RAM—nor of programming as a mathematical activity. At ICME II, held in Exeter in 1972, Seymour Papert arrived with a considerable amount of bulky hardware from MIT to talk of the promise of the computer language Logo. But the hints are there of arguably the main force pulling on the head of the fish of mathematics—and mathematics education. As I mentioned earlier, it was only in 1976 that the computer-assisted proof of the four colour theorem appeared (to significant controversy). I still have an envelope from the mathematics department at the University of Illinois at Champaign-Urbana (where the provers of the theorem, Kenneth Appel and Wolfgang Haken, worked), franked in red capitals with the assertion that "Four Colors Suffice," which is the title of Robin Wilson's (2002) book on the history of the problem.

With regard to greater association and connectivity between mathematics educators of various stripes, I point obviously to the CAMT article, but also to the range of authors cited in all the articles—touchstone figures who emerged in greater numbers during this decade as mathematics education became more systematic and visible (not least in terms of journals, organisations, conferences, and doctoral programmes). It was May 1968 that saw the first publication of *Educational Studies in Mathematics,* and January 1970 when the first issue of NCTM's *Journal for Research in Mathematics Education* hit the stands. The first International Congress on Mathematical Education (which I think of as the Olympic Games, also held in the same years, apart from the first one) was held in 1969 in Lyon, France, and the first annual Psychology of Mathematics Education conference was held in 1977 (the group was created at the ICME III conference, held in Karlsruhe, Germany, in 1976).[12] In 1980, David Wheeler, based at Concordia University in Montreal, published the first issue of *For the Learning of Mathematics,* an international journal based in Canada. The three journals mentioned here provided the main face of Anglophone, academic mathematics education in Europe and North America into the 1980s.

Wheeler was also intimately involved in the creation of the Canadian Mathematics Education Study Group, a significant organisation of both university-based mathematicians and mathematics educators that meets annually to this day.[13] With regard to the goals and aims of CAMT, the closest embodiment today is perhaps the Canadian Mathematics Education Forum,[14] in the sense of the range of Canadian educators concerned with the teaching of mathematics at all levels meeting and talking together. It may just be my personal view (as it was the decade in which I moved from mathematics to mathematics education), but I see the 1970s as the time when mathematics education got itself organised, gained strength (like a tropical storm over the Gulf of Mexico?), and launched itself into and onto the world.

My third theme is the most obvious but is, nevertheless, core: the centrality of curriculum and the associated question of who gets to decide. These issues played out through the articles collected here and continue to be played out in Alberta (and elsewhere) today. Should the mathematics curriculum be broad or narrow in terms of its focus and aspirations? What should be the balance (if that is the right word) between pure and applied mathematics, and how does the affordable presence of technology that can handle "real" (that is, messy and unstructured) data influence this? New math and old math; modern math and postmodern math; the "basics" and their antipode (which I find myself unable to name succinctly).

The question of basics brings to mind a U.K. TV series started in the seventies called *Leapfrog*, which was scripted by a group of five innovative mathematics educators (including Dick Tahta) called *Leapfrogs*. The first series (polarising within the profession) was dynamic, inclusive, and broad-ranging in terms of what it considered appropriate televised mathematical experiences for upper elementary students. (One of my first pieces of educational research was to produce a 50-page report on the series to present at ICME IV, held in Berkeley, California, in 1980.) When the second series was being developed, the question of its name came up: one broad observation was that people would not know the series was about mathematics. The group decided to take the bull by the horns and reclaim ground—they renamed the series *Basic Maths*.

At bottom, any curriculum really concerns lessons,[15] and below that, the tasks that students engage with.[16] In 1961, the Russian educator Daniil El-konin observed: "The basic unit (cell) of educational activity is the educational task.... An educational task differs fundamentally from other types of problems in that its goal and its result consist of a change in the acting subject himself, not in a change in the objects on which the subject acts" (quoted in Davydov and Markova, 1983, pp. 60–61).

So if curriculum designers thought in terms of tasks and their rationales (what changes in the "acting subject"? how? why?), as well as the efficacy of a particular task in achieving the goals inherent in the rationale (whether explicitly stated or not), a task-based mathematics curriculum (such as the partial one Kieren was offering in his pieces) could prove a delight to behold (rather than the more widely spread exhortative, assertive, or moralistic forms—the student *will*, the student *should*, the student *must*). But, realistically, I know that it is always possible to defeat the intentions of any curriculum embodied in tasks by misusing the tasks themselves, either intentionally or not, by not using them to attend to the mathematics to which they are intended to provide access.

Looking back with the hindsight of more than 35 years is always a luxury. When I agreed to write these comments, I had no idea what a personal journey it would be, nor how prophetic I would find some of the pieces (as well as the generally unchanging nature of the arguments and points of contention). I do see mathematics education generally as a pop-up subject—but, just as with analysis, that does not mean it is not worthy of engaging with. And I am grateful to have had the career that I have had, which started during a period of rapid, intricate, and continually problematic growth. I feel I have managed to live through interesting times.

NOTES

1. Sandy Dawson's (1969) doctoral dissertation at the U of A in part attempted to relate Lakatos's work (much of which had been published in a series of academic articles in a philosophy journal in the early sixties) to questions of the school teaching of mathematics.
2. See www.corestandards.org/Math.
3. At the Second International Congress on Mathematical Education (ICME), held in England in 1972, plenary speaker mathematician René Thom (1973) made similarly trenchant yet fascinating remarks about modern mathematics and the significance of Euclidean geometry (as opposed to Euclid *tout court*) as a source of rich problems and clear meaning.
4. Valerie Walkerdine (1988), in her book *The Mastery of Reason,* takes a close look at the lexical pseudo-pair *more* and *less* and its implications for a variety of subtle mathematics education issues, both inside and outside school.
5. Martin Hughes's (1986) book *Children and Number* documents grades 1 and 2 children's attempts to depict/symbolise subtraction.
6. One such author is American schoolteacher and university professor Eric Gutstein. See his book *Reading and Writing the World with Mathematics* (2005) or the reader he co-edited with Robert Peterson, *Rethinking Mathematics* (2005).
7. See, for instance, http://creatingbalanceconference.org/.
8. See, for instance, Marilyn Frankenstein's book *Relearning Mathematics* (1990).
9. See, for instance, Higginson (2006).
10. Nick Jackiw has drawn my attention to the likelihood that *computer* was first used as a job title in the 19th century for the Great Trigonometric Survey (of British India).
11. For a history of such devices used in the United States over the past two centuries, see Kidwell, Ackerberg-Hastings, and Roberts (2008), *Tools of American Mathematics Teaching, 1800–2000.*
12. See www.icmihistory.unito.it/pme.php.
13. See http://publish.edu.uwo.ca/cmesg/.
14. See http://cms.math.ca/Community/Canada/.
15. Alan Bishop talks about the need for teachers to "lessonise" the curriculum, a task that many elementary textbooks have sought to carry out for teachers, by means of the design layout known as the two-page spread.
16. There is an important and often ignored distinction between *task* and *activity.* The former is usually under the teacher's control (even if selected from a text or the Internet). The latter is the student's and occurs in response to the task provided. See Love (1989).

REFERENCES

Ainley, J. (1991). Is there any mathematics in measurement? In D. Pimm & E. Love (Eds.), *Teaching and learning school mathematics* (pp. 69–76). London, UK: Hodder and Stoughton.

Chazan, D., Callis, S., & Lehman, M. (Eds.). (2007). *Embracing reason: Egalitarian ideals and high school mathematics teaching.* New York, NY: Routledge.

Davis, B., & Simmt, E. (2006). Mathematics-for-teaching: An ongoing investigation of the mathematics that teachers (need to) know. *Educational Studies in Mathematics, 61,* 293–319.

Davydov, V., & Markova, A. (1983). A concept of educational activity for schoolchildren. *Soviet Psychology, 21*(2), 50–76.

Dawson, A. (1969). *The implications of the work of Popper, Polya and Lakatos for a model or mathematics instruction* (Unpublished doctoral dissertation). University of Alberta, Edmonton, AB.

Frankenstein, M. (1990). *Relearning mathematics: A different third R: Radical mathematics.* London, UK: Free Association Books.

Gray, J. (2008). *Plato's ghost: The modernist transformation of mathematics.* Princeton, NJ: Princeton University Press.

Gutstein, E. (2005). *Reading and writing the world with mathematics: Toward a pedagogy for social justice.* New York, NY: Routledge.

Gutstein, E., & Peterson, R. (Eds.) (2005). *Rethinking mathematics: Teaching social justice by the numbers.* Milwaukee, WI: Rethinking Schools.

Higginson, W. (2006). Mathematics, aesthetics and being human. In N. Sinclair, D. Pimm, & W. Higginson (Eds.), *Mathematics and the aesthetic: New approaches to an ancient affinity* (pp. 126–142). New York, NY: Springer.

Hughes, M. (1986). *Children and number: Difficulties in learning mathematics.* Oxford, U.K.: Blackwell.

Kidwell, P., Ackerberg-Hastings, A., & Roberts, D. (2008). *Tools of American mathematics teaching, 1800–2000.* Baltimore, MD: Johns Hopkins University Press.

Lakatos, I. (1976). *Proofs and refutations: The logic of mathematical discovery.* Cambridge, UK: Cambridge University Press.

Love, E. (1989). Evaluating mathematical activity. In D. Pimm (Ed.), *Mathematics, teachers and children* (pp. 240–262). Sevenoaks, UK: Hodder and Stoughton.

National Council of Teachers of Mathematics (NCTM) (1970). *A history of mathematics education in the United States and Canada* (32nd Yearbook). Washington, DC: NCTM.

Oesterle, S. (2011). *Diverse perspectives for teaching math for teachers: Living the tensions.* Unpublished doctoral dissertation, Simon Fraser University, Burnaby, BC.

Paine, L., Fang, Y., & Wilson, S. (2003). Entering a culture of teaching: Teacher induction in Shanghai. In E. Britton, L. Paine, D. Pimm, & S. Raizen (Eds.), *Comprehensive teacher induction: Systems for early career learning* (pp. 20–82). Dordrecht, NL: Kluwer Academic Publishers.

Pimm, D. (1995). *Symbols and meanings in school mathematics.* London, UK: Routledge.

Pirie, S. (1982). *Deficiencies in basic mathematical skills among nurses: Development and evaluation of methods of detection and treatment.* (Unpublished doctoral dissertation). University of Nottingham, Nottingham, UK.

Pirie, S. (1987). *Nurses and mathematics.* London, UK: Royal College of Nursing.

Rotman, B. (2008). *Becoming beside ourselves: The alphabet, ghosts, and distributed human being.* Durham, NC: Duke University Press.

Skemp, R. (1964–1970). *Understanding mathematics* (Books 1, 2, 3, 4M, and 5M). London, UK: University of London Press.

Thom, R. (1973). Modern mathematics: Does it exist? In A. Howson (Ed.), *Developments in mathematical education* (pp. 194–209). Cambridge, UK: Cambridge University Press.

Walkerdine, V. (1988). *The mastery of reason: Cognitive development and the production of rationality.* London, UK: Routledge.

Whitehead, A. (1925). *Science in the modern world.* New York, NY: Macmillan.

Wilson, R. (2002). *Four colors suffice: How the map problem was solved.* London, UK: Penguin.

Wilson, S. (2002). *California dreaming: Reforming mathematics education.* New Haven, CT: Yale University Press.

MATHEMATICS EDUCATION IN ALBERTA IN THE 1980s

Len Bonifacio

The 1980s produced a great deal of change in mathematics education in Alberta, both at the department level as well as in classrooms across the province. Some of this change occurred because of a wave of general education reform around the world, but some was a result of local economic, political, and social factors here in Alberta.

Globally, industry was changing in nature; there was a greater demand for workers in high-skilled positions and less need for low-skilled workers. The world was becoming more information-based and less industrial-based.

Provincially, many of the changes were related to assessment. In 1982, the Department of Education introduced the Achievement Testing Program for grades 3, 6, and 9 to assess the core subjects of language arts, mathematics, science, and social studies. As well, in 1984 the government brought back grade 12 examinations, formerly called departmental exams, now known as diploma exams. The English 30 or 33 exam became a requirement for high school graduation, and exams were developed for the Math 30 and 33 courses. Because of all these new assessment tools, the

Selected Writings from the Journal of the Mathematics Council of the Alberta Teachers' Association, pages 155–159

government developed a new student information system to provide accurate analysis of student data and information to policymakers.

Also at the provincial level, teacher mathematics organizations became more robust and assertive; they didn't just deliver information about curriculum and assessment, they pushed for more input into all issues related to mathematics education.

Many jurisdictions in Alberta moved away from set textbooks from major publishing firms in the early 1980s for junior high mathematics courses to a softcover locally produced, consumable workbook as their main teaching and assignment source. These workbooks contained very little in terms of instruction and virtually no context for mathematics concepts; instead, they emphasized repeated practice of procedural skills. The desire of school districts to save money on resources might have been the main force behind this. Later in the decade, the move to provide more practical applications of mathematics concepts grew and led to the government producing application and problem-solving related booklets in the early 1990s.

The 1980s still saw many students dropping out of mathematics courses. Not many high school students signed up for advanced mathematics courses like calculus or analytical geometry, and the vast majority of those who did were male.

Demographic shifts and population changes made school closures a large concern in many areas of the province during the 1980s. To counter reducing student numbers, many school districts began to introduce focus programs, such as the International Baccalaureate program at the high school level. This was successful in attracting students from outside school catchment areas and increasing numbers to a viable level. This trend later progressed to include a great variety of focus programs and sports academies, and students were not bound to their neighborhood schools anymore. This model was first introduced by Edmonton Public Schools and was later copied by many jurisdictions in North America, and was a result of the move to a site-based management style, which was piloted in the late 1970s and fully incorporated in the early 1980s. As Joan Cowling, former board chairperson of Edmonton Public Schools, said in 1995: "It was around 1984 or 1985 when the concept of school-based management seemed to become institutionalized and have taken on a life and philosophy of its own" (cited in Delaney, 1995).

The 1980s also saw huge changes in technology in education with the development of microcomputers at prices that schools could afford. The excitement of this new tool led many school districts to begin bringing computers such as the Commodore 64, Apple IIe, and Vic 20 into their classrooms in the early eighties. In mathematics classes, these computers were primarily used for interactive programs written in BASIC language to deal with concepts related to number. Some progressed further to look at geometry concepts through the use of the LOGO program with its iconic turtle on the screen. This program was very successful, especially with young children and physically disabled children, and was also popular with teachers. Dr. Dale Burnett, of the University of Lethbridge, said in his article that the reasons for this were threefold: synthesis (the necessity to build on previous knowledge), self-control (the user had some control in the setting of tasks and how to approach them), and sharing (students helping students and feeling good about it, as opposed to working in isolation on a task). Floppy disks became the norm for program and data storage, although many teachers battled with cassette drives for this purpose. Technology for education changed rapidly and, later in the decade, IBM made its computers more affordable and applicable to education. Computer processing classes became widespread as did BASIC programming and word processing. Spreadsheet programs such as Lotus 1-2-3 became a popular area of study, and spreadsheet work was incorporated into mainstream mathematics curriculum later in the century. The universities around the province provided many in-servicing and professional development courses for teachers to learn how to use microcomputers in their mathematics classrooms. Dr. Milt Petruk at the University of Alberta was one of the leaders in this area.

The 1980s also saw the rising importance of statistics in secondary mathematics curriculum. Historically, this branch of mathematics had always been treated poorly, and its proponents had to fight for it to be treated the same as other branches in the mathematics field. With the growth of the consumer society, research companies, and an information-based global focus, the greater presence of market surveys and opinion polls in our lives pushed the need for population study and sample surveying as mathematics topics in schools. An indication of the new respect for statistical topics is evidenced in this comment from Dennis Haack, in the article entitled "Statistics in the High School": "But statistics has become more than a research tool. Statistics has become a language in its own right. We are bombarded by numbers, but what do the numbers mean?" (p. 7). The interpretation and understanding of statistics were seen as more important than the actual calculation of statistical measures.

A great deal of attention in the 1980s was directed at trying to make mathematics more relevant to girls. Many educators attempted to dispel the myth that females are innately poorer than males in mathematical ability. Conferences, NCTM initiatives, and much research were devoted to increasing female participation in all aspects of mathematics, including employment in scientific and mathematics-related careers. This movement was quite successful as evidenced by the increase in the number of females in mathematics classes and pursuing related postsecondary programs and careers through the end of the 20th century and beyond.

The search to understand more about how people learn and how to make mathematics teaching more relevant spurred the constructivist approach to teaching. Professors like Dr. Sol Sigurdson, at the University of Alberta, tried to help teachers understand the logic behind the constructivist approach to learning. As Sigurdson wrote in an article entitled "A Constructivist Approach to Teaching Mathematics," "The main tenet of this view is that all learners actively construct theories, no matter how minor, about what is appropriate action for responding to any particular situation" (p. 8). This was an indication that the processes of learning and teaching were being looked at from a more scientific point of view: recognizing relationships among all mathematics topics, acknowledging the many-faceted aspects of even the simplest of mathematical concepts, and paying more attention to what the learner brings to the situation.

By the end of the decade, mathematics curriculum had shifted to a three-pronged focus: intended, implemented, and attained. This approach came out of the Second International Mathematics Study of 1989. "Intended" referred to the curriculum developers, "implemented" to how the curriculum was realized in classrooms, and "attained" to how the curriculum was demonstrated by student achievement and attitudes.

In conclusion, the decade of the 1980s saw a great deal of change affecting mathematics education in Alberta, some of that change on the global stage, and much of it more local, at the provincial and district level. Great progress in mathematical resources occurred late in this decade. Technology use in mathematics classes made great inroads and was definitely here to stay. Overall, I believe that great strides were made in this decade to make mathematics education more interesting to a wider population and more relevant to the world around us.

Len Bonifacio's current position is Secondary Mathematics Consultant for Edmonton Catholic Schools. He has been involved in mathematics education for 39 years, at the junior high and senior high levels, as well as a mathematics examiner at the provincial level, and now at the district level. He is especially interested in instructional strategies and assessment in mathematics.

REFERENCES

Delaney, J. G. (1995). The development of school-based management in the Edmonton Public School District. Retrieved from http://www.mun.ca/educ/faculty/mwatch/vol1/delaney.html.

Haack, D. G. (1980). Statistics in the high school. *delta-K, 19*(4), 6–9.

Sigurdson, S. E. (1986). A constructivist approach to teaching mathematics. *delta-K, 26*(1), 8–13.

BIBLIOGRAPHY

Alton, E. V., & Gersting, J. L. (1980). An alternative course for the 'I hate math and I've never been good at it' student. *delta-K, 20*(1), 4–6.

Burnett, J. D. (1988). Logo: An opportunity for synthesis, self-control and sharing. *delta-K, 27*(1), 25–31.

McEwen, N. (1995). Educational accountability in Alberta. Retrieved from http://www.csse-scee.ca/CJE/Articles/FullText/CJE20-1/CJE20-1-04McEwen2.pdf

O'Shea, T. (2003). The Canadian mathematics curriculum from new math to the NCTM standards. Retrieved from http://smc.math.ca/Reunions/FCEM2003/panel/oshea.pdf

CHAPTER 21

STATISTICS IN THE HIGH SCHOOL

Dennis G. Haack

Editor's Note: The author is the section leader of the Biostatistics and Epidemiology Section of the Tobacco and Health Research Institute at the University of Kentucky. As a member of the National Council of Teachers of English (NCTE) Committee on Public Doublespeak, Dr. Haack writes and lectures on the misuse of the language of statistics.

The high school mathematics curriculum is continually changing. One of the more recent changes has been the inclusion of a course in statistics (see Pieters, 1976). As to the specific makeup of a high school statistics course, there is not likely to be agreement. As to the primary objective of such a course, there should be agreement. The purpose of this paper will be to look at the objective of a high school statistics course.

The key to the development of any course in statistics is deciding what statistics is. Statistics has, since the publication of R. A. Fisher's *Statistical Methods for the Research Worker* in 1925, been thought of as a set of research tools. In this regard statistics is the investigation of a population. The population of interest may exist or may be created by the researcher.

The study of an existing population is by a sample survey. A part of or sample from a population is selected and studied. Examples of sample surveys include opinion polls, marketing research surveys, TV-viewing and radio-listening surveys, and pre-election polls. A 100 percent sample is referred to as a census. Of interest in the study of sample survey techniques is how a survey is designed as well as how to analyze and interpret survey data.

On the other hand, a researcher may wish to study a population which he creates. For example, an agricultural researcher might test a fertilizer on a crop which he has planted on a test plot. The researcher is stimulating the use of the fertilizer by farmers; that is, he tries to create a population that would exist if farmers used the fertilizer on their crops.

Another example of the investigation of a created population involves research on the effects of a drug. A population is created in the laboratory which would simulate use of the drug if the drug were put on the market.

As with the study of an existing population, the study of a created population involves a researcher with the design of his experiment as well as with the analysis and interpretation of experimental data. So we see that statistics is the study of a population which exists or is created. Statistics provide a set of tools which are required by an investigator for the design of a population study and the analysis and interpretation of the data generated by the study.

Traditionally, statistics courses at all levels have been an attempt to teach statistics as the study of a population. Distinction between experimental and sample survey investigations may or may not be made.

But statistics has become more than a research tool. Listen to the news this evening, or read a newspaper or a news magazine. Listen to public officials and advertisers. Statistics has become a language in its own right. This language pervades the media making it nearly impossible to understand a newscast without being quite familiar with the language of statistics. What are these words we hear—"estimates," "significance," "projections," "averages," et cetera? We are bombarded by numbers. But what do the numbers mean?

This is what statistics is to most Americans: a language which is very often used and too often misused. Statisticians have, for the most part, not taught about the language of statistics. Even students who have completed a traditional course in statistics cannot usually understand this language.

Statistics can be thought of as the tools required for the study of a population, or statistics can be thought of as a language. We must decide which type of statistics we are to teach our students.

A first course in statistics should not try to teach statistics as a research tool. There are two main reasons for this. First, the study of statistics, a research tool, requires students to memorize the use of formulas, if not to memorize the formulas themselves. Students become so involved with learning to calculate statistics that they fail to learn what the statistics mean. Retention of the manipulative skills is minimal, causing students to have little if any knowledge of statistics after a course of this type is completed.

A second reason why a first course in statistics should not teach statistics as a research tool is that students, after taking a traditional statistics course, are no better able to understand the statistics they'll encounter in the media than they were before the course started. The better students might be able to run a t-test, but they are not likely to have a feeling for what is involved with a determination that, say, significantly more animals in a treatment group developed cancer than did animals in a control group. Some of the students might be able to calculate the probability of selecting a red ball from an urn, but they may not know how to interpret the statement, "The probability of rain is 20 percent today." That is, the most we can hope of a student is that he or she will become a manipulative "whiz." A student might become quite good at "plugging and chugging": plugging numbers into a formula and chugging until a number results. Yet our students are not likely to be able to interpret the statistics they might have learned to calculate.

Statistics should be taught as a language rather than as a research tool. Students should first be taught how to interpret statistics. A student will be much better off being able to understand statistics than only able to calculate them.

Statistics can be taught as a language. It is being done at the University of Kentucky (see Haack, 1976). The idea behind the course is to down-play the calculation of statistics while concentrating on how to interpret statistics. In fact, students do not calculate any statistics in the course. There is, therefore, no need for mathematical formulas. The course is conducted in a strictly verbal, nonsymbolic manner. Examples used in the course come from the media. Ideally, students will be able to apply the principles they learn to statistics they will encounter, or have encountered in other areas.

One of the major drawbacks with a nonsymbolic statistics course has been the lack of a text, requiring a large amount of work by the teacher. Texts are now becoming available (see, for example, Haack, 1979).

One of the more interesting aspects of teaching statistics as a language is that students become genuinely excited about being able to detect misuses of statistics. When I started this experiment in teaching a few years ago, I did not look forward to trying to find examples of the misuse of statistics. Such examples are, of course, very instructive. As I began looking for cases of the misuse of statistics, I became awed by how easy examples were to find. I became more and more convinced that a course of this type was needed. Students also relish catching advertisers and public officials misusing statistics—that is, detecting doublespeak.

Doublespeak is the "involved, inflated, and often deliberately ambiguous use of language" (Webster's New Collegiate Dictionary) (see Rank, 1974, and Dieterich, 1976). The misuse of the language of statistics is statistical doublespeak. Statistical doublespeak can be avoided if statistics are properly understood (see Haack, 1977). This is the objective of the course I propose.

It is possible to teach statistics as a language. It is a challenging, yet rewarding undertaking. As you contemplate offering a course of this type, you might want to look at some of the books that can be used as reference material. There are a few good, readable books that may help you teach about statistics, the language.

With emphasis on sample surveys there are:

1. Gallup, G. *The Sophisticated Poll-Watchers Guide.* Princeton Opinion Press, 1972.
2. Roll, C. W., Jr. and A. H. Cantril. *Polls: Their Use and Misuse in Politics.* Basic Books, 1972.
3. Wheeler, M. Statistics. *Lies, Damn Lies, and Statistics,* 1976.

These books lack adequate discussion of the science of studying an existing population but do give a good discussion of the "art" of sample surveying. On the general topic of statistics and statistical doublespeak consider:

1. Bross, I. D. J. *Scientific Strategies in Human Affairs: To Tell the Truth.* Exposition Press, 1957.
2. Campbell, S. *Flaws and Fallacies in Statistical Thinking.* Prentice Hall, 1974.
3. Federer, W. T. *Statistics and Society.* Dekker, 1973.
4. Hauser, P. M. *Social Statistics in Use.* Russell-Sage Foundation, 1975.
5. Huff, D. *How to Lie with Statistics.* Norton, 1954.
6. Messick, B. M. *Mathematical Thinking in Behavioral Sciences.* Readings from *Scientific American.* Freeman, 1968.
7. Mosteller, F. (editor). *Statistics By Example.* Addison-Wesley, 1973.
8. Reichard, R. *The Figure Finaglers.* McGraw-Hill, 1974.
9. Taner, J. (editor) *Statistics: A Guide to the Unknown.* Holden-Day, 1972.

You will find these books to be very interesting. Taner's collection of essays is an excellent source for the statistics course I propose. The essays are on the application of statistics in just about any area that students might have an interest.

REFERENCES

Dieterich, D. (Ed.). (1976). *Teaching public doublespeak.* Urbana, IL: NCTE/Citation Press.

Haack, D. G. (1976). A nonsymbolic statistics course. *Communications in Statistics,* A(5) 10, 943–947.

Haack, D. G. (1977, Fall). Statistical doublespeak. *Kentucky Council of Teachers of English Bulletin, 27,* 1

Haack, D. G. (1979). *Statistical Literacy: A Guide to Interpretation.* North Scituate, MA: Duxbury Press.

Pieters, R. S. (1976). Statistics in the high school curriculum. *American Statistician, 30,* 134–139.

Rank, H. (Ed.). (1974). *Language and public policy.* Urbana, IL: NCTE/Citation Press.

This chapter was originally published as:

Alton, E. V., Gersting, J. L. (1980). An alternative course of the "I hate math and I've never been any good at it" student. *delta-K: Journal of the Mathematics Council of the Alberta Teachers' Association, 20*(1), 4–6.

CHAPTER 22

AN ALTERNATIVE COURSE FOR THE "I HATE MATH AND I'VE NEVER BEEN ANY GOOD AT IT" STUDENT

Elaine V. Alton and Judith L. Gersting

We've all encountered this student. Probably his (or her) dislike of mathematics comes from his previous lack of success with mathematics courses. His understanding of what constitutes mathematics is usually limited to visions of very complex arithmetic manipulations, with perhaps some x's and y's scattered throughout.

A high school student such as this is one whose career goal probably does not include college, or, if it does, not in a discipline where any knowledge of algebra is required. This student is either unprepared for or uninterested in first-year algebra; perhaps even unprepared for or uninterested in a general mathematics course. Many states have a graduation requirement of one year of mathematics. What opportunities can we offer this student to meet such a requirement? Should he be "retreaded" in a course containing the type of mathematics with which he has already had difficulty and which aims to prepare him for a level of mathematics he will probably not use?

Some schools are trying an alternative approach by offering a mathematics course designed to awaken the student's appreciation of mathematics in his world. Facility with computation is not one of the objectives of such a course, but hopefully the student will have a better understanding of mathematical thinking in a wider sense and be aware of uses of mathematics that affect him as a citizen, regardless of his occupation.

At the college level, similar problems exist with a large number of students. Why should an English major learn how to solve a quadratic equation? Can't we offer something more appropriate to satisfy a graduation requirement? In this article, we will discuss an alternative mathematics course which we developed at Indiana University–Purdue University at Indianapolis (IUPUI). There is nothing in the course, either in content or prerequisites, which makes it necessarily a college-level course. In fact, almost the same course is offered in several local high schools. If you are interested in developing an alternative mathematics course for your own school, the following discussion may prove helpful.

In all phases of setting up our course at IUPUI, that is, in the formulation of objectives, selection of text and topics, choice of methods of presentation, et cetera, we attempted to keep in mind the special needs and attitudes of the group of students who would take the course. In particular, we anticipated that many students would dislike or fear mathematics due to past failures; would have little awareness that mathematics enters into aspects of life where, as educated citizens in a technological society, they should strive to make informed decisions; and would think of mathematics as pretty dry and uninteresting stuff. Many of our students did indeed fit this picture, but we were pleasantly surprised at the number who really did not dislike mathematics—they simply had no need for traditional mathematics in their course of study.

At the beginning of the course, we shared with the students three broad objectives which had been formulated, namely:

1. to give you an insight into mathematics as a way of thinking, as an area of human endeavor, an art and a science that has been both useful and interesting to mankind throughout history
2. to help make you aware of the ways in which mathematics touches your everyday life
3. to show you that mathematics can be fun

As part of the course evaluation at the end of the semester, we gave the students another copy of the objectives and asked how well they thought each had been met. There were surprisingly positive comments; in fact, this was overall the most highly rated item in the course evaluation.

Choices that we made in each of the following four categories were influenced by our desire to achieve the course objectives.

TEXTBOOK

We used *Mathematics, A Human Endeavor* by Harold Jacobs, W.H. Freeman and Company, 1970. This is a visually appealing book in an easy-to-read format. It does not overwhelm the student, but at the same time has a number of interesting problems that do require thinking. In addition, there is an excellent teacher's guide to accompany the text, which is filled with ideas for additional discussion, demonstration, et cetera. This guide, unlike many others, is a truly practical supplement.

TOPICS

We covered the following topics from the textbook: inductive and deductive reasoning, number sequences, counting, probability, statistics, topology. In addition, we included a unit on metric measurement and one on computer science. These were done in the same spirit as the material found in the textbook. The text contains enough material to be used for a one-year high school course.

COURSE REQUIREMENTS

Grades in the course were based on homework problems, a series of short tests, a brief term paper, and class participation. Often, the class broke into small groups and the members of each group worked together on homework problems. There was a test after each topic. The tests grew progressively harder during the course of the semester, but the students did not seem to notice, or at any rate did not object.

We provided three topics for the term paper—Fibonacci numbers, magic squares, and the four-color problem. Each of these is a topic that is easy to understand, interesting, and about which there is a lot to say; of course, there is more now to say about the four-color problem than there was when we first taught this course! We provided a short list of references for each topic, but a number of students found additional references and wrote quite good papers. A student could also write a paper on an approved topic of his own choice. Class participation was a component of our grading scheme because it was essential to the success of the method of presentation we used.

METHOD OF PRESENTATION

This was probably the most unique feature of the course. We decided that mathematics did not need to be presented as a spectator sport, and we attempted to involve each student in an active learning experience. We never gave a single lecture in the course. Instead, the students participated in activities or experiments designed to lead them to ask questions and formulate and test conclusions. Some of the activities were individual, some were done as demonstrations by volunteers from the class, but most were done within the small groups mentioned earlier. The membership of the groups developed quite naturally, and as the semester progressed there was a lot of mutual help and support going on in the individual groups.

We used a lot of instructional aids—posters, models, overhead transparencies, handouts, demonstrations, readings or pictures from other books— to stimulate discussions. The teacher's guide to the text provided many excellent suggestions, such as the "paper cutting race," to introduce the Moebius band and the topology unit. At the very least, it was impossible to sit passively in this class, and we certainly kept everyone awake!

There was one more unusual feature to this course. We submitted a course proposal, which was approved a year before the class was scheduled. The semester before the course was offered, both of us were involved in planning the course objectives, content, and requirements, as well as in developing and collecting instructional materials.

When the class actually began, we used what might be called "cooperative teaching." Each of us attended every class, and while one person was responsible for directing the class sessions during a given topic, the other person knew what was going to happen each day and could aid in guiding the discussion, setting up for a demonstration, or working with the groups. The short tests for each topic were designed jointly; we conferred on assignment of final grades. Because this is not a cost-effective way to teach any but a large class, we have not repeated this arrangement, but it was a good way to initiate a course of this nature and was an interesting experiment. You and another teacher might want to try this arrangement if you have the opportunity to combine several classes.

How successful has this course been? From all reports, the high school course has been well received and has apparently reached some previously turned-off students. Our own course at IUPUI has received high student evaluations and has grown slowly, in spite of the fact that mathematics is not a specific graduation requirement for many programs.

If you are interested in beginning such a course at your own school, you may encounter some opposition from faculty or administrators who feel the course is not sufficiently mathematical (read computational). One reply to this is that the student has not benefited much from his past computational courses; at least in an alternative course such as this he has a chance to learn some new mathematical ideas and develop a new and more positive attitude toward mathematics. One of our students said, while working with geometric sequences, "I was never any good at mathematics, but *this is fun.*"

This chapter was originally published as:

Nicol, G. (1984). Expecting girls to be poor in math: Alternatively, chance for a new start. *delta-K: Journal of the Mathematics Council of the Alberta Teachers' Association, 24*(1), 26–27.

CHAPTER 23

EXPECTING GIRLS TO BE POOR IN MATH

Alternatively, Chance for a New Start

Gordon Nicol

Janet Ferguson, federal Minister of Science and Technology, told a conference recently that "half-baked myths about their lack of mathematical ability are responsible for girls dropping out of sciences in high school." Women from across Alberta heard the statement during a three-day conference sponsored by a university organization, Women in Scholarship, Engineering, Sciences, and Technology.

The minister said that during her travels through offices and campuses in Canada, she found "shockingly few women in professional and management positions" and that "girls need more opportunities to play with tools and explore their environments." More than ever, women must understand the impact of technology. "We need to examine our behavior and think about the ideas we're giving to our daughters.... Today, every human is living a life that is drastically altered every minute by science and technology," Ferguson said.

In light of these statements and in light of new federal and provincial initiatives aimed at producing realistic and lasting change, this old topic deserves a fresh look. So important a matter is this to the National Council of Teachers of Mathematics that it produced an official position statement on the subject, as quoted below in its entirety:

THE MATHEMATICS EDUCATION OF GIRLS AND WOMEN

The National Council of Teachers of Mathematics is committed to the principle that girls and women should be full participants in all aspects of mathematics, both as students and as teachers.

Often employment opportunities and continuing educational progress are closed to many young women because of powerful social influences that discourage them from continuing their study of mathematics beyond that required by school policy. Mathematics educators, therefore, must make individual and organizational commitments to eliminate psychological as well as institutional barriers to women in mathematics. Innovative ways must be explored to convince both students and parents of the vital importance of continuing to take mathematics courses so as to keep open both educational and career options.

Each school or school system that does not have an equal proportion of the sexes in its most advanced mathematics classes should examine both its program and its faculty for influences that lead to "math avoidance" by girls and young women.

The teacher is in a key role to stimulate and encourage students to continue the study of mathematics. Teachers at all educational levels must take positive steps and use appropriate learning materials and experiences to overcome the mistaken notion that mathematics is a male domain.

Suitable programs, adequately financed, must be developed to promote the mathematical education of females. Both simple justice and future economic productivity require that we do so without further delay (April 1980).

It's really not enough for us to admit there is a problem, unless we are pre pared to commit a significant amount of resources to remedy the situation. NCTM has, in fact, led the way for us by preparing a program called *Multiplying Options and Subtracting Bias*, consisting of videotapes and workshops.

Shirley Hill, former NCTM President, says that two essential conditions for the solution are (a) an increased awareness of the realities and options on the part of young people (both male and female), parents, teachers, counselors, and others who advise and influence them, and (b) intervention with systematic, deliberate programs to change fallacious beliefs and remove barriers to free choice.

The program preface indicates that it is not enough merely to *tell* females about the importance of mathematics in keeping career and life goals open, nor is it sufficient to *ask* females to change their behavior without changing the complex and embedded societal beliefs and forces operating on them. Rather, the handbook points out that the educational environment, consisting of the significant groups of math teachers, counselors, parents, and male and female students, needs changing. *Multiplying Options and Subtracting Bias* "was designed and developed to change these significant groups' beliefs about women and mathematics as well as to change each group's behavior."

The 192-page handbook introduces the preparation, rationale, objectives, and format of the workshops, qualifications and selection of facilitators, and the evaluation of workshops, format guides, and facilitator resources. These resources include a validation report of the videotape *Multiplying Options and Subtracting Bias* and selected reprints, including *Sexual Stereotyping and Mathematics Learning* (Fennema and Sherman), *Multiple Levels for Change* (Sells), *Math Anxiety* (Zanca), *Counselling the Math Anxious* (Tobias and Donady), and *The Power of the Raised Eyebrow* (Burton). Also included are an annotated bibliography, *Mathematics/Science Intervention Strategies fop Female Students* (Fennema, Caretta, and Pedro), and evaluation instruments.

The program includes a student workshop, a teacher workshop, a counselor workshop, and a parent workshop, each of which includes an overview, activities, frequently asked questions or comments with suggested responses, overhead masters, handout masters, and a 30-minute videotape available in four formats.

Now is an opportune time to provide a full hearing of the subject. Anything less would be unfair, inconsiderate, and perhaps negligent. After all, mathematics is everybody's business.

This chapter was originally published as:

Sigurdson, S. E. (1986). A constructivist approach to teaching mathematics. *delta-K: Journal of the Mathematics Council of the Alberta Teachers' Association, 26*(1), 8–13.

CHAPTER 24

A CONSTRUCTIVIST APPROACH TO TEACHING MATHEMATICS

Sol E. Sigurdson

Editor's Note: During the school year 1985–1986, Dr. Sol E. Sigurdson was on sabbatical leave from the University of Alberta, where he taught methods and graduate courses in mathematics education. His interests focus on classroom change brought about by inservice and curriculum change.

Over the last few years, psychologists and educators have been interested in going beyond behavioristic and Piagetian views to new conceptualizations of learning, especially in using the computer as a "model" of how we think and learn. One of the new conceptualizations has been "information processing." Proponents of this view claim that when we think, we basically process information; it's as simple as that. This, however, leads to a further question: How do we manage this processing? To account for the management of processing, it is suggested that the learner engages other processes called metacognitive processes. But, still we might ask: What manages the metacognitive processes? Although this is not a trivial question, most proponents presently do not differentiate between levels of management,

Selected Writings from the Journal of the Mathematics Council of the Alberta Teachers' Association, pages 177–184
Copyright © 2014 by Information Age Publishing

simply naming all those processes above the cognitive level metacognitive processes. In fact, the difference between cognitive and metacognitive is not always clear. For the time being, let us say that strictly mathematics propositions, procedures, and processes are called cognitive, while management decisions about such matters as when to use them, in what order, and with what degree of confidence are called metacognitive processes.

Another related view of learning has been called a theory of "personal constructs." The main tenet of this view is that all learners actively construct theories, no matter how minor, about what is appropriate action for responding to any particular situation. If a particular theory leads to inappropriate action, we revise the theory. This view, like information processing, also utilizes the notion of "metacognitive processes" managing our theory development. According to the personal constructs view, learners of differing capabilities exist because both our cognitive capacities and our metacognitive (management) capabilities differ. Another explanation, which goes beyond differences in cognitive or metacognitive components, is that some learners' perceptions are blinded (by emotion, say), so that they are unable to differentiate between appropriate and inappropriate action and, consequently, construct poor theories.

What relevance do these new conceptions have to the mathematics classroom? The one outstanding impression that the personal constructs view leaves is that our classrooms consist of 25 or so finely tuned, sensitive, self-initiating, theory-generating, learning "beings." The metacognitive aspect, on the other hand, leads us to question how much of a commitment we teachers have in attending to the development of metacognitive processes. The information processing aspect begs the question of how to present information for efficient storage and easy access. Psychologists and educators are still exploring answers to these questions and will be for many years. In the meantime, what aspect of these theories can be useful to teachers in dealing with the complex world of classroom instruction?

In order to make these ideas more available for teacher use, I will combine the three notions—information processing, personal constructs, and metacognitive processes into one "constructivist" view of learning. In this article, I will describe constructivist principles of learning and further derive from them constructivist guidelines for classroom teaching of mathematics. Mathematics teachers are encouraged to think about, and use, these ideas to improve their classroom instruction. Psychologists and educators, who are continually striving for new insight into the learning process, would surely appreciate feedback from the most significant learning laboratory of all, the classroom. Curriculum examples will not be used to describe this view because these new conceptions of learning are equally relevant to all grade levels. The word *constructivist* has been around for many years. I am not concerned that my usage may be slightly different than that of others.

CONSTRUCTIVIST PRINCIPLES OF LEARNING

1. Purposeful Constructions

Students construct their own theories for responding to a given situation, and, as they see their knowledge leading them to inappropriate action, they revise their theories. Learning proceeds from the current conceptions or theories of knowledge that the learner possesses. "Tuning," that is, modifying or adjusting, is an important learning process. Appropriate theories are best constructed in the light of some acknowledged purpose.

2. Learning How to Learn

Learners' awareness of their knowledge (mathematical content and processes, and metacognitive processes) at any time aids learning. Metacognitive processes (management of cognitive knowledge) are especially important, and these may be a major source of individual differences between slow learners and others.

3. Confidence

Because learning means taking risks and experimenting with new cognitive constructions, the atmosphere for learning must be familiar and full of trust. Inaccurate perceptions can be caused by either strong positive or negative emotions.

4. Framework for Information

Learning occurs in a context that provides a framework for the organization of information. The most appropriate context is one which is most applicable to the future situation in which the knowledge will be used. A framework for mathematical knowledge can consist of mathematical, everyday, and scientific elements.

5. Structure of Knowledge

All mathematical knowledge consists of propositional (conceptual and relational) structures and procedural (algorithmic and methods) structures. The process through which we understand and manipulate mathematical situations is grounded in specific content structures.

6. Complexity of Concepts

Propositional structures and procedural structures are complex content structures, a fact which is often disguised through rote learning and teaching. Although, traditionally, we teach through analyzing and breaking down knowledge, the constructivist sees "building up" as an equally valid learning process. Procedural structures (algorithms) are linked in important ways to propositional structures (concepts).

7. Transfer of Knowledge

As we learn, we learn context, as well as content and process. Transfer of knowledge must not be assumed; it occurs only as a new context is seen as the learned one.

Although a deeper understanding would require considerable elaboration on all of these principles, perhaps we can employ a constructivist teaching tactic, and let the reader come to understand the principles as they are used to develop the "guidelines for classroom teaching." Classifying something as complex as human learning in "seven principles" seems to be an utterly futile undertaking. However, I would like to elaborate slightly on the structure and complexity principles. Recognized in the structure principle, first of all, is the importance of relationships among all mathematical concepts and that any understanding of mathematics is a matter of recognizing all these relationships. Also implied in the structure principle is that all mathematical activity, such as problem solving, is highly dependent on these structures. The complexity principle, while acknowledging the many-faceted aspect of even apparently simple concepts such as multiplication, stresses that understanding and use of knowledge must take into account all, or most, of these facets.

Of course, these learning principles can be applied to the teaching of any subject, but our concern here is what this might mean for the teaching of mathematics. In deriving these guidelines for classroom teaching, it became apparent that several possible interpretations would be valid. Once again, I have opted for seven, knowing that these can only serve as general suggestions.

CONSTRUCTIVIST GUIDELINES
FOR CLASSROOM TEACHING

1. Unit Context

Mathematics should be taught in the context of a three- to four-week unit constructed around a mathematical, everyday, or scientific application of the content. Students should feel comfortable and familiar with this application context.

Rationale: The purposeful constructions and the framework principles are satisfied by this. The actual application context would not only be a function of the content, but also of the grade level of the class, the characteristics of the students, and the school environment.

2. Curriculum Tasks

The tasks which comprise the unit should be conducted with a view to the students engaging their current conceptions, mastering task, and learning from it. The focus of the task should be central to the unit application.

Rationale: The learning how to learn and the confidence principles suggest that the task be a manageable part of the unit. The structure principle suggests that relevant mathematics knowledge be an integrated part of the task.

3. Managing the Task

All students should be given assistance in dealing with the task—determining task difficulty, monitoring their understanding of it, apportioning time for it, and predicting how well they can perform it. The teacher should pay special attention to the students' perception of the task. Individual differences should be noted and provided for in this aspect.

Rationale: The purposeful constructions and learning how to learn principles are important here, especially in helping students become aware of their knowledge and knowledge processes. This guideline is the core of the instructional process.

4. Task Variety

Tasks should include a range of learning activities, such as direct examples, reviewing, textbook use, note taking, concrete materials, understanding, amplification of basic concepts, problem solving, self-inquiry, practice exercises, group activities, discussion, and questioning.

Rationale: The purposeful constructions principle does not imply that student learning should be of a discovery nature, but only that learning should have some purpose. The complexity principle not only suggests that a considerable amount of guidance, even direct examples, is appropriate, but also that a variety of approaches is necessary to achieve an understanding of a mathematical topic.

5. Assessment Tasks

Assessment should be carried out primarily within the context of the unit.

Rationale: The transfer principle suggests that we should first apply learning to the context of the unit. If we do testing beyond the context of the unit, we should be conscious of how the new context relates to the learned one. In actual (real-life) use of mathematics, contexts that are important to the student are most often familiar ones.

6. Mathematical Learning

(a) Readiness

Readiness for content learning must be noted, but only in the context of the learning task. What does the learner bring to the situation? Students' awareness of their own readiness is also important.

Rationale: Purposeful constructions are derived from previous "theories" that the student has. This is the central premise of the constructivist view. The learning how to learn principle suggests a self-awareness of these previous theories.

(b) Concepts

Concepts, the pivotal ingredients of mathematics learning, must be constructed from the student's prior knowledge. Learning of complex subject matter is achieved through many different propositional structures. Specific instructional devices, such as concept maps and structured apparatus, should be employed.

Rationale: The framework, structure, and complexity principles all indicate the necessity of a thorough conceptual basis for mathematics learning.

(c) Skills

Skill development, as it relates to the curriculum unit, is important. Care should be taken in selecting the application context for curriculum units. Skills and algorithms (procedural structures) are founded upon certain propositional structures. Skills should be learned as broader "method" approaches.

Rationale: Although our principles do not address the matter of skills directly, the structure principle advocates a solid basis for all procedures, while purposeful constructions implies that all skill learning be in context.

(d) Applications

All applications occur in the context of the unit. They should be dealt with as an indication of the use, and usefulness, of mathematics, and also as a way of relating the real world to the development of mathematics.

Rationale: The framework principle means that applications can be an important contribution to the framework for learning mathematics. The purposeful constructions principle suggests applications as a primary reason for studying mathematics. Lastly, the teacher must be constantly aware of transfer and the problem of the context of learning.

(e) Problem Solving

Problem solving should be approached through a study of the particular kinds of problems in each unit. Problem solving is a particular way of knowing content.

Rationale: The structure principle suggests that all mathematics is dependent on specific knowledge. The metacognitive processes of the learning how to learn principle manage only cognitive knowledge. A constructivist view does not support broad generalizable problem solving strategies.

7. Goals of Mathematics Learning

The major goals of mathematics teaching are that students gain understanding of complex areas of mathematical knowledge, use this knowledge in relevant situations, and understand their own processes and capabilities for functioning in a mathematical environment.

Rationale: The constructivist view not only provides new insight into how mathematics should be taught, but also implies a somewhat revised goal for mathematics teaching; "practice, feedback, and coaching" are not enough. Although the view expands upon what understanding means, one of the more interesting issues it raises is how teachers should regard their efforts toward improving students' capabilities for learning how to learn.

The strongest message of a constructivist approach is the desirability that teachers make clear to themselves and to students the purpose of learning mathematics. Making clear the purpose, without trivializing it, will be of great benefit in improving mathematics teaching. At this writing, I believe the weakest part of these guidelines is the matter of "context" and, therefore, the matter of what a sensible unit context might be. It seems essential that the context include, but go beyond the bounds of, mathematics itself. It certainly need not be confined to students' interests. Plausibility to the student might be a better guideline. Clearly, the broader the context, the more mathematics it will subsume. However, the greater breadth might tend to lose focus. Also, the notion of curriculum task and its position between the unit context and mathematics to be learned is somewhat problematic. An appropriate resolution of these weaknesses will need to be worked out in light of both the proposed principles of learning and the other guidelines.

Obviously, this interpretation of the constructivist perspective leaves many gaps. If a teacher were to conduct lessons solely on the basis of this statement (even assuming the availability of a textbook), I would predict chaos. The statement can only be seen as an attempt to modify already competent practice. Certainly, these are not prescriptions for teaching. Rather, I see them as interesting guidelines that can be tried, discussed, revised, and reinterpreted. A constructivist would see a teacher interpreting these guidelines on the basis of the teacher's existing "theories," and then, perhaps, rejecting them as invalid or "tuning" existing theories, using them, and then revising or discarding them.

At the very least, these guidelines should provide the basis for an interesting curriculum unit which would go far in explicating the guidelines. This would provide an opportunity for psychologists to say that their views have been misread or misinterpreted, which would be very useful. It might even serve to have them rethink their ideas in the light of feedback given by teachers. Whatever happens, teachers of mathematics are obligated to begin investigating ways that these new conceptualizations of learning can benefit them. Teachers certainly owe it to themselves and, in some sense, they owe it to psychologists and educators who are searching for new insight into the very important but, too often, frustrating process of learning mathematics.

BIBLIOGRAPHY

Claxton, G. (1984). *Live and let live: an introduction to the psychology of growth and change in everyday life.* London, UK: Harper and Row.

Cobb, P., & Steffe, L. P. (1983). The constructivist researcher as teacher and model builder. *Journal for Research in Mathematics Education, 14*(2), 83–94.

Frederiksen, N. (1984). Implications of cognitive theory for instruction in problem solving. *Review of Educational Research, 54*(3), 363–407.

Magoon, A. J. (1977). Constructivist approaches in educational research. *Review of Educational Research, 47*(4), 651–693.

Posner, G. (1982). A cognitive science conception of curriculum and instruction. *Journal of Curriculum Studies, 14*, 343–351.

Shavelson, R. J. (1981). Teaching mathematics contributions of cognitive research. *Educational Psychologist, 16*(1), 23–44.

Wagner, R. K., & Sternberg, R. J. (1984). Alternative conceptions of intelligence and their implications for education. *Review of Educational Research, 54*(2), 179–223.

This chapter was originally published as:

Le Maistre, K. (1988). If this is television, shouldn't my intelligence by insulted? *delta-K: Journal of the Mathematics Council of the Alberta Teachers' Association, 27*(1), 5–6.

CHAPTER 25

IF THIS IS TELEVISION, SHOULDN'T MY INTELLIGENCE BE INSULTED?

Kate Le Maistre

Editor's Note: Kate Le Maistre is mathematics and science consultant with Jerome-Le Royer School Commission, English Educational Services, Ville d'Anjou, Quebec. Le Maistre presented a paper to the NCTM Canadian Conference held in Edmonton in 1986.

THE FREEDOM MACHINE

Those of us who have the responsibility of meeting a class of children up to 200 times a year and teaching them a subject like mathematics need help from time to time. My favorite aid is the videocassette recorder (VCR).

The VCR can free us from complicated technology and from network schedules. Even the manufacturers of VCRs have been surprised by the popularity of the flat box that so many of us store underneath the television set. Two factors that have contributed to this popularity are the convenience and ease of use of the VCR. We can "time-shift" or watch programs at a time convenient to us, not when the networks feel that we should watch them. This is important to a classroom teacher, who can show some or all of a program when it fits into the lesson plan, not necessarily when it is being transmitted. If you have not yet bought a VCR for your home, the total in-service time needed to train the most unmechanical person to insert, play, rewind and eject a tape is about five minutes. In seven minutes, you can find how to fast forward, pause, put a transparency on the screen, trace the picture for later use on an overhead projector and probably come up with several innovative ideas of your own.

If all else fails, ask one of the children in your class to show the tape; after all, your students are growing up with VCRs as we did with radios.

Compare the ease of operating a VCR with the agony we used to go through in setting up a 16 mm movie. I thought life would be easy when my school bought a self-threading projector, but, although I didn't tear the sprocket holes from as many films, when I stopped the film to ask a question or to explain a point, someone near the projector was bound to call out: "Miss! I can smell something burning!"

HOW CAN I COMPETE WITH "MIAMI VICE"?

A recent study found that "young people between the ages of six and 11 watch, on the average, 27 hours of television a week" (Merrow, 1985). As teachers, we cannot control what children watch at home—and there are times when we all like to spend an hour or two looking uncritically at television. While our students are spending so much time using television as recreation, they are not reading books, yet our teaching methods remain textbook-based. I am not suggesting that because children do not read books for entertainment, we should not use books at school. Quite the contrary; part of our role is to encourage students to use as many sources of information as possible, and any good teacher will use as many strategies as possible to get through to students. We are hiding our heads in the sand if we do not use a strategy that is both familiar and attractive to students.

"Talk" and "chalk," closely followed by worksheets, seem to be the basics of the teacher's arsenal. Apart from the occasional overhead projector, the tools in many classrooms have not changed appreciably since the days of the slate and pencil. Yet we are trying to use these archaic tools with a generation of students who are often more technologically sophisticated than we are.

The most recent mathematics programs available on instructional television contain what the professionals call "production values," or what we might call "razzle-dazzle," attention-grabbers or—shades of teachers' college—motivators. These programs include action shots of children of the same age as the intended audience, animation, outside locations,

simulations and computer-generated graphics. If anything is going to compete with the glitzy productions of the big networks, these programs stand a good chance.

I THOUGHT WE WERE SUPPOSED TO ENCOURAGE ACTIVE PARTICIPATION

Here is a typical scenario for using videotapes in a mathematics class: A grade 5 teacher finds that her classes on the comparison of decimals have not been as successful as usual. She looks at the teachers' guide to Math Works and decides that the title of program 15 is promising. The program summary describes a sequence in which two girls are running a race timed to hundredths of a second; an animation sequence involving the Three Bears; another short story in which two boys discover that 0.5 kg of meat is more than 0.33 kg; a gemologist explaining why decimals are important in his work; and a summary sequence in which two girls apply their knowledge of decimals.

The teacher then previews the tape to decide where to place it in her lesson plans. Should she use the whole program? Some scenes only? Should she do the suggested activities before showing the tape? At some point during the program? As a follow-up? Should she try to fit the entire 15-minute program into one class? Are there enough concepts to make several lessons?

Next, she prepares the program activities in the teachers' guide as well as any others that are appropriate, copies any assignment sheets and collects necessary materials. Again the teachers' guide is useful because it provides a black-line master and lists of simple materials used in the activities.

At this point, the only thing left is to show the tape and watch the children's faces. If you've never done this before, be prepared for a pleasant surprise.

I am not suggesting that playing videotapes should be the only, or even the major, activity in mathematics classes. To suggest as much would mean simply replacing one provider of information (the teacher) with another (the "box"). Rather, I am suggesting that a videotape can act as a motivator, as an introduction to a topic, as a unit review, as one more version of the same information, as a tutorial for an absent student—in fact, as one more tool in the teacher's toolbox.

After all, when you're a teacher in front of a class, isn't it reassuring to have a variety of tools in the toolbox?

SUGGESTED MATHEMATICS PROGRAMS

"Two Plus You"
"Math Patrol"
"Math Patrol 2"
"Math Patrol 3"
"It Figures"
"Landscape of Geometry"

Check the ACCESS listings for other mathematics programs available on instructional television.

REFERENCE

Merrow, J. (1985, November). Children and television: Natural partners. *Kappan*, p. 212.

CHAPTER 26

LOGO

An Opportunity for Synthesis, Self-Control and Sharing

J. Dale Burnett

Editor's Note: J. Dale Burnett is an associate professor in the Faculty of Education, the University of Lethbridge. Dr. Burnett teaches courses on applying computers in education. This paper was presented at the Seventh National Congress of the Council for Exceptional Children held in Regina, Saskatchewan, in October 1986.

INTRODUCTION

Logo is a computer language specifically designed for children. I am not implying that it is a "kiddie language" (Logo is suitable for graduate work in computing science) but rather that the syntax and the domains of inquiry are readily accessible to young children. Logo is successful with very young children (Lawlor, 1985), with physically disabled students (Goldenberg, 1979), with students with learning disabilities (Weir & Watt, 1981), as well as with students in regular classrooms and in gifted programs (Carmichael et al., 1985).

Selected Writings from the Journal of the Mathematics Council of the Alberta Teachers' Association, pages 189–199

What are some of the reasons for Logo's success? *Synthesis, self-control,* and *sharing* (the three *Ss*), plus the teacher, are key factors in Logo's success.

Synthesis refers to the natural necessity to build on one's previous knowledge (the Piagetian concept of constructivism), using both real world knowledge as well as a growing understanding of the rules of the Logo language. Self-control flags the value of permitting the learner to have a substantial degree of autonomy in what tasks are set and in the method of approaching them. Sharing refers to the social context in which much Logo activity occurs. Students helping students and feeling good about it (and about themselves) are common features of many Logo settings (Carmichael et al., 1985).

SYNTHESIS

I will discuss the nature of synthesis at two levels. After reviewing how the concept of synthesis fits into current psychological theory, I will show how synthesis can be applied to the situation of an individual facing his or her first exposure to Logo. The second subsection then shows how this theory might apply to the situation of an individual learner, faced with their first exposure to Logo.

Synthesis and Psychological Theory

The educational community owes an enormous debt of gratitude to a self-proclaimed non-educator: the Swiss psychologist-epistemologist Jean Piaget. When one hears Piaget's name, one immediately thinks of children and of stages. The first association is a good one, the latter misleading. His substantial contribution to current psychological perspectives was not the idea of stages (which suggests that development is discrete rather than continuous) but rather that of development. Development implies change and growth. The purpose of education is to facilitate development. Change and growth are our mandate.

Equilibration is the term used to describe the underlying process of mental development by which individuals organize their ideas into noncontradictory wholes. This process occurs through the complementary subprocesses of assimilation and accommodation. Because individuals draw heavily upon what they already know (their present cognitive structure), the label *constructivism* can be used to describe theoretical perspective.

Traditionally, developmental literature has treated only a particular subset of the total picture: cognitive development, physical development, social development, moral development, and emotional development are all familiar terms. The thesis is that such exclusions or restrictions, while well intentioned ("let us control for all possible sources of variation except one, and then observe the effect of this one remaining factor"), are fundamentally misguided. The resulting information is misleading because individuals never find themselves in such controlled situations outside of the research environment. Classroom practices or curriculum guidelines that fail to take this natural complexity into account are inappropriate.

Figure 26.1 illustrates the interaction between mental development and other developmental factors. The arrows indicate a posited causal effect. Thus an increase in emotional development causes an increase in mental development. Similarly, an increase in mental development causes an increase in emotional development. The positive signs beside each arrowhead indicate that the relationship between the two nodes is in the same direction (for example, an increase in one causes an increase in the other, or a decrease in one causes a decrease in the other). Second, a distinction is made between mental development and cognitive development. The intention is to distinguish between overall mental development, which might include feelings and intuitions, and the more restrictive conceptual domain of cognitive development. The charting conventions follow those outlined by Roberts et al. (1983).

The psychological literature of the last decade has increasingly focused on cognitive approaches and the literature on learning emphasized individual's building upon their previous knowledge and experience. The cognitive emphasis has also expanded to encompass not only strict rational and logical perspectives but also emotional, affective, and social components. Psychology is becoming both more holistic and more philosophic (for example, is knowledge constructed or discovered?) as professionals (for example, Solomon, 1986) begin to reflect on the conceptual underpinnings of many of their ideas.

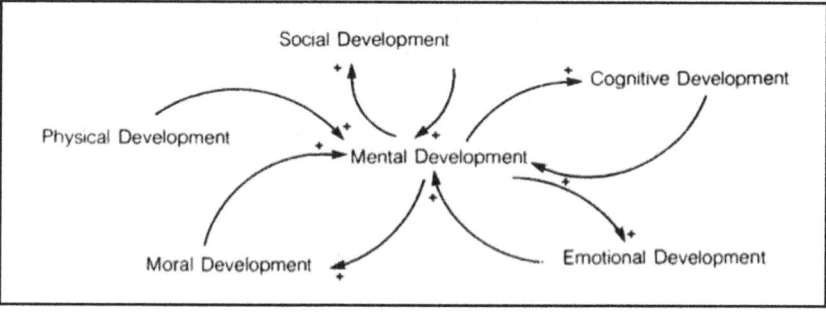

Figure 26.1 Casual loop.

In addition to receiving contributions from philosophy, cognitive science (as the new discipline is called) has been strengthened by ideas from computing science. At first glance, this development seems remarkable, since one field is concerned with human ideas and nature while the other is ostensibly interested in machines and electricity. The term "artificial intelligence" is familiar to most people (Winston, 1977; Haugeland, 1985), and the term "expert system" is beginning to appear in the educational literature (Hayes-Roth et al., 1983; Van Horn, 1986). However, lest the novice become enamored too quickly with these new ideas, cautionary notes have also appeared (Dreyfus & Dreyfus, 1986; Weizenbaum, 1976).

Thus, cognitive science is practising what it preaches: The discipline itself is synthetic, building on any relevant bit of knowledge. One branch of computing, system dynamics, has taken the idea of modeling and simulation, combined it with the biological concepts of feedback and used the idea to construct conceptual as well as computer-based models of phenomena. One of the first applications of this approach to reach the public's attention was the Club of Rome's famous publication Limits to Growth (Meadows et al., 1972), which attempted to construct a model of the world showing interactions among population, agriculture, industry, pollution, and natural resources. The same approach clarified the complexities of mental development. Diagrams and models should not be viewed as right or wrong but as appropriate or inappropriate for a given purpose or function. As the function changes, so may the model. Thus the previous model is explicated to reveal various kinds and interactions of development.

Synthesis in the Individual Learner

Figures 26.2, 26.3, and 26.4 illustrate a particular perspective on learning.

Figure 26.2 shows how a person's knowledge is positively related to the individual's receptivity to new events, which in turn is positively related to

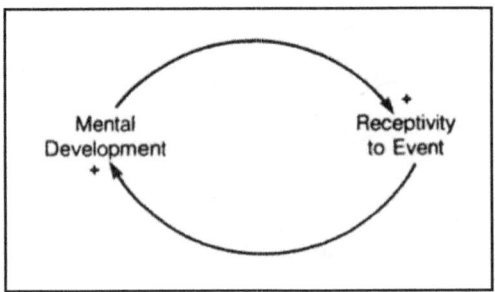

Figure 26.2

their knowledge. The net effect is a "constructive circle" in which learning begets more learning. "Receptivity to an event" is closely related to Vygotsky's (1962, 1978) concept of "zone of proximal development."

Both nodes clearly require amplification, and new nodes and causal arrows need to be identified. One might assume that knowledge consists of knowledge about using computers and about the specific language of Logo, as well as "other knowledge" that may have a bearing on the present situation. The other knowledge may be very important. Existing knowledge about Logo may be zero: the individual may never have heard or seen it before. Existing knowledge about computers is not likely to be zero (most people have at least heard of them and have seen pictures of them), but it may be very limited (the person may not have actually touched one or watched someone else use one). (See Figure 26.3.)

We now have three knowledge nodes or "containers," plus one receptivity node. Thus the Logo node contains the amount of Logo knowledge that the user brings to the task (assumed to be zero). We now insert a brief exogenous variable: the instructional event. This event consists of a brief demonstration of the Logo commands "FORWARD" and "RIGHT."

An alternative representation may place more focus on the centrality of the Logo experiences. Consider Figure 26.4.

The effectiveness of this instruction depends upon (1) what the student already knows, (2) the student's attitude toward Logo, and (3) the student's generic ability to learn. All three factors are "within the student." External factors include (1) the teacher's attitude toward Logo, (2) the teacher's attitude toward the student, (3) the teacher's understanding of Logo, and (4) the actual instructional sequence. The addition of these nodes further

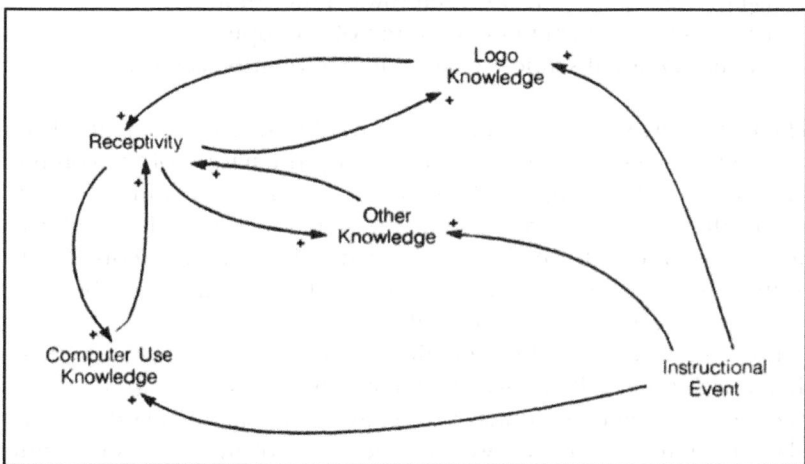

Figure 26.3

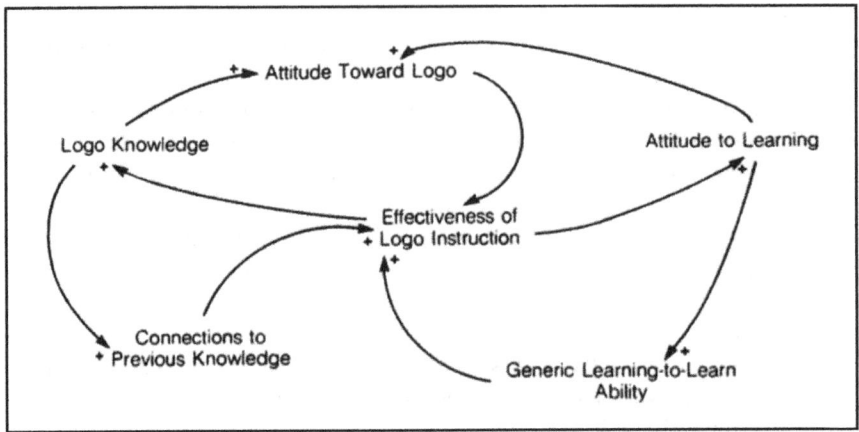

Figure 26.4

complicates the situation, but the nodes may be important. Constructing a pleasing diagram is less important than constructing an adequate
explanation.

Clearly, the effectiveness of instruction should not be viewed as a simple
topic. We immediately realize that instruction is enhanced when

1. students can relate the instruction to what they already know
2. students are positively disposed toward the topic
3. students are positively disposed toward learning
4. teachers are positively disposed toward the topic
5. teachers are positively disposed toward the student
6. teachers have a firm understanding of the topic
7. the instructional sequence takes the above into account

The preceding summary is important not because it is particularly novel
or complex but because it permits us to grasp the nature of the complexity
"at a glance." Many people will consider the model incomplete, but additional information can be added. Another difficulty with the above approach is the ease with which we can construct alternative representations
with little basis for choosing among them. Then again, it may be that the
alternative representations are equally appropriate.

Let us consider some of the details that might indicate that the first experience with Logo will be successful. First, the chances are quite good that
novices (preschoolers or adults) will relate the commands FORWARD and
RIGHT to their existing real-world experience of moving about. Indeed,
this approach is no accident and was specifically built into the design of the
original Logo. Drawing is also an early experience for virtually all children.

The particular terminology and certainly the syntax may be new, but the general context should strike a responsive chord in most learners. Thus the first condition is likely to be met, at least to some degree. However, the student's initial attitude toward Logo is more difficult to estimate and is likely related to whatever attitudes the student may possess regarding computers. Attitudes may be positive, negative, or neutral. A strongly negative attitude may well affect the outcome. The student's attitude toward learning is also important. A positive attitude ("learning something new is fun") is a substantial asset; on the other hand, a negative attitude ("school is boring") is a handicap. With young children, all three factors are often positive, perhaps explaining the level of success of introducing young children to Logo.

Three teacher factors were also identified. The teacher's attitude toward Logo is important. A skeptical or negative rating may be a powerful determiner of the outcome. We should not assume that the computer or Logo is a positive enough factor to compensate for a teacher who does not believe in using it. Similarly, teacher attitudes toward particular students should not be ignored. Sometimes teachers may feel that they are teaching the whole class, but this is not the case. Students filter the information as though it were directed at them. If previous events indicate that the teacher does not respect or value a particular student or students, then the teacher's impact is diminished, if not eliminated. A lesson that "looks good on videotape" may be entirely negated by an event that occurred two weeks earlier in the classroom.

The teacher's understanding of Logo is also important. Teaching a subject that one does not understand is indeed difficult; this applies to Logo. Finally, the particular instructional sequence is important and will be discussed later.

SELF-CONTROL

I have already acknowledged the contributions of Jean Piaget: I will now do the same for Carl Rogers. Although Rogers has written many books, I will quote from one, *Freedom to Learn for the Eighties*. The introduction contains this statement:

> It [the book] appears in a peculiar time in our history when many are saying that we must teach only the "basics," that we must tell children what is right and wrong, that we must teach them to obey and follow.... They hold that students are in school to be taught, not to discuss problems or make choices. (Rogers, 1983, pp. 1–2)

Rogers (1983, p. 18) states that "the primary task of the teacher is to permit the student to learn." He then distinguishes between meaningful learning and learning that has no personal meaning and only occurs "from the neck up." Rogers says that meaningful learning has five characteristics: It has a quality of personal involvement, is self-initiated, is pervasive, is evaluated by the learner, and has meaning as its essence.

Other authors have noted this distinction between meaningful and meaningless learning. I wish to highlight for a moment the second characterization—that of self-initiation. Noss (1984) focused on the related issue of ownership. Either by design or by default, many students engaged in Logo activities have had opportunities to ask their own questions, to set their own tasks and to explore their own ideas. Such events are a sad reflection of our present educational system. As a result, we have little information on what occurs in such situations. However, the findings of a number of studies in which this was allowed to occur (Carmichael et al., 1985; Lawlor, 1985; Noss, 1984; Watt, 1979) all point in a positive direction.

SHARING

Maslow (1970) acknowledged the importance of sharing in his hierarchy of motives by placing it just after basic physiological and safety needs. Yet, my experience shows that most educators view Maslow's hierarchy as something to be memorized for a psychology exam rather than as something to consider in designing the curriculum. The Logo community may, in part, be responsible for resurrecting the idea of sharing. As a result, many of the exciting events surrounding Logo experiences have a highly social flavor to them. Researchers gathering data on this dimension are impressed by its richness (Carmichael, 1985). Others have failed to look for it, have not noticed it or have set up an environment to prevent it from happening (since it might contaminate the results).

The issue has broader implications. What is the role of sharing in the school environment? What are the relationships between sharing and individualized instruction and cheating? How much of the school curriculum explicitly gives students an opportunity to share? If this number is low, then why would we expect our graduates to be adept at sharing or working together? The potential for sharing occurs at many levels as well: There can be sharing among classmates working on the same task. There can be sharing among students of different classes or grade levels. For example, grade 6 students could work with grade 3 students, or gifted students could work with students with learning disabilities. It is naive to assume that the primary domain of learning is at the level of the subject matter. Finally, there is the sharing between student and teacher. One of my favorite anecdotes from Papert's *Mindstorms* is that of a student who, working with his teacher on a problem, suddenly says, "You mean you really don't know!"

TEACHER

What is the primary function of education and what is the role of the teacher in facilitating this function? The first question is dangerous because it appears to imply a single answer. Perhaps a more appropriate question is "What are some of the principal functions of education?" This question at least leaves the door open for new ideas that may have been missed in an earlier formulation. One such idea is "learning how to learn." Novak and Gowin (1984) asked, "How can we help individuals to reflect upon their experiences and to construct new, more powerful meanings?" (p. xi). They go on to say:

> Whereas training programs can lead to desired behaviors such as answering math problems or spelling correctly, educational programs should provide learners with the basis for understanding why and how new knowledge is related to what they already know.... (p. xi)

Perhaps we have passed through an era when American behavioristic perspectives have held sway (What can students do? What are your behavioral objectives? What are the scores on standardized tests?) and are entering an era, also with a strong American flavor, when we ask, "What do students understand? The difference in perspectives is fundamental. Unfortunately, much of our current practice is based on a perspective that may be outmoded philosophically, psychologically, and educationally. The shift toward understanding is not as simple as learning a new instructional technique. What is learning? How do students learn? What is the proper relationship of teacher to student? Alternative teacher roles vie for attention: distributor of facts, organizer of drill programs and seatwork, facilitator, encourager, and fellow explorer.

The teacher has an important role to play. For example, it is an error to assume that the synthesis within the mind of a student first exposed to Logo occurs naturally and spontaneously. The teacher can facilitate learning by bringing some of these potential connections into explicit awareness. Thus FD 100 may be related to FD 50 or to BK 100 or to FD-100. Relationships to movement commands in English or to another language may be observed. How would you tell someone from Japan to go forward? How would you tell Logo turtle to go forward? Exploring relationships among numbers may provide an excellent introduction to mathematics and the relative magnitudes of different numbers. Turtle steps and metric units may be viewed as analogous. How would you tell a robot to go forward? Now the class can discuss robotics for a while.

Connections abound. The secret is to look for them. Another example of looking for connections occurs at the meta level of problem solving when the teacher suggests that a student "play turtle" in order to figure out how to draw a particular figure with a sequence of Logo commands. Other suggestions such as breaking a complex problem into a number of simpler subproblems or developing an overall structure to the solution can be related to other nonLogo activities such as writing a term paper, baking a cake, or studying for a history exam.

The teacher should be aware of numerous potential connections: the relating of Logo commands to one another, the relating of Logo to other non-Logo environments, the relating of Logo problem solving to generic problem solving. Everyone should be alert for connections between the specific situation and other knowledge. As a result, the student should see that learning Logo is much like learning anything else. The synthesis should include not only low-level activities such as learning how to use Logo language, but higher meta-level activities such as debugging, planning, organizing, problem solving, attitude awareness, communicating, and sharing approaches and strategies. The basics of education may be at the other end of the continuum from where we have been looking. It may be very difficult to show some of these connections empirically. That does not necessarily mean they do not exist but that our current research procedures are at fault.

Education also benefits from a little faith. On the other hand, researchers must continue their efforts to provide further insights into our understanding of the learning process.

REFERENCES

Carmichael, H. W., Burnett, J. D., Higginson, W. C., Moore, B. G., & Pollard, P. J. (1985). *Computers, children and classrooms: A multisite evaluation of the creative use of microcomputers by elementary school children.* Toronto, ON: Ontario Ministry of Education.

Dreyfus, H. L., & Dreyfus, S. E. (1985). *Mind over machine: The power of human intuition and expertise in the era of the computer.* New York, NY: Free Press.

Goldenberg, E. P. (1979). *Special technology for special children.* Baltimore, MD: University Park Press.

Haugeland, J. (1985). *Artificial intelligence: The very idea.* Cambridge, MA: MIT Press.

Hayes-Roth, F., Waterman, D. A., & Lenat, D. B. (Eds.). (1983). *Building expert systems.* Reading, MA: Addison Wesley.

Lawlor, R. W. (1985). *Computer experience and cognitive development.* Chichester, UK: Ellis Horwood.

Maslow, A. H. (1970). *Motivation and personality* (2nd ed.). New York, NY: Harper and Row.

Meadows, D. H., Meadows, D.L. Randers, J., &Behrens, W.W. (1974). *Limits to growth.* London, UK: Pan Books.

Noss. R. (1984). Children Learning Logo Programming. Chiltem Logo Project. Interim report No. 2.

Novak, J. D., & Gowin, D. B. (1984). *Learning how to learn.* New York, NY: Cambridge University Press.

Papert, S. (1980). *Mindstorms: Children, computers and powerful ideas.* New York, NY: Basic Books.

Roberts, N., Anderson, D., Deal, R., Garet, M., & Shaffer, W. (1983). *Introduction to computer simulation.* Reading, MA: Addison-Wesley.

Rogers, C. (1983). *Freedom to learn for the eighties.* Columbus, OH: Charles E. Merrill.

Solomon, C. (1986). *Computer environments for children.* Cambridge, MA: MIT Press.

VanHorn, M. (1986). *Understanding expert systems.* Toronto, ON: Bantam Books.

Vygotsky, L. S. (1962). *Thought and language.* Cambridge, MA: MIT Press. (Original work published in Russian in 1934.)

Vygotsky, L. S. (1978). *Mind in society.* Cambridge, MA: Harvard University Press.

Watt, D. (1979). *Final Report of the Brookline Logo Project Part III: Profiles of Individual Students' Work.* MIT Logo Memo No. 54.

Weir, S., & Watt, D. (1981). Logo: A computer environment for learning-disabled students. *The Computing Teacher,* pp. 11–19.

Weisenbaum, J. (1976). *Computer power and human reason.* San Francisco, CA: W.H. Freeman.

Winston, P. H. (1977). *Artificial intelligence.* Reading, MA: Addison-Wesley.

CHAPTER 27

READABILITY

A Factor in Textbook Evaluation

Yvette M. d'Entremont

Editor's Note: Yvette d'Entremont is a doctoral candidate in the Department of Secondary Education at the University of Alberta.

Selecting an appropriate text for a school mathematics program is an important task and should not be taken lightly. Because mathematics has a language of its own, one of the factors that should be considered in textbook selection is readability.

Mathematics teachers are aware that reading mathematics material is different from reading other subject matter. Reading in mathematics is more than reading words. It involves decoding the words; decoding and interpreting the various mathematical symbols; and being able to interpret, comprehend, and solve mathematical sentences and phrases. The mathematics textbook serves as an aid in developing language in mathematics.

One way in which students acquire skills and knowledge is by reading instructional materials; therefore, they must have textbooks that are easy to read and comprehend. Progress in learning mathematics and the language of mathematics will be achieved if the reading level of the textbooks is appropriate to the grade or course for which the texts are intended.

Selected Writings from the Journal of the Mathematics Council of the Alberta Teachers' Association, pages 201–206

Departments of education and teachers are faced with an ever-increasing flood of printed materials, which differ widely in content, style and difficulty, and from which selections have to be made. In this situation, readability formulas may help by providing teachers with an additional guide for selecting suitable material. The textbook that is most effective is the one in which the author, through his or her writing style and vocabulary, produces a text with a reading level that is matched with the reading level of the student (Kennedy, 1974).

In preparing new textbooks, the readability level of the text is considered by some publishers. In 1982, the readability level of *Starting Points in Math 10* was analyzed. J.E. Freeman, associate program manager for Ginn and Company (publishers of *Starting Points*), stated that it is the company's policy to establish readability levels of textbooks by the Fry Graph but that provincial departments of education that purchase books do not inquire about the readability level of particular texts.

When considering whether to adopt a new mathematics textbook, Alberta Education does not apply readability formulas but pilots or field tests the book. Publishers are invited to submit textbooks that they feel will fit the scope and sequence of the curriculum sent to them by Alberta Education. A committee of teachers from across the province then evaluates the textbooks submitted by the publishers and chooses a number of them to field test. After piloting the textbooks, the teachers come together to discuss the books and the program and how they fit together. They select the books that will be listed as "basic resources." Schools then select the books that they wish to use from the list of basic resources (Jim Neilsen, 1988).

Many students experience difficulties in comprehending the explanations and problems found in mathematics textbooks. Concern about this has led me to assess the readability level of the text referred to earlier, *Starting Points in Math 10*, which I use with my grade 10 students.

The readability level of particular textbooks can be determined by using readability formulas. Applying formulas usually involves selecting a sample from a text, counting some easily identifiable characteristics such as the average number of words per sentence or the proportion of polysyllabic words in the sample, and performing a calculation to produce a score (Gilliland, 1972). Thus, readability formulas are based on correlational data, the correlation between sentence length and passage reading difficulty.

My objectives were to ascertain the readability level of the text by using two readability formulas, the Fry Readability Graph and McLaughlin's SMOG Grading Formula, and by administering cloze tests. A readability formula is a formula that is intended to provide quantitative objective estimates of the difficulty of reading (Klare, 1963).

Three passages were selected at random from the text:

1. Finding the Equation of a Line, Given Two Points (p. 36)
2. Adding and Subtracting Rational Expressions (p. 288)
3. The Pythagorean Theorem (p. 257).

The Fry Readability Graph, Figure 27.1, (Fry, 1968) and McLaughlin's SMOG Grading Formula were applied on each passage selected.

The Fry Readability Graph uses two factors to determine reading level: the average number of sentences per 100 words and the average number of syllables per 100 words. The intersection point of these two factors on the Fry Graph gives the grade level.

The McLaughlin SMOG Grading Formula, developed in 1969 by G. Harry McLaughlin, is based on only one factor: the number of words having three or more syllables in 30 selected sentences. The grade level is calculated by adding "3" to the nearest appropriate square root of the polysyllabic word count.

The Fry Readability Graph and McLaughlin's SMOG Grading Formula were not designed for use with mathematics materials, but they have been modified to measure the readability of a variety of mathematics books. In applying the formulas to mathematics textbooks, the samples selected should include only sentences. Non-sentence material such as pure computation, equation-solving, geometric proofs, titles of chapters, and illustrative problems are not part of the content examined (Johnson, 1957).

Readability scores were calculated on the text in question using the above readability formulas. Analysis of the selected passages by use of the Fry Readability Graph produces 163.7 as the average number of syllables per 100 words, and 4.9 as the average number of sentences per 100 words. Plotting the average number of syllables per 100 words and the average

Grade	Average Number of Sentences per 100 words	Average Number of Syllables per 100 words
1	14.3	120
3	8.6	123
6	5.8	129
9	4.5	149
12	4.0	162

Extracted from Fry's Readability Graph. [From ''Reading Level Determination for Selected Texts'' by K. Kennedy, *The Science Teacher* 41 (March 1974): 26.]

Figure 27.1

number of sentences per 100 words on the Readability Graph results in an average reading level of grade 12 for the text, *Starting Points in Math 10.*

Analysis of the same selected passages by use of the SMOG Grading Formula produces a polysyllabic word count of 105 in 30 selected sentences. Calculating the square root of 100 (the nearest appropriate figure to 105) and adding 3 to the square root gives a figure of 13. Therefore, according to the SMOG Grading Formula, the reading level of the text in question is grade 13.

When discussing these results, one must consider the grade level of the intended user of the text as well as the accuracy of the formulas involved. Fry (1968) states that the Readability Graph results are accurate to "probably within a grade level" (p. 514). For the SMOG Grading Formula, the standard error is about 1.5, providing a range of three years (McLaughlin, 1969).

The Fry Readability Graph and McLaughlin's SMOG Grading Formula are based on different prediction criteria. The Fry Readability Graph predicts the reading level that a student must have to be able to read the text with 50 to 75 percent comprehension. The SMOG Grading Formula attempts to predict the reading level necessary to read with 90 to 100 percent comprehension.

The results obtained predict that a grade 12 reading level is required to read *Starting Points in Math 10* with 50 to 75 percent comprehension and that a grade 13 reading level is required to read the text with 90 to 100 percent comprehension. The results suggest that a reading level of grade 13, which is three levels above the intended grade level of the user, is required to be able to read and fully understand the text.

The text was further tested by evaluating individual students' comprehension. The students were divided into three groups and randomly assigned a cloze test on the same passages selected for the readability formulas. A cloze test is a mutilated passage in which every fifth word or symbol from the passage has been deleted and replaced with a blank. In constructing cloze tests for mathematics texts, not only words but symbols such as >, % and 5 may be deleted. The student is then required to fill each blank with the exact word or symbol according to the original text material. The cloze procedure allows readers to use their knowledge of language patterns and their ability to respond to contextual clues (Malo, 1978).

The cloze procedure is advocated as a measure of the readability of "mathematical English" by Hater and Kane (1975). In 1970, they conducted a study to adapt the cloze procedure to the language of mathematics and to assess its behavior as a measure in that language.

The scores obtained by the students on each test were separated into three categories:

0%–43% correct—frustration level
44%–57% correct—instruction level
58%–100% correct—independent level

Average Number of Syllables per 100 Words

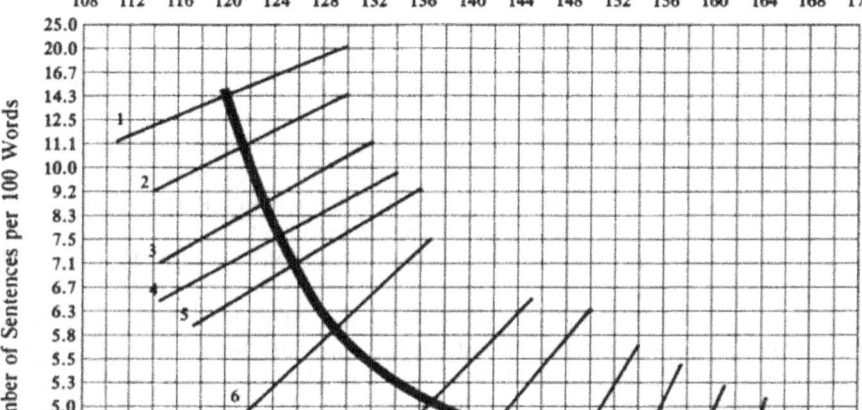

Figure 27.2 Fry's Readability Graph. *Source:* "Reading Level Determination for Selected Texts" by K. Kennedy, *The Science Teacher,* 41 (March 1974): 26.

Bormuth (1968) found that a score of 75 percent on conventional comprehension tests is comparable to a score of 44 percent on a cloze readability test made from the same passage (Figure 27.2). The three levels listed above have been accepted as a standard when interpreting cloze test results.

The percentage score means achieved on passages from *Starting Points in Math 10* were 57.38, 55.52, and 71.50 for passages 1, 2, and 3, respectively. These scores suggest that passage 3 may have been easier than passages 1 and 2, perhaps due to familiarity with the topic (Pythagorean Theorem). Averaging the percentage score mean of each passage provides a text mean of 61.47. The text mean of 61.47 falls into the independent level, but is only slightly above the instructional level of 44 to 57 percent. Because students were familiar with the content of the cloze tests (which may have affected the scores), one can conclude with a degree of certainty that the text assessed is suitable to be used if instructional support is provided.

One must always keep in mind that readability formulas and cloze tests are tools that can be used to assess the readability level of texts. Readability scores give an approximate grade level for materials and should be used as guides rather than absolute values. Knowing the readability level of a particular text can influence whether one will adopt it for a group of students.

REFERENCES

Bonnuth, J. R. (1968, April). The Cloze Readability Procedure. *Elementary English*, pp. 429–436.

Burmeister, L. E. (1974). *Reading strategies for secondary school teachers*. Reading, MA: Addison-Wesley.

d'Entremont, Y. (1984). *The readability of two high school mathematics texts*. Unpublished master's thesis, Mount Saint Vincent University, Halifax, NS.

Dunlop, W. P., & J. Strope, G. J. (1982). Reading mathematics: Review of literature. *Focus on Learning Problems in Mathematics, 4*(1), 39–49.

Earle, R. A. (1976). *Teaching reading and mathematics*. Newark, DE: International Reading Association.

Freeman, J. E. (1982, December). Associate Program Manager, Mathematics. Ginn and Company, Educational Publishers. Scarborough, Ontario. December 1982.

Fry, E. (1968). A readability formula that saves time. *Journal of Reading, 11*(7), 513–516, 575–578.

Gilliland, J. (1972). *Readability*. London, UK: University of London Press.

Hater, M. A., & Kane, R. B. (1975). The Cloze Procedure as a measure of mathematical English. *Journal of Research in Mathematics Education, 6*(2), 121–127.

Hater, M. A., & Kane, R. B. (1970). *The Cloze Procedure as a Measure of the Reading Comprehensibility and Difficulty of Mathematical English,"* EDRS Ace. No. Ed 0400881.

Hittleman, D. R. (1978). Readability formulas and cloze: Selecting instructional materials. *Journal of Reading, 22*, 117–122.

Johnson, D. A. (1957). The readability of mathematics books. *Mathematics Teacher, 50*(2), 105–110.

Kane, R. B. (1970). The readability of mathematics textbooks revisited. *Excellence in Mathematics Education—For All, 7*(63), 579–581.

Kennedy, K. (1974, March). Reading level determination for selected texts. *The Science Teacher, 41*, 26–27.

Klare, G. R. (1963). *The measurement of readability*. Ames, IA: Iowa State University Press.

Malo, D. G. (1978). *Analysis of the grammar and logic of 'if' in grade 6 science textbooks*. Unpublished master's thesis, University of Alberta.

McLaughlin, G. H. (1969). SMOG grading—A new readability formula. *Journal of Reading, 12*(8), 639–646.

National Council of Teachers of Mathematics. (1982). *How to evaluate mathematics textbooks*. Reston, VA: Author.

Neilsen, J. (1988). Program Manager, Senior High Math. Alberta Education, Edmonton, Alberta. February 1988.

This chapter was originally published as:

Yvon, B. R., & Zaitz, J. (1988). Combining literature and math. *delta-K: Journal of the Mathematics Council of the Alberta Teachers' Association, 27*(2), 20–24.

CHAPTER 28

COMBINING LITERATURE AND MATHEMATICS

Making Math Books and Finding Math Concepts in Books

Bernard R. Yvon and Jane Zaitz

Editor's Note: Dr. Bernard Yvon is a professor of mathematics education and child development at the University of Maine. He was a speaker at the 1986 NCTM Canadian Conference in Edmonton. Dr. Yvon was a contributor to Mathematics in the Early Childhood Classroom. Jane Zaitz is preparing to return to teaching. She was a student in one of Dr. Yvon's classes.

There is more to teaching math than one textbook or many worksheets. Counting books, geometric shape books, and many other math concept books can be made and shared by children. Hundreds of library and classroom books contain math concepts to be discussed, written, and even acted out. Factual books can be used to make graphs and charts. Poetry books about numbers can be read, and similar ones can be written by children themselves. Numerous activities and projects, including the making of books, can be undertaken to motivate math students and involve them in the everyday world of numbers and literature.

MAKING MATH BOOKS

Books come in all sizes; the "big book" is becoming popular for shared reading experiences in whole language programs. Large groups of children can read in unison from the pages of these books, which are printed with large type. Such reading is fun and noncompetitive. However, instead of buying expensive copies of such books, why not make them? Have each child make one page. The making of a book requires skills in language arts, creative writing, art and, if it is a counting book, math.

A grade 2 class in one school enjoyed making a book that they entitled "The Colorful Counting Book" as a gift for the kindergarten and grade 1 classes. The grade 2 students cooperated and shared ideas while compiling the book. In the first lesson, each child selected a number and object(s) such as balls, trains, lollipops, gifts, and balloons, which would attract young children. The students did their drawings on an 8" × 12" (20 cm × 30 cm) paper that had a ruled line across the bottom for the copy. In big books, the lettering should be as large as possible—½ inch (1 cm) for lower case letters and 1 inch (2 cm) for upper case letters-so that a group of children can read the book from a distance. Some children may need to practise making big letters on separate sheets of paper. After practising lettering, their sentences should be checked for errors in spelling and punctuation.

After students have completed their sentences, they can begin their drawings. Drawings should initially be done on smaller paper, then transferred to the larger sheets that make up the big book. If the book is a counting book, numerals should be written in large squares on the top right corner of each page. Once the drawings and lettering are done, the students can assist the teacher in putting the book together.

Pages should be glued back-to-back; that is, page 1 is glued to page 2, page 3 is glued to page 4, and so on. Each page becomes sturdier and more durable when reinforced with another sheet of paper. If possible, the pages should be laminated to protect them from wear and tear. The front and back covers can be made by students who have finished their pages. Holes should be punched in the sheets, then two large rings inserted to hold the book together.

The fun continues when the children see their finished product and read it among themselves. If they've made the book for another class, they can perform an oral presentation when giving the book. Each of the writers can read his or her own page. To do so will provide each with a sense of accomplishment. The students who receive the book will be very excited. To know the authors and even, at times, have played with them in the schoolyard is exciting.

Counting books are only one of the many different kinds of math books that a class can make. A colorful geometric shape book with two- and three-dimensional shapes was made by students in a grade 8 class and left as a gift to their successors when they graduated into high school. The procedure described above was used. Students referred to their math text and their teacher while making the more difficult geometric shapes.

A simple sequence book like Maurice Sendak's *Seven Little Monsters,* which shows seven monsters doing different things, could be tried. Any child can create a favorite character and make several pictures of that character performing different, humorous acts. The potential for creativity with a simple math concept is great and can be used to turn young math geniuses loose.

A poetry book like the classic *Over in the Meadow* by Olive Wadsworth creates wonderful sets of 10 directions between mother animals and their offspring in an easily reproduced rhythm pattern. Creative students in grades 4 through 8 could write similar books of directions from parents to children, or teachers to students. The potential for fantastic illustrations and humor are enormous. The finished product could remain in the classroom or be shared with or acted out for other classes. The whole school will be inspired to make other poetry or math-related books.

A dictionary of difficult mathematical terms could be made, illustrated, and shared by upper class students. It would serve as a handy reference book and would help to develop students' math vocabulary. A small binder to which pages are easily added could be used to make this dictionary.

A book showing parts of favorite desserts, cut up into servings, could be helpful in teaching students about fractions. A story about a growing family in which the parents keep dividing food into smaller and smaller portions could place fractions into a humorous account of family life. Metric units and decimals could be explained in a book in which each child selects a unit or term and draws a picture that illustrates the length or size of the unit. Graphing and charting activities can also be a part of this creative process of making math relevant to members of the class. After they start producing books, reading the books of others and making graphs on particular units of study, students will come up with more and more ideas in brainstorming sessions.

NONFICTION

Books of facts such as the *Guinness Book of World Records, The Book of Lists* (for the middle and upper grades), and *Do You Know? One Hundred Fascinating Facts* by Random House (for younger children) list intriguing data for children to study and make into graphs and charts. For example, a child who is interested in the speeds that animals can travel can make a graph that illustrates animals, from slowest to fastest moving. Older children can chart or graph information about buildings, sports, populations, speeds of vehicles, or other subjects that interest them.

Logic, order, and planning all go into making a graph or chart. Handsome finished products can be displayed, shared, and discussed by young mathematicians. Follow-up activities can include the construction of intriguing problems and questions that require the interpretation of data from the charts and graphs. Each student is an authority on his or her graph and can verify others' results with his or her expertise. Positive classroom dynamics are at work; each child is king of his own castle and knight at his neighbor's castle.

The variety of factual material is overwhelming. Teachers can search libraries for books related to science, social studies, history, geography, and other areas of interest.

ACTING OUT STORIES WITH MATH PROBLEMS

Many stories and nursery rhymes have number and math concepts in them. "The Three Little Pigs," "The Three Bears" and "The Three Little Kittens" all have sets of three. Instead of just reading these stories to younger children, the teacher can organize a reading group to read them or find volunteers to act them out. Puppets are a good way for shy children to begin experiencing drama. In the book *The Teacher Who Could Not Count*, by Craig McKee and Margaret Holland, students teach their teacher to count by acting out each number with their bodies. Games in which numbers are acted out and guessed can be great rainy day activities for children. Remember, Roman numerals need extra cooperation and teamwork.

In many books, such as *It Could Always Be Worse* by Margot Zemach and *Mushroom in the Rain* by Mirra Ginsberg, people or animals are added to the original set. In the Yiddish folktale, *It Could Always Be Worse*, a poor family that lives in an overcrowded house keeps adding, on the advice of the rabbi, more inhabitants (animals) to the house. After the father can no longer stand the overcrowded conditions, the rabbi advises him to return the animals to the shed. Life seems peaceful and pleasant after the animals depart. The equation that corresponds to the story is 6 (children) + 2 (parents) + 1 (grandmother) + 3 (fowl) + 1 (cow) − 4 (animals) = 9 (the original number of family members).

Good listening and math skills are demanded of the children so that they can write an equation on the board after listening to a story. With practice, the children can reverse the process and tell a story from a simple addition or subtraction sentence. Their original stories can be acted out as well.

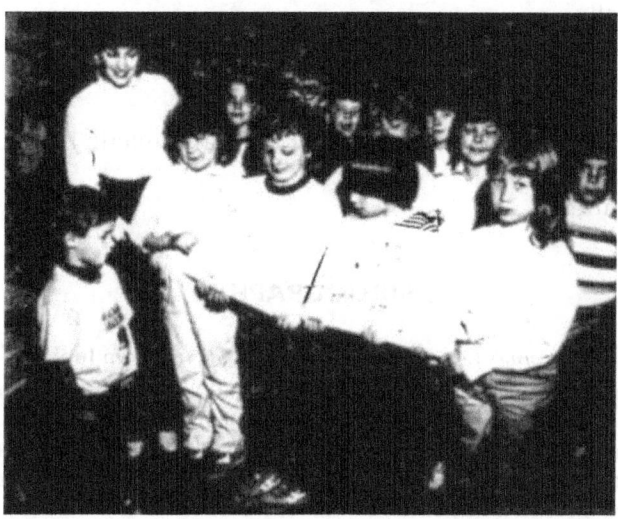

Math and books...books and math equal a fun learning experience. With a little imagination and a big desire to relate math to other areas of the curriculum, every teacher and class can make books that excite and motivate. Likewise, abstracting mathematical equations, activities and concepts from books may take some time and planning, but will make math the most exciting class of the day. The bibliography that follows will help you start an integrated math program that combines the world of numbers with the world of children's literature.

PROCEDURE FOR MAKING A BIG OR LITTLE BOOK WITH YOUR CLASS

1. Have a planning session in which children select a number or a math concept.
2. Have them draw a picture and write a sentence describing the picture, which illustrates the number or math concept. This should be done on draft paper.
3. Have the students practice large printing, if necessary.
4. On large sheets of paper (construction paper works well) draw a box in the top right-hand corner, where a number will be written.
5. Check students' lettering and spelling, and make corrections or additions before allowing the students to start their final copy.
6. Make front and back covers. On the front cover, indicate the class, the year, and the names of the authors.
7. Laminate or use clear contact paper.
8. Punch holes and place reinforcements around the holes for strength.
9. Use two 2-inch rings to hold the book together.
10. Read and share the book with other classes.
11. Hang the book on a coat hook to store it.
12. Take photographs of the book for the class bulletin board, class journal or newspaper.
13. Have fun!

BIBLIOGRAPHY

Anno, Mitsumasa. *Anno's Counting Book.* New York, NY: Crown Junior Books, 1977.
A counting book that is beautifully illustrated with country scenes. On the first page, zero is indicated with an empty landscape; the next page has one piece of scenery. As the number increases, so does the number of objects filling the landscape. A fun, natural way to count. The last page talks about early number systems and one-to-one correspondence.

Charlip, Remy, and Terry Joyner. *Thirteen.* New York, NY: Parents' Magazine Press, 1975.

A wordless concept book consisting of 13 picture sequences in which shapes evolve into new forms. Good for developing observation skills among children of all ages. Needs introduction.

Emberley, Ed. *Ed Emberley's Drawing Book of Animals.* Boston, MA: Little Brown, 1970.

A wonderful book that teaches how to draw animals with simple lines, squares, triangles, and angles. Has fun art lessons for all ages.

Hillman, Priscilla. *The Merry Mouse Counting and Colors Book.* New York, NY: Doubleday, 1983.

A small, square cardboard book with colorful drawings of mice. Counting up to 10. Written in verse.

Hoban, Tana. *Circles, Triangles and Squares.* New York, NY: Macmillan, 1974.

Beautiful black-and-white photographs of city scenes and everyday objects show the three most common geometric shapes for children to identify. Other books by Hoban include *Shape and Things*; *Look Again*; *Push-Pull, Empty Full: A Book of Opposites*; *Count and See*; and *Round and Round and Round.* Each could be used to promote discussion or provide follow-up activities for young children.

Hutchins, P. *One Hunter.* New York, NY: Greenwillow, 1982.

One hunter meets up with 10 African animals hidden in the forest and walks by each camouflaged set. A humorous account that allows for guessing and counting. (Preschool to Grade 2)

Mathews, Louise. *The Great Take-Away.* New York, NY: Dodd, Mead and Co., 1980.

One hog in a town of pigs steals. In rhyme, with five subtraction problems to solve. (Grades 1 to 3)

McKee, Craig, and Margaret Holland. *The Teacher Who Could Not Count.* School Book Fairs, Inc., 1981.

A story about a mixed-up teacher who makes mistakes in learning to count. Her students act out the numbers with their bodies to teach her properly. Great for rainy days or for number games in physical education.

Merriam, E. *Project 1,2,3.* New York, NY: McGraw-Hill, 1971.

A fascinating book for urban or rural children to learn about life in a huge complex. Has eight pages at the end for observation and counting.

Oxenbury, Helen. Helen *Oxenbury's Numbers of Things.* New York, NY: Heinemann, 1967.

A counting book about a lion. Simply but humorously illustrated. Depicts numbers 1 to 50.

Pienkowski, Jan. *Numbers.* New York, NY: Harvey House, 1975.

Numbers is similar to Pienkowski's books *Colors, Sizes,* and *Shapes.* The numbers one through 10 are illustrated with objects in a natural setting. On the opposite page, an abacus shows combinations of 10. (For two- to six-year-olds)

Random House. *Do You Know? One Hundred Fascinating Facts.* New York, NY: Author, 1979.

Lots of facts about things smallest to largest, from animals to vehicles. Ideal for graph and chart making.

Scarry, Richard. *Richard Scarry's Best Counting Book Ever.* New York, NY: Random House, 1975.

A counting book in which Willy Bunny counts everything he sees. Goes to 100. Ideal for playing such games as "You Find It." Ask a child such questions as "How many firemen have green mops?"

Sendak, Maurice. *Seven little Monsters.* New York, NY: Harper and Row, 1975.

A simple, short account of seven monsters who get into trouble. The book could provide inspiration to children for making their own sequence books.

Shapiro, A. *Mr. Cuckoo's Clock Shop*, Los Angeles, CA: Intervisual Communications, 1978.

A rhyming story about a clock shop with a large clock that has movable hands. The reader moves the time ahead one hour per page.

Silverstein, S. *The Missing Piece Meets the Big O.* New York, NY: Harper and Row, 1981.

A triangle searches for his whole and meets many disappointments until the Big O tells him to wear off his edges and become a circle. A good introduction for young children to various shapes.

Wadsworth, Olive. *Over in the Meadow: A Counting Out Rhyme.* New York, NY: Viking-Kestrel, 1985.

A counting book of the numbers one to 10. Ideal for ideas when asking a class to make their own books.

Warren, Cathy. *The Ten-Alarm Camp-Out.* New York, NY: Lothrop, Lee and Shepard Books, 1983.

A story about a mother armadillo and her nine babies who like even numbers. They have a strange camping adventure. An enjoyable story with counting practice. (Preschool to Grade 2)

This chapter was originally published as:

Ediger, M. (1989). Psychology in teaching mathematics. *delta-K: Journal of the Mathematics Council of the Alberta Teachers' Association, 27*(4), 20–23.

CHAPTER 29

PSYCHOLOGY IN TEACHING MATHEMATICS

Marlow Ediger

Editor's Note: Marlow Ediger is professor of Mathematics Education, Northeast Missouri State University, Kirksville, Missouri. Dr. Ediger is a frequent contributor to *delta-K.*

Numerous reputable psychologies are provided to assist mathematics teachers in guiding students to achieve optimally. The teacher of mathematics should study the diverse psychologies of education to implement the best teaching strategy possible. The lay public focuses on student achievement in the three Rs or the basics. Mathematics represents a highly salient basic. Students need to do well in mathematics to do well in school and in society. Teachers need to select objectives, learning opportunities, and appraisal procedures that assist learners to achieve as well as is possible.

BEHAVIORISM IN THE MATHEMATICS CURRICULUM

Precise, measurably stated objectives and their use is the heart of behaviorism. Objectives are selected prior to being implemented in the classroom. Generally, no student participation has been emphasized in selecting these goals. Behaviorism can be emphasized with state-mandated objectives in terms of core competencies and key skills. At the state level, precise measurably stated objectives have been chosen. The department of education of each state selects a cross section of educators to agree upon the stated ends. The mathematics teacher then plans learning opportunities to help students attain each objective.

A second example of behaviorism emphasizes instructional management systems (IMS) at the district level. The central office selects a cross section of teachers within the district to select salient objectives in mathematics. Again, the classroom teacher must emphasize each objective in teaching–learning situations.

The mathematics teacher, without stated mandated objectives or IMS, may write and implement specific ends for student attainment.

An early pioneer in measurably stated objectives and their use was B. F. Skinner emphasized programmed learning in either textbook or software form. The ingredients of programmed learning include

1. sequential items of small amounts of information acquired by students in each step of learning
2. students responding to a test item, such as a multiple-choice question based on information presented in book or software form
3. learners receiving feedback based on the response made
4. reinforcement being rather common with high frequency of correct responses made

Behaviorism, in its diverse manifestations, emphasizes that a student either does or does not achieve an objective as a result of instruction. If an objective is not attained, the mathematics teacher needs to try a different teaching strategy.

Behaviorism appears to be a dominant psychology of education emphasized in the teaching of mathematics. With the popularity of behaviorism, the writer recommends

1. that each objective in mathematics be carefully selected in terms of being useful in school as well as in society
2. that students achieve success in attaining sequential objectives
3. that a variety of challenging learning opportunities be provided for learners to attain each end
4. that students experience meaning, interest and purpose in achieving desired ends
5. that critical and creative thinking, as well as problem solving, receive ample attention in the mathematics curriculum
6. that appraisal procedures be varied, valid, and reliable to evaluate learner progress

HUMANISM IN THE MATH CURRICULUM

Humanism, as a psychology of learning, emphasizes students being heavily involved in determining objectives, learning opportunities and evaluation procedures. Each student is guided to attain selfrealization. The late A. H. Maslow (1954), humanist psychologist, listed five sequential levels where individuals need assistance to achieve realization of self:

1. assisting students to meet physiological needs, such as adequate food, clothing, and proper shelter
2. helping learners to feel safe and secure in their environment
3. guiding students in meeting love and belonging needs
4. developing situations in which esteem needs of students are being met
5. assisting learners to achieve self-actualization

Only after the above sequential needs of students have been met can students achieve optimally, according to humanism as a psychology of learning. Meeting the needs of students to increase achievement behooves any school system.

Input from learners in selecting objectives, learning opportunities and appraisal procedures is highly important. There are several excellent ways of emphasizing humanism in the mathematics curriculum. One plan is to utilize learning centres. More tasks than any one student can complete would be at the diverse centres. Students learn to make decisions. They choose, sequentially, which tasks to complete and which to omit. Each learner then selects what is perceived to be of interest, meaning, and purpose. Sequence in selecting ordered tasks resides within the student. A psychological curriculum is then evident. Internally, the student makes choices in terms of tasks to pursue.

A second plan of humanism as a psychology of education is to use a contract system. In a contract, the students and their teachers together plan specific learning opportunities for the former to complete. There must be considerable input from the students in the contract for humanistic psychology to be evident. The due date of the contract is indicated with the students' and the teachers' signatures.

A third plan of humanism is when the teacher lists, for example, 10 activities for students to consider to complete in mathematics. Each student may choose 5 or more to complete. The student here has input as to what to pursue and what to omit.

Humanism emphasizes a humane mathematics curriculum. Humaneness is defined as students being able to decide from among alternatives which learning activities have value and need to be completed satisfactorily.

The writer, in evaluating humanism in teaching mathematics, recommends that

1. worthwhile tasks be developed for students to pursue sequentially (trivia is to be omitted for learners to pursue)
2. students be guided to stay on task and not digress from achieving relevant objectives
3. tasks be written on diverse levels of achievement to challenge each student to achieve as much as possible

THE STRUCTURE OF KNOWLEDGE

During the 1960s and early 1970s, much emphasis was placed upon mathematicians on the higher education level identifying structural ideas for public school students to attain. The structure of knowledge emphasized underlying principles that provided a framework for an academic discipline. Thus, in the academic discipline of mathematics, selected broad generalizations provided a structure for students in ongoing lessons and units. The key ideas included the commutative property of addition and multiplication, the distributive property of multiplication over addition, the property of closure, and the identity elements.

The structure of knowledge approach, as identified by Jerome Bruner of Harvard University and his associates (1960), emphasized that public school students utilize methods of learning used by mathematicians on the higher education level. An inductive procedure is then evident. Students are guided by the teacher to learn by discovery in moving from the specific to the general to achieve structural ideas. Materials to use in teaching students to acquire content inductively include inactive (manipulative items), iconic (pictures, drawings, slides and filmstrips emphasizing main ideas), and symbolic (abstract content such as printed words and numerals).

The structure of knowledge approach has much to recommend itself. The writer recommends that

1. teachers emphasize structural ideas in a spiral curriculum. However, the spiral curriculum should not be excessively repetitious. The structure has a built-in review when these key generalizations receive attention at more complex levels in mathematics curriculum.
2. induction receive adequate attention in teaching–learning situations. However, continued use of inductive methods is time-consuming. The mathematics teacher needs to inject meaningful explanations also at definite points in ongoing lessons and units.
3. creative teaching in using diverse methodologies be emphasized thoroughly. Methods and subject matter have to be adjusted to the present achievement level of each student as students differ in interests, purposes, and present levels of achievement.

DIAGNOSIS IN MATHEMATICS

Mathematics teachers must utilize the concept of diagnosis in teaching–learning situations. To diagnose means to pinpoint specific difficulties students experience in computation, concept development, and problem solving. Students need assistance to overcome errors made.

Robert Gagne (1985) advocates a hierarchy of objectives be stated in measurable terms for student attainment. If a learner cannot achieve a specific end, the teacher needs to move to an easier sequential objective. Reversing to easier ends is necessary until the student's present attainment level is found. The last three levels of Gagne's hierarchy are especially important to know when teachers diagnose difficulties students experience in mathematics. The three in sequence are concept learning, rule learning and problem solving. Thus, if students cannot solve a problem in mathematics, the teachers need to assist the former to determine if they understand the involved rules. For example, if the problem involves finding the volume of a cylinder, the students must understand the involved formula: $\pi r^2 h$. If the learners do not understand the rule to determine the volume of a cylinder, they need assistance in attaching meaning to concepts. The separate concepts are radius, radius times radius, pi, and height.

Diagnosis is involved when the mathematics teacher assists the student to pinpoint specific weaknesses in a lesson or unit. Gagne provides a quality model for mathematics teachers to follow in helping learners to progress sequentially.

In using diagnostic-remediation procedures in the teaching of mathematics, the writer recommends that

1. students attach meaning to each sequential step of learning
2. learners be assisted to perceive holism and sequence in the subject matter learned. Diagnosis is available if a student fails to attach meaning to ongoing rules (generalizations) and concepts to solve problems in mathematics.

CONCLUSION

Relevant principles of learning from the psychology of education need to be implemented in teaching–learning situations. The teacher of mathematics must assist each student to attain in an optimal manner.

Four schools of thought were discussed in the psychology of education: behaviorism, humanism, the structure of knowledge, and diagnosis based on a hierarchy of objectives.

The writer recommends that teachers of mathematics

1. implement tenets of behaviorism with its measurably stated objectives. Higher levels of cognition must not be hindered with the use of behaviorism in teaching-learning situations.
2. provide ample opportunities for students to engage in decision making. Learners need to have chances to select sequential learning opportunities as advocated by humanism.
3. stress the structure of knowledge so that students may perceive that subject matter is related.
4. adequately diagnose and remediate students' problems in lessons and units. Students need to perceive mathematics as being holistic and not isolated specifics in diagnostic/remediation situations.

REFERENCES

Ashlock, R. B. (1982). *Error patterns in computation*. Columbus, OH: Charles E. Merrill.

Bley, N. S., & Thornton, C. A. (1981). *Teaching mathematics to the learning disabled*. Rockville, MD: Aspen.

Ballew, H. (1973). *Teaching children mathematics*. Columbus, OH: Charles E. Merrill Publishing.

Bruner, J. S. (1960). *The process of education*. New York, NY: Vintage Books.

Cawley, J. F. (1985). *Cognitive strategies and mathematics for the learning disabled*. Rockville, MD: Aspen.

Fehr, H. F., & McKeeby Phillips, J. (1972). *Teaching modern mathematics in the elementary school* (2nd ed.). Reading, MA: Addison-Wesley.

Gagne, R. (1985). *The conditions of learning*. New York, NY: Holt, Rinehart and Winston.

Higley, J. (1983). *Activities deskbook for teaching arithmetic skills*. West Nyack, NY: Parker Publishing Company.

Jenson, R. (1973). *Exploring mathematical concepts and skills in the elementary school*. Columbus, OH: Charles E. Merrill.

Kennedy, L. G. (1984). *Guiding children's learning of mathematics* (4th ed.). Belmont, CA: Wadsworth Publishing Company. Chapter One.

Lesh, R., & Landau, M. (Eds.). (1983). *Acquisition of mathematics concepts and processes*. New York, NY: Academic Press.

Maslow, A.H. (1954). *Motivation and personality*. New York, NY: Harper and Row,.

Stem, C., & Stem, M. B. (1971). *Children discover arithmetic*. New York, NY: Harper and Row.

This chapter was originally published as:

Percevault, J. B. (1984). The development of problem-solving skills: Some suggested activities (Part II). *delta-K: Journal of the Mathematics Council of the Alberta Teachers' Association, 24*(1), 11–16.

CHAPTER 30

THE DEVELOPMENT OF PROBLEM-SOLVING SKILLS

Some Suggested Activities (Part II)

John B. Percevault

Editor's Note: John B. Percevault is an Associate Professor at the University of Lethbridge. During 1982–1983, while on administrative leave, he worked with grade 3–6 teachers in Lethbridge elementary schools. This article presents some of the problemsolving skills and strategies that were used in the schools. Activity 1 ("Reading in Mathematics") and Activity 2 ("Developing Models") were given in Part I (see *delta-K,* 23, no. 2 [May 1984]).

ACTIVITY 3: COLLECT AND RECORD DATA

The spatial visualization activities may also be used to develop the skill of collecting and recording data. Choose a particular configuration and record data. The data from Activity 2 are presented in tabular form below.

At the point where the partial table includes the first three sets of data, children may mark the rows of three and count the objects. Challenge the children to complete the next two rows. Have them predict the number of rows and the number of objects that would be required. The recognition of the patterns (counting numbers, and odd numbers starting with 3) is

Selected Writings from the Journal of the Mathematics Council of the Alberta Teachers' Association, pages 221–227
Copyright © 2014 by Information Age Publishing

Diagram	Number of Rows of 3 Objects	Number of Objects
000	1	3
0 000 0	2	5
0 0 0 0 0 0 0 0 0	3	7
0 0 0 0 0 0 0 0 0 0	———	———
	———	———

Activity 2

a skill. Extending the pattern to answer a question such as the following begins to develop into a problem-solving strategy: "If 21 objects were arranged in this pattern, how many rows each containing three objects would there be?" Children are challenged to find a short way of determining the number of objects used or number of rows required, in at least two ways. At this level, they are using a problem-solving strategy. For further applications, consider:

Use real materials, make diagrams, collect and record, search for patterns, extend, look for relations, and predict and generalize.

1.   5 objects per row   --5 objects
                                                       --9 objects
                                                       --13 objects
                                                       --___ objects
                                                       --___ objects
                                                       --___ objects

2.   •   •   •   Conditions?
   • •   • •   • •   What comes next?
      • • •   • • •
         • • • •

3.   • •   • • •   • • • •   Conditions?
   • •   •   •   •   •   What comes next?
      • • •   •   •
         • • •   • • •

ACTIVITY 4: RECORDING AND LISTING

A second set of collecting and recording activities follows: activities relating to addition facts. Skills that are involved are listing, organizing data, and diagramming. The problem-solving strategy that may evolve is generalizing a rule or relationship.

Have children list all the sets of two addends that yield a sum of five. The listing would include: $1 + 4$, $3 + 2$, $0 + 5$, $2 + 3$, $5 + 0$, and $4 + 1$ (not necessarily in this order).

Challenge the students to organize the list. Give the cue to let one addend increase or decrease. The organized list is evident:

```
0 + 5    Commutative patterns
1 + _    (for example, 0 + 5 and 5 + 0)
2 + _
3 + _
4 + _
5 + _
```

Repeat the activity with the sum of 6, 7, 8, . . .

Encourage students to diagram the facts. A diagram of the addition facts for a sum of 5 is shown below.

```
0 + 5 = 5
1 + 4 = 5
2 + 3 = 5
3 + 2 = 5
4 + 1 = 5
5 + 0 = 5
```

Explore the following after the completion of listing and/or diagramming: How do odd and even numbers differ? If one addend is increased by two, what happens to the other addend? [The compensation principle may be expressed thus: If $a + b = c$, then $(a + k) + (b - k) = c$.] How does the number of 2 addend facts for a given sum compare to the sum $(S + 1)$? These questions will always produce a finite list with $S + 1$ members, where "S" is the sum.

ACTIVITY 5: LISTING-SUBTRACTION

Ask students to list two numbers that produce a difference of two. Accept any and all correct responses, such as:

```
   4            15          2002
  -2           -13         -2000
 ───          ────        ──────
   2            2            2
```

Many students will realize intuitively that the list is infinite. The teacher may wish to develop an organized list, such as:

```
   2        3        4        5
  -0       -1       -2       -3
 ───      ───      ───      ───
   2        2        2        2
```

Have students examine the list. Ask them if they can state a relationship. [The compensation principle in subtraction may be expressed thus: If a – b = c, then (a + k) – (b + k) = c.] Have the students use the generalization of compensation in addition.

$$\text{If } 2 - 0 = 2, \text{ then is } (2 + 10) - (0 + 10) = 2?$$

$$\text{If } 2 - 0 = 2, \text{ then is } (2 + 20) - (0 + 20) = 2?$$

Further examples may be explored, such as: If $32 - 30 = 2$, then is $42 - 40 = 2$? Is the answer to $34 - 17$ the same as the answer to $37 - 20$? Which subtraction is easier to perform?

An activity such as the one above can be used to introduce the "equal addition algorithm" for subtraction.

ACTIVITY 6: ORGANIZED LISTS IN A PROBLEM

Problem: I am thinking of two numbers. The sum of the two numbers is 13. The difference of the two numbers is 3. The numbers are ___ and ___. Obviously, the elementary student could guess and check. However, a student who has a more organized approach could use the skills developed above. Ask such questions as:

What are we to find?
  Answer: two numbers.
How are they related? (What is the condition?)
  Answer: They have a sum of 13.
Is there another condition?
  Answer: Yes, they have a difference of 3.
How many two-addend addition facts are there for a sum of 13?
  Answer: 14.

Can you develop an organized list?
   Answer: Hopefully, yes.
How do you want to organize?
   Answer: Answers may vary.

The following table may be developed with the students:

Sum of 13	Difference	
13 + 0	13	
12 + 1	11	Look for patterns in addends
11 + 2	9	and in difference columns.
10 + 3	7	
_____		
_____		How does the difference vary?
7 + 6	1	(By 2). Why?
6 + 7	1	
_____		
_____		Do we need to do the rest of
_____		the table?
_____		
_____		
0 + 13		

Assume absolute value only at this stage.

Encourage discussion on how the problem was solved. Answers that indicate that students realize that one condition (sum of 13) was satisfied first and then checked in the second condition (difference of 3) show the beginning of a problem-solving strategy, because the problem has been divided into two or more problems.

Vary the conditions. For example,

Sum of 14	Difference of 6
Sum of 14	Difference of 3 (This is impossible. Why?)
Sum of __	Difference of __

Have students work in pairs to make up problems for their classmates to solve.

ACTIVITY 7: COUNTING PATTERNS

Use the hundred board. Have students count and color every second, third, and so on, square. After the first three rows are completed, encourage the students to look for patterns and to use the patterns to complete the coloring of appropriate squares. Counting by twos and fives is suggested as a starting point. The partial pattern for each is shown.

Counting by twos pattern:

1		3		5		7		9	
11									
21				25					
						47			
								69	
				95				99	

Counting by fives pattern:

1	2	3	4		6	7	8	9	
11	12	13	14		16	17	18	19	
21	22	23	24		26	27	28	29	
				65					
				75					
				85					
				95					

Explore the counting by threes pattern:

1	2		4	5		7	8		10
11				16					
				26					
31									40
41				46					50
				56					
61									70
71				76					80
				86					
91									100

The numerals 3, 6, 9, 12, . . . may be arranged in the format:

3	6	9
12	15	18
21	—	—
—	—	—

After students have extended the format, regular patterns may be explored. Each column increases by nine. The sum of the digits in each column is three, six, or nine. This may not be evident for a numeral such as 39, where the sum of digits is 12. However, U is found in the column that sums to three.

Further questions may include:

Is 47 a multiple of three (included in the counting by threes table)? Justify your answer.

Can you place 81, 42, and 96 in the appropriate column? Justify your answer.

Find the "counting by nines" on the hundred board and in the table.

The counting by eights pattern is given below:

								9	10
11	12	13	14	15					20
21									
31									
						97	98	99	100
101	102	103	104	105	106	107	108	109	110

Again, encourage students to explore alternate arrangements, such as the following:

```
 8,    16,    24,    32,    40,
48,    56,    64,    etc.
```

COMMENTARY

1980s

An Agenda in Action, A Decade of Change

A. Craig Loewen

In 1980 the National Council of Teachers of Mathematics (NCTM) published a document entitled *An Agenda for Action*. The eight recommendations in that document became one of the most significant collections of ideas for mathematics instruction of the decade, and have since shaped both research directions and thinking about mathematics instruction specifically and mathematics education generally. The agenda was widely adopted across North America, and the recommendations wound themselves inextricably into mathematics classrooms and academia alike.

The eight recommendations can be found on the NCTM website (nctm. org). Since 1980 they have been further developed in the NCTM journals and the 1983 NCTM yearbook, *The Agenda in Action*. At the close of the decade, the NCTM publication *Curriculum and Evaluation Standards for School Mathematics* (1989) defined new standards for mathematics instruction. The original eight recommendations (copied from www.nctm.org/standards) are that

Selected Writings from the Journal of the Mathematics Council of the Alberta Teachers' Association, pages 229–233
Copyright © 2014 by Information Age Publishing

1. problem solving be the focus of school mathematics in the 1980s
2. basic skills in mathematics be defined to encompass more than computational facility
3. mathematics programs take full advantage of the power of calculators and computers at all grade levels
4. stringent standards of both effectiveness and efficiency be applied to the teaching of mathematics
5. the success of mathematics programs and student learning be evaluated by a wider range of measures than conventional testing
6. more mathematics study be required for all students and a flexible curriculum with a greater range of options be designed to accommodate the diverse needs of the student population
7. mathematics teachers demand of themselves and their colleagues a high level of professionalism
8. public support for mathematics instruction be raised to a level commensurate with the importance of mathematical understanding to individuals and society

The ten articles selected by the editors to represent the 1980s for this commemorative issue of *delta-K* publication address one or more of these recommendations.

PROBLEM SOLVING

The first item of the agenda was the recommendation to "let problem solving be the focus of school mathematics for the 1980s." The recommendation stems from the idea that effective problem solving is ultimately the purpose of mathematics instruction, and that problem solving is also a process whereby mathematics can be learned. It is in engaging the exploratory experiences of solving problems that the applications of mathematics come to life and students come to see purpose in their learning. And with purpose comes enjoyment. The inherent importance of problem solving both as a goal and a process implies that the mathematics curriculum itself should be organized around it. In the article by Sigurdson, "A Constructivist Approach to Teaching Mathematics," the theories of constructivist principles of teaching and learning are applied to mathematics classrooms. Sigurdson presents problem solving as a way of "knowing content." D'Entremont's article, "Readability: A Factor in Textbook Evaluation," contests that the readability of classroom textbooks may limit students' understanding of content and their ability to solve problems. She argues that readability should be considered in the selection of classroom resources and she provides an example of an Alberta-approved grade 10 classroom text written at the grade 12 or early postsecondary reading level. Percevault's article "The Development of Problem-Solving Skills: Some Suggested Activities" (the second of two parts) directly addresses problem solving as it provides a range of exploratory activities that encourage the development of certain problem-solving skills and strategies while teaching addition, subtraction, and commutativity.

UNDERSTANDING

The second of the eight recommendations of *An Agenda for Action* is that "basic skills in mathematics be defined to encompass more than computational facility." This recommendation addresses the common misconception that mathematics is ultimately about formulas, numbers, and the ability to find the right answer consistently, immediately, and efficiently. But what does it mean when we say a concept is *understood*? There are many dimensions to understanding that include the ability to connect new knowledge to existing knowledge, mastery of the language about a concept, the ability to create and interpret representations of the concept, the ability to apply new knowledge, and so forth. Each of these dimensions of understanding needs to be part of a robust mathematics curriculum and evident in the acquired abilities of our students. In "Statistics in the High School," Haack addresses the question of understanding statistics, the abuses of statistics, and the misuses of the language of statistics. He also discusses the structure of knowledge, the interconnection of ideas, the role of student awareness (metacognition) in learning, and the importance of knowledge transfer as indicators of learning. Yvon and Zaitz, in "Combining Literature and Mathematics," suggest that children's literature provides a context for the learning of mathematics as students see characters engaging in mathematical tasks. Finally, Ediger, in "Psychology in Teaching Mathematics" discusses the structure of mathematical knowledge and the role of engaging behaviorist and humanist principles in the development of effective learning and teaching environments.

TECHNOLOGY

The third NCTM recommendation is that "mathematics programs take full advantage of the power of calculators and computers at all grade levels." It is simply evident that technology has changed dramatically since the 1980s. The availability of technology, range of applications, software, and cost have all changed beyond anything that could have been imagined in 1980, and those changes brought profound implications for classrooms, such as the introduction of virtual classrooms and virtual schools. Two of the selected articles reflect the technology of the time. Le Maitre, in her article entitled "If this Is Television, Shouldn't My Intelligence Be Insulted?," speaks of the need to teach in the manner in which our students best learn (the enduring thesis) and extols the value of viewing videotapes to enhance lessons. An amusing moment in her article occurs when she encourages teachers concerned about their skills in working a videocassette recorder to "ask one of the children in your class to show the tape; after all, your students are growing up with VCRs as we did with radios." Burnett's article, "Logo: An Opportunity for Synthesis, Self-Control, and Sharing," addresses the opportunities for mathematical exploration inherent within the Logo environment. He argues that Logo matches well with what we know about how students learn.

MEETING THE NEEDS OF LEARNERS

The sixth agenda item speaks of the need for a more complete and flexible mathematics program that addresses the diverse needs of the variety of learners in our classrooms. Programs are needed that reach out to students, involve them, and impress upon them the importance of mathematics in society and in their lives. The article by Alton and Gersting, "An Alternative Course for the 'I Hate Math and I've Never Been Any Good At It Student,'" describes a course they developed and delivered at Indiana University–Purdue University at Indianapolis (IUPUI) for students who had negative perceptions of mathematics and saw it only as a mystical collection of formulas and numbers. In this course students were encouraged to see math as applicable to life, a way of interpreting and understanding the world, and even as something to be enjoyed. The article by Nicol, "Expecting Girls to be Poor in Math: Alternatively, Chance for a New Start," explores the NCTM position statement on gender and mathematics learning and recommends some useful materials for encouraging girls to study math.

As one reads these articles one cannot help but be struck by how they illustrate the character of *delta-K* and of the Mathematics Council of the Alberta Teachers' Association (MCATA) itself. The council has reached out across both sides of the border with its many international submissions. *Delta-K* has included articles written by students, professors, and classroom teachers. The articles themselves have both addressed big-picture topics and provided specific classroom activities for both upper and lower grades; the articles have included research results, theoretical musings, and practical activities. But each, at its heart, has this goal as its foundation: to gain and maintain improvement in mathematics teaching and learning. This is ultimately the purpose and intent of *delta-K* and its parent organization, the Mathematics Council of the Alberta Teachers' Association.

These ten articles provide a snapshot of our society at a time of great change and development. We see genuine efforts being made to focus on problem solving, address gender bias in classrooms, recognize learning as an act of meaning building, teach in a manner responsive to student needs, and incorporate technology fully into the learning/teaching process. Collectively, these changes have altered the nature of mathematics learning and teaching, as well as the content of the curriculum. A decade of change . . . an agenda in action.

Dr. Craig Loewen is currently serving as the Dean of Education at the University of Lethbridge (Alberta, Canada). His area of teaching and research is secondary mathematics education, a position he has held for over 25 years. He is particularly interested in alternative ways of teaching mathematics that inspire and influence understanding and problem solving behavior. Dr. Loewen previously served as editor of *delta-K* and has published more than a dozen articles with the journal.

REFERENCES

Krulik, S., & Reys, R. E. (Eds.). (1980). *Problem solving in school mathematics: 1980 yearbook.* Reston, VA: National Council of Teachers of Mathematics.

National Council of Teachers of Mathematics (NCTM). (1980). *An agenda for action: Recommendations for school mathematics of the 1980s.* Reston, VA: Author.

National Council of Teachers of Mathematics (NCTM). (1989). *Curriculum and evaluation standards for school mathematics.* Reston, VA: Author.

Shufelt, G., & Smart, J. R. (Eds.). (1983). *The agenda in action: 1983 yearbook.* Reston, VA: National Council of Teachers of Mathematics.

THE DAWN OF THE INFORMATION AGE

Darryl Smith

The political, social, physical, and educational worlds are not mutually exclusive; they coexist and affect each other, sometimes in reactionary ways. The launching of the Soviet Union's and the world's first artificial *Sputnik* satellite on October 4, 1957, is a case in point. It showed that North America was lagging behind the Soviets in the space race and triggered the "New Math" movement of the 1960s with its emphasis on mathematical structure, set theory, and number bases other than 10. Can you imagine what our world of mathematics would be like if humans were all born with, say, four digits on each limb? Surely we wouldn't have a base-8 number system!

It seems to me that often mathematics education is trying to catch up to unfolding world events. We are all familiar with the question about what came first—the chicken or the egg? However, what if the question was, "What came first—science or mathematics"? In the early 19th century, the famous German mathematician Carl Friedrich Gauss referred to mathematics as the "queen of the sciences." Historically, mathematics has been successful at uncovering and explaining the nature of physical reality; the symbiotic successes of mathematics and science continued into the 1990s.

Selected Writings from the Journal of the Mathematics Council of the Alberta Teachers' Association, pages 235–238
Copyright © 2014 by Information Age Publishing
235

Many consider the 1990s to be the dawn of the information age. That decade saw the beginning of widespread use of the internet by the public, DOS-based computers had evolved to Microsoft's Windows 3, and the Mac-Plus of 1986 was replaced by the MacClassic and the MacII. The original Intel Pentium processor was introduced in 1993, and Windows 95 and 98 arrived, along with the Apple iMac. In 1999, the Apple PowerMac G4 boasted in excess of one billion floating point operations per second!

In 1994, Sir Andrew Wiles, a Royal Society research professor at Cambridge University, proved a theorem that had been postulated circa1665 by Pierre de Fermat. Fermat's Last Theorem, which states that $x^n + y^n = z^n$ has no integer solutions for $n > 2$ and $x, y, z \neq 0$, has the distinction of being the theorem with the largest number of published false "proofs," since in excess of 1,000 were published between 1908 and 1912. In 1998, Sir Wiles received a silver plaque from the International Mathematical Union in place of the Fields Medal to recognize his achievements (the Fields Medal is restricted to those under 40 and Sir Wiles was 41 at the time of his discovery). Sir Wiles has an asteroid (9999 Wiles) named after him, and he was appointed Knight Commander of the Order of the British Empire in 2000.

In science and space exploration, the Galileo probe was launched on October 18, 1989, and inserted into orbit around Jupiter on December 8, 1995. The Human Genome Project began in 1990, a year that also saw the launching of the Hubble Space Telescope (now proposed to be in service until 2014). In 1992, the first Canadian woman in space, Dr. Roberta Bondar, flew aboard NASA's space shuttle Discovery. The Pathfinder mission to Mars was launched on December 4, 1996, and landed on the Martian surface on July 4, 1997. The existence of extra-solar planets was discovered, Dolly the sheep was cloned, genetically engineered crops were invented, GPS became fully operational, and HIV/AIDS mortality was reduced through the use of highly active anti-retroviral therapy (HAART).

In my opinion, the most exciting development in mathematics education during the 1990s was the introduction of the graphing calculator. Casio marketed its fx-7000G in 1985; Texas Instruments introduced its Ti-81 in 1990, followed by many others in the TI-family. There were brief stints with Casio, Hewlett-Packard, and Sharp, but Texas Instruments emerged as the leader in mathematics education. Paired with an overhead palette and perhaps a CBL Ranger, the graphing calculator became a teaching tool of the highest order. The graphing calculator not only did away with the tedium of calculation; it allowed more time to spend on the mathematics of a situation. More importantly, students used the graphing calculator to explore and even discover mathematics. Established programs such as advanced placement and international baccalaureate both shifted their objectives to require the use of a graphing calculator in some sections of their examinations.

The calculus reform that was initiated in the 1990s emphasized that a particular problem could be solved numerically, algebraically, or graphically (the Rule of Three), and that these seemingly disparate techniques were actually complementary. This in turn facilitated more communication among students of mathematics and other fields in the physical and social sciences. The modern calculus texts of the 1990s, complete with excellent multicolour graphics and calculator and computer screenshots, saw the inclusion of more real-world and nonstandard problems, which, in turn, increased the demand for more conceptual understanding. Students seemed to take a great deal of interest in, for instance, related rate or max/min problems that had their origins in the fields of medicine and physiology.

Specialized examinations and surveys such as the Trends in International Mathematics and Science Study (TIMMS), initiated in 1995 by the International Association for the Evaluation of Educational Achievement (IEA), compared student proficiency in mathematics and science among participating countries throughout the world at the grades 4 and 8 levels. The summary report dated September, 2000, claims that "Canadian (grade 8) students performed relatively well in mathematics and science!" It goes on to state that only six countries had achievement results that were "significantly higher" than those of Canada in mathematics, and only five countries were "significantly higher" in science. The Program for International Student Assessment (PISA) Study began in 2000 and compares 15-year-old students in reading, mathematics, and science; Canada scored near or at the top in all three disciplines when compared to G8 countries. In mathematics, only 11 percent of Canadians scored below Level 2 (the best result among G8 countries), and 18 percent of Canadians scored above Level 4 (second only to Japan).

In a paper entitled *The Canadian Mathematics Curriculum from New Math to the NCTM Standards*,[1] Dr. Thomas O'Shea of British Columbia's Simon Fraser University puts the 1990s in perspective by reflecting that changes in Canadian mathematics curricula paralleled curriculum movements worldwide. According to Dr. O'Shea, Canadians experienced the euphoria of the "New Math," and that period was followed by a back-to-basics movement. The National Council of Teachers of Mathematics (NCTM) *Agenda for Action* of the 1980s made problem solving central to the mathematics curriculum of that decade, and the NCTM Standards shaped the curriculum of the 1990s. Dr. O'Shea characterized Canadian curriculum swings in education as "muted," saying that they occurred later in Canada then they did in the United States, ostensibly because of the provincially based educational decision making process of those years. The ministers of education of the western provinces (British Columbia, Alberta, Saskatchewan, and Manitoba) as well as the Northwest and Yukon Territories signed the Western Canadian Protocol for Collaboration in Basic Education (WCP) in December of 1993

(Nunavut joined in 2000). The WCP saw the development of common curriculum frameworks with learning outcomes in language arts, mathematics, and international language; the first common curriculum framework in mathematics (K–9) was released in both official languages in June, 1995. I believe that the agreements made in the WCP were a major development in mathematics education.

During the 1990s, mathematics education experienced significant change not only in Alberta schools, but across Canada and North America. This change was due not only to the implementation of technologies such as graphing calculators and computers in classrooms, but also to the publication of vastly improved textbooks. Perhaps, however, even more important was the most enjoyable trend towards collegiality and communication that developed among mathematics educators in the 1990s; educators were more than willing to share exercises, exams, teaching ideas, and knowledge on a large scale, thanks in part to the internet. Mathematics conferences, item writing, and marking sessions as well as august groups such as the Mathematics Council of the Alberta Teachers' Association (MCATA) and various mathematics learning consortia throughout the province all had many positive effects on mathematics education in Alberta that continue to be enjoyed to the present day.

NOTE

1. O'Shea, Thomas, Simon Fraser University, "The Canadian Mathematics Curriculum from New Math to the NCTM Standards." Excerpts from the third draft of Chapter 18 of the NCTM's Mathematics History Volume, 2003, http://cms.math.ca/Events/CSMF2003/panel/oshea.pdf

Darryl Smith retired in 2002 after having taught for 34 years with Edmonton Catholic, 30 of which were spent at Austin O'Brien High School. Since his retirement, he has continued to tutor many students in mathematics, and that role has required him to keep abreast of curriculum developments. He looks forward to many more years of mathematics education.

This chapter was originally published as:

Hubber, D. (1990). It's all Greek to me: Mathematics anxiety. *delta-K: Journal of the Mathematics Council of the Alberta Teachers' Association, 28*(2), 13–15.

CHAPTER 31

IT'S ALL GREEK TO ME

Math Anxiety

Darlene Hubber

Editor's Note: Darlene Hubber teaches special education in Brooks, Alberta. She has 12 mentally handicapped and learning disabled students in grades 4 to 6.

"I'm not a numbers person." "Don't ask me; I never could understand math." "I just don't have a mathematical mind." "What, me do math? I can barely add two and three." "It's all Greek to me!"

Sound familiar? If you have heard your students make such proclamations, then you have already had some experience with math anxiety. Tobias defines math anxiety as the feeling of panic, helplessness, paralysis, and mental disorganization that arises among some people when they are required to solve a mathematical problem. (Tobias & Weissbrod, 1980)

The prevalence of the problem is reported to be as high as 68 percent among students enrolled in math classes (Lindbeck & Dambrot, 1986). Some anxiety is desirable. Tension can serve as a motivator, but too much anxiety, as in the case of math anxiety, can inhibit learning.

Several factors contribute to the development of math anxiety (Greenwood 1984; Martinez 1987; Tobias, 1980). The first is the "math as a gift, not as a set of learned and practiced skills" misconception. Math anxious learners think only those born with a mathematical mind can fully comprehend complex numerical operations. They assume that competent math students arrive at solutions instantly, and they have little or no faith in their own ability as a result. Even when they are able to come up with a solution to a problem, they lack confidence in their answers and assume they couldn't possibly have figured it out correctly.

Math anxiety appears to be a uniquely North American phenomenon. An examination of the math attitudes of Asians and Americans revealed some interesting discrepancies. The Asians thought math ability was fairly evenly distributed and that skills were developed through study, persistence, and hard work. Americans, on the other hand, viewed mathematical ability as a very rare and uncommon talent (Tobias, 1987).

Early math experiences can contribute to the development of math anxiety. Many of us painfully recall being summoned to the blackboard to solve a problem, only to make an embarrassing blunder in front of a room full of witnesses. Some math anxious students recall negative experiences with particular teachers. They remember the stress and confusion caused by timed tests, the emphasis on one and only one correct answer and on the "right" way to arrive at solutions to problems. They remember pages and pages of drill and practice.

Some researchers believe "the principal cause of math anxiety lies in the teaching method used to convey basic mathematical skills" (Greenwood, 1984). They suggest the explain-practice-memorize paradigm isolates facts from the problem solving process of which they are a fundamental part. Students who cannot understand the thought processes that underlie a problem's solution begin to perceive math as an incomprehensible mystery.

Teachers can be math anxious too. Such teachers contribute to the anxiety levels of their students by relying primarily on text explanations and do-theproblems/ correct-the-problems assignments. They often refer authoritatively to the teacher's guide and seldom work out problems with students in front of the class.

Inadequate out-of-class experience with math can contribute to anxiety development as well. Students who are never given an opportunity to solve real-life problems fail to see how the skills they have learned are applicable to situations outside the four walls of the classroom. They lack motivation because the meaningfulness of the skills is not apparent to them.

The language of mathematics can be complicated and ambiguous. For students who never become proficient *readers* of math, this can create confusion and anxiety. Words such as attitude and root have very different mathematical connotations. One researcher reports interviewing a student who assumed that "least common denominator" meant "most unusual" (Tobias,

1978). Based on her understanding of the term, she produced unique, but obviously incorrect responses.

Gender may play a role in the development of math anxiety. Studies show that although girls account for 49 percent of the secondary school population in the United States, they comprise only 20 percent of those taking math beyond geometry (Tobias, 1978). This discrepancy is not a result of differing levels of ability so much as it is a reflection of societal perceptions. Parents, teachers, and peers are more likely to excuse poor math performance by girls. As a result, girls may perceive math as a primarily masculine domain and lack the confidence and motivation to develop their own skills in this area. Girls may also feel pressure to appear "dumb" in order to conform to these perceptions.

Girls may find math more difficult because they lack math experience. Few of the playthings commonly provided for girls promote the development of mathematical understandings. The toys generally provided for boys, however, are of the "take apart and put together" variety that do develop these understandings. In addition, many mathematical concepts are used in sports, such as hockey, football, and baseball, which traditionally have much higher levels of male participation.

Role conflict, negative math experiences, inadequate instructional emphasis, lack of understanding of the language of math, and misconceptions about the nature of math learning all contribute to the development of math anxiety. Math anxiety leads to math avoidance by students who feel they cannot and will never be able to experience success in the mathematics classroom. It may also contribute to lower achievement levels by otherwise able students who, for whatever reason, panic when asked to perform mathematical computations. Helping these students overcome their anxiety and, ideally, preventing it from afflicting the generations to come is now a critical issue. Math competence has become undeniably necessary for many technological careers. Students without a firm grounding in mathematics greatly reduce their career options.

Programs have been developed at the college level for the math anxious members of the population. Most programs include counseling and education components. Participants are involved in some form of group or individual counseling where their anxieties are examined. They are also provided with direct instruction in mathematics with an emphasis on meaningful problem solving. Although these programs are beneficial, they treat the symptoms not the cause. Preventing the development of math anxiety is the responsibility of educators.

An anxiety-free learning environment can be created by removing tension and competitiveness; students must not be afraid to ask questions and make mistakes. A nonthreatening learning environment is a necessary component of such a classroom. Allowing students to correct their own work, learn through trial and error, offer answers without fear of humiliation, and solve problems as committees helps create this kind of environment and nurture student confidence.

A change in emphasis from the traditional explain-practice-memorize paradigm is also required. Math learning experiences that are creative and encourage active participation by students can contribute to the development of more positive attitudes toward mathematics.

Instruction matched to the cognitive levels of the student results in lower anxiety levels. Concepts must be understood in their concrete forms before more abstract applications are introduced. To help students visualize concepts, manipulatives can be used at all levels.

A lower level of anxiety is also evident in students for whom numbers have real-life significance. Students must be aware of the personal usefulness of math; it must be meaningful from the students' perspective.

Systematic instruction in problem solving helps reduce the levels of anxiety in math classrooms. Understanding a variety of problem solving strategies gives students the tools to attack problems successfully outside the classroom. Students must have experience in looking for solution patterns, substituting simpler numbers and diagramming or sketching problems in order to use these strategies successfully.

Some attention must be paid to the unique reading demands of math. The language of math must be clearly understood for students to succeed. Strategies such as substituting more easily understood synonyms, underlining and discussing problem words, and rereading for clarification, if employed effectively by math students, will reduce anxiety levels.

"Most people leave school as failures at math or at least feeling like failures" (Tobias, 1978). They are defeated before they even begin by their lack of confidence in their own abilities. As teachers, we can help reduce anxiety levels by changing the way we teach math. We must demonstrate the importance of math, provide students with a multitude of experimental opportunities to test mathematical concepts, and teach students the strategies necessary for effective problem solving. As our world becomes increasingly technological, we must develop more effective methods of teaching mathematics. Numeracy has become as important as literacy. We must dispel the myth that math is a secret code comprehensible only to an elite few.

REFERENCES

Adwere-Boamah, J. et al. (1986, Spring). Factor validity of the Aiken-Dreger mathematics attitude scale for urban children. *Educational and Psychological Measurement, 46*, 233–241.

Greenwood, J. (1984, December). My anxieties about math anxiety. *Mathematics Teacher, 77*, 622–623.

Lindbeck, J, S., & Dambrot, F. (1986, November). Measurement and reduction of math and computer anxiety. *School Science and Mathematics, 86,* 567–577.

Maninez, J. G. R. (1987, November). Preventing math anxiety: A prescription. *Academic Therapy, 23,* 117–125.

Tobias, S. (1980). Math anxiety: What you can do about it. *Today's Education, 69,* 26E–29E.

Tobias, S. (1978). *Overcoming math anxiety.* New York, NY: Norton and Company.

Tobias, S. (1987). *Succeed with math: Every student's guide to conquering math anxiety.* New York, NY: College Board.

Tobias, S., & Weissbrod, C. (1980, February). Anxiety and math: An update. *Harvard Educational Review, 50,* 63–70.

This chapter was originally published as:

Hauk, M., & Quinn, B. (1992). Moving out of the comfort zone. *delta-K: Journal of the Mathematics Council of the Alberta Teachers' Association, 30*(2), 4–9.

CHAPTER 32

MOVING OUT OF THE COMFORT ZONE

Marie Hauk and Bryan Quinn

The project had just begun, and already Bryan was experiencing self-doubt. Why was he rocking his boat and entering uncharted waters? He is a good math teacher; he works conscientiously toward having his students enjoy math and experience success. So why was he choosing to depart from the safety of his familiar, traditional classroom structure? Why was he now standing in his classroom directing traffic as his grade 7 students attempted to rearrange their desks into groups of four?

To begin with, the students had come to class unprepared. None had submitted lists for proposed group members, so class time was required to do this. Few had brought new folders or duotangs as requested the previous day. To top it off, class time had been shortened due to a school activity. Bryan had planned to have his students spend the whole period collaborating on an introductory poster activity in which they would identify, discuss, and display the purposes of group work. Not only was there less time for this activity but also it had become evident that to keep the students on task, they needed more specific direction than anticipated.

Bryan capitulated. After all, it was Friday afternoon! Perhaps if he supplied the folders and duotangs for the students, they could get organized properly on Monday. Besides, he knew Marie would be there to help. Bryan and Marie were collaborating on all aspects of the planning and teaching of this fractions unit in which students would make extensive use of concrete materials within a small-group learning setup.

What motivates successful teachers to make significant changes in their teaching style? In other words, *if it ain't broke, why fix it?* Good teachers continue to be learners throughout their careers. Not only do they reflect on what they are doing or have done but also they seek alternatives. Current research and literature offer new directions as theories regarding teaching and learning change. For Bryan and Marie, it was a matter of making classroom practices consistent with their philosophical stance. They believe that students must have opportunities to be involved responsibly and actively in their own learning. They also believe that for students to truly understand and appropriately apply mathematical concepts and skills, they must have concrete experiences in personally meaningful problem solving contexts.

The topic of fractions is notorious for being abstract and difficult. Many students who appear to have achieved success through traditional chalk-and-talk methods have often developed only a superficial understanding. For this reason, it seemed to be an appropriate area in which to introduce an alternative approach. While Bryan's previous use of concrete materials in teaching math was primarily for teacher demonstrations, he perceived that the manipulatives had facilitated his students' understanding. Ongoing use of concrete materials within a small-group setting for a unit in mathematics was a new experience for Bryan and the students.

One month was allocated for this unit. During this time, a major theme, *Out of the Comfort Zone*, emerged. While Bryan used this phrase explicitly on a number of occasions, it was descriptive of both the students' and teachers' experiences. The comfort zone was defined by a number of interwoven beliefs and practices regarding teaching and learning. The path for moving out of the comfort zone was neither smooth nor one-way; the urge to return remained strong.

The students' experience of moving away from what was familiar and comfortable centred on day-to-day personal gain and may be described as *Seeking Immediate Rewards*. From the teachers' perspective, however, concern with short-term personal effects was only one factor. As the innovators, they were accountable for moving themselves and the students out of the comfort zone. This demanded daily self-evaluation of both the practical and the philosophical aspects of the experience. Their movement was influenced by time and confidence, and may be categorized as *Coping with Time Constraints* and *Looking for Support*.

THE COMFORT ZONE

Most teachers and students operate within a multifaceted comfort zone without being explicitly aware of its existence. Consciousness of this phenomenon occurs when any boundaries are crossed. This project revealed the following comfort zone fronts: curriculum and pedagogy, classroom organization, control of learning, learning resources, and evaluation of learning. Movement requires a departure from traditional practices.

A particular curriculum focus tends to give rise to a compatible pedagogical style. When curriculum focuses on product-oriented objectives, pedagogy that produces easily measured short-term results may appear to be successful and is often viewed as desirable. A move toward emphasizing process over product, where the results are less tangible and often long-term, is viewed with suspicion by traditionalists. Classroom organization sets the tone for learning. The traditional arrangement of desks in rows fosters a nonsocial approach to learning. A small-group arrangement creates a different atmosphere: No longer is talking among students taboo, it is required! Traditionalists view control of learning to be the prerogative of the teacher, who presents and explains the concepts and skills to the students. However, having students work in groups allows them to assume ownership of the learning process. They are held more accountable both for their own learning and for assisting others within their group.

The textbook is often taken for granted as a learning resource. Because of the security it can offer to teachers, students, and parents alike, it may be the only resource used. In fact, too often, the quality of teaching is judged on the basis of what page the class is on. The textbook *becomes* the course even though its content may not be congruent with the prescribed curriculum. While a good textbook can be of value, ideally, it should be viewed as just one of many resources used in the classroom. Manipulatives, calculators, and other print materials can enrich learning.

Evaluating mathematics learning traditionally has been done through measuring concepts and skills in an objective manner, usually by means of paper-and-pencil tests. However, learning is also subjective. Assessment strategies that focus on process and include the use of manipulatives and oral reasoning as well as written work result in more valid assessments.

SEEKING IMMEDIATE REWARDS

The second day of the unit began with a pleasant surprise for Bryan. Waiting to greet his students, he braced himself for impending confusion. However, the students entered the classroom, and, without waiting for instructions, quickly maneuvered their desks into their respective groups of four.

The novelty of the situation, having two teachers in the classroom and the opportunity for social interaction, seemed to be motivational; certainly the students were eager to try the new setup for learning.

Another novel aspect for the students was keeping two-part logbooks. The first section was used for in-class activities and homework assignments, the second as a journal for personal writing (about group work, the use of concrete materials and sundry comments). Many journal entries indicated students' discomfort with the writing process, both with expression of ideas and spelling and grammar. Despite their difficulties with writing, many students made honest observations and offered insightful comments. To present the experience realistically from the students' point of view, examples of journal entries have not been edited. The students' real names have been changed to maintain their anonymity.

To capitalize on the students' desire to seek immediate rewards, a prize draw was instituted very early in the program. Students earned stamps for their group during the course of a week. A group received an entry for each five stamps earned. On the basis of the accumulated entries from the previous week, a draw was conducted each Monday. Stamps were given rather liberally for a wide variety of student behaviors—literally anything that contributed in a positive way to the functioning of the class (for example, arranging desks quickly and quietly, taking group attendance, working as a group rather than individually, being polite, assisting others, writing in logbooks). Despite the fact that the group prizes were mere token gifts, earning stamps proved to be an overwhelmingly powerful motivator for individual students while strengthening group dynamics.

The students' generally positive attitude toward the group arrangement remained throughout the unit. Their opinions fluctuated, however, depending on perceived personal benefits from day to day. Early on, Bruce commented in his journal, "I think every thing going prity good. Exept some times people act up. I like this more than the ragular class." At the end of the unit, Blair expressed a common opinion, "Today we are finished off your fractions and group work. I thought the group work was good and I hope that we have it next year."

Most students were pleased with their selection of group members. In early journal entries, Mike commented, "Today Rob was the leader. He was very nice," and Flora wrote, "I do like Wally in my group." Bonding among group members became evident when a change was made early in the unit. When one girl in Group 4 left the class, Bryan replaced her with a girl who had been unhappy in Group 1, and he transferred one boy from Group 2, the only group of five, to Group 1. Bryan gave the five boys in the group the responsibility for deciding which one of them should move. They found the decision difficult and ended up putting their names in for a draw. Mike's concern was evident when, later that day, he wrote, "Darcy had to move to group #1. He is so far coping with the group fine." Another day, he commented further, "Darcy is doing fine with group #1."

Using manipulatives for mathematics was a new experience for most of the students. As a result, these students used play time to become familiar with each of the materials. A certain amount of free play had to be tolerated before the students would use the manipulatives for the intended purposes. The students were inclined to handle and explore the colorful materials rather naturally. They often described the work with manipulatives as being fun. Lois wrote, "Well, now I learned about cuisenaire rods. Their kinda fun!" Helen similarly reflected, "I learned that Cuisunaire block are used for fractions. They are used as telling sizes. It's easy and fun too."

An unspoken question seemed to develop: "If we're having fun, can we be learning?" (Is it OK to have fun while learning?) Some dealt with this uncertainty about the legitimacy of *playing* in math class by differentiating between meaningful and nonproductive types of play. This was expressed by Helen who wrote, "It is fun to play with shapes and I just learned never to give Amanda any kind of blocks to play with because that is one of her favorite kinds of toys, building blocks." While having fun was important, Gloria's comment, "I enjoy having to do this kind of work," suggests that the playing was goal-directed.

Having fun was important, but students gained more appreciation for the manipulatives as they became aware that the latter helped them develop greater understanding of the mathematics being explored. A number of journal entries reflected this attitude. Howard wrote, "When we used the manipulatives it made the problems easier because it helped us understand them. The fraction strips was probably the most helpful manipulative." That some students were initially dubious was reflected by Wayne: "Today we learned about what maded 1/3, whole, 1/2, & how to compare other rods. The thing that was fasinating was using the Cuisenaire rods to make it easier to understand. The thing I was surprised about was to talk with my group members and understand fractions much better." The visual aspect was appreciated by Lois, who wrote, "Dividing fractions is pretty easy with cuisenaire rods. I guess it is better to see it than to think it in your head." The need for tactile experience was expressed by Shawn: "Also we got to use little blocks cuisenaire rods bar graphs and stuff like that. Using thing you can Feel is better than just thinking about it because you can see them and Feel them and ther wright there in front of you bye your self you can only think about it."

Students who previously have had only superficial exposure to procedures and algorithms may become frustrated when they are confronted with the more lofty goal of attaining a deeper understanding of concepts. Evidence of this was indicated by Melanie: "Right know we are working on fraction bars. I don't understand so I'm behind the group. I don't find the directions straight forward and I just don't get it." The need for patience in this respect was recognized by Gloria, who wrote, "Understanding fractions is a little bit hard. I used to know a lot about fractions but now it is getting harder. One thing I like about groups is that theres lots of things to do. We also worked with colored cubes. Its kind of neat because you get to do all sorts of things with shapes. I learned that its easy once you know what your doing. Not everyone thinks it easy but I think they should give it a try. At first this was hard for me but once you give it a try it can be easy. I just hope everyone gives everything a try."

Probably the most motivating aspect for the students was the opportunity for social interaction. This enthusiasm was highlighted by Lois, who wrote, "Boy this was fun! Today I got a double period! Woa dude! This is heavy! I like working in a group. Its totally dense!" As for the concrete materials, fun was a common descriptor of group work, but there was also a sincere desire to learn, as expressed by Judy: "I think group work is really fun and eciting. I honestly belive that this group work thing will really work and I'm sure that we will get a lot of work done.... This will probably be the best math classes that we ever had." She went on to address explicitly the importance of interpersonal relations by writing, "I love this working in group. I mean this is really fun. I find that working in groups you get to know people better. I think that math class is better now that we are working in groups too bad we cant stay like this forever." Douglas expressed similar thoughts: "Today I am going to write how I like groups better than working by myself. Its better I find it because you can shar ideas and meet new people and become better frends or learn more about them." The benefits of collaborating and sharing the responsibility of learning were often mentioned by the students. For example, Shawn wrote, "When we were working in groups I found it Fun. It was better than working buy yourself because when you do a question in a group you get to hear what everybody has to say and you can discuss it." Similarly, Howard said, "I think it was a good experience, working in groups. It made the work easier and I understood more that we worked together. It helped some people with their listening." Gloria agreed, "Groups are fun to work in because you can help people or they can help you."

The most common complaints about group work were noise and lack of cooperation. This was indicated by Gloria, who noted, "Some people hate groups because its destracting when some group members keep talking." However, this was not necessarily perceived as a reason to quit group work. Lois acknowledged the distractions, but was not discouraged: "I guess it was pretty noisy. The work was a lot easier and we learned to cooperate with each other. I think in the future, we should get into more groups!" Some of the students were unsure about the value of group work in terms of their own learning. This feeling was expressed by Billy, who wrote, "In this section I feel that I did not lern as much as I would if we were not in groups

In other words I feel I would have lerned more by myself." Even Judy, who enjoyed the experience, said, "I think learning how to use fractions in a group was fun and I really dont know if I could have done better or worse if I would have been by myself."

At the end of the unit, the students generally spoke favorably about group work and indicated that they would prefer to continue learning that way. Cheryl wrote, "The fractions unit is over and I really enjoyed working in groups. We are now doing ratios and rates and now we work individually. I enjoyed sharing the load of work with others." Endorsement was offered by Mike, in his comment, "Today we are back to normal in math. BORING! I like group work way better. I learn better that way. The group is more better way to learn math."

Before the project, in terms of performance, most students in this class were either high achievers or low achievers. Although this pattern was not altered significantly by the project, it seemed as though the high achievers felt more challenged while the low achievers, despite low marks, felt as though they benefited in other ways. Not surprisingly, the bottom line for most students is marks. Those who did not achieve well on the tests expressed disappointment. Melanie wrote, "I used to think math was easy but ever since the first term my marks are slipping majorly! I don't know why I just don't understand THIS." Shawn related, "We had a test on fractions I thought that I would do O.K. but I didint I only got 48% Not to good at all. I try to study more and do better." Even some students who did well, relatively speaking, indicated that they had greater selfexpectations. Rachel noted, "On February 18 we got our tests back. I got 72%. That's okay but is not so great. I don't understand how to multiply and divide fractions yet." Similarly, Lois commented, "Well, I didn't do very well on my fractions test. I got 72% Pretty bad, huh? I hate it. I'm a 70%–80% person. I'm so bad. Why can't I do any better. I guess I have to 'Apply' myself to it."

The disappointment, however, was not directed at the group setup. Gloria lamented, "On the test I didn't do very well. I use to be good at fractions but is differnt now. Some how its alot harder than the years before. I think I need some help with fractions." Shawn, in fact, supported group work in spite of his own performance: "We had a test on Fractions I thought that I would do pretty good on the test but I didint I only got 33% thats not to good but at least I tried. It was a Fun experienc being in groups I hope are class will get to do it again."

On the other hand, some students who were pleased with their marks did attribute their success to working in groups. Blair wrote, "... I liked the group work and we should have it again. I liked the fraction unit because I did good on the tests and I found it easy." Helen was most pleased with her results on both tests. After the first test, she enthused, "...all I can tell you is I GOT 79% ON MY MATH TEST!!!... I think I work much better in groups. I usually fail. I know I'm one of few people who are better in groups and I know we won't stay like this but I wish we could." Following the final test, she reported, "We don't have to go in groups anymore. I really enjoyed it though. I obviously did better in groups though I actually passed my 2 tests!!! I am happy, my mom is happy and if I get 80% on my whole report card I get a new bike!!!!!"

COPING WITH TIME CONSTRAINTS

Throughout the project, time was a significant factor. Both unit and daily planning required many hours. The planning began with consideration of the, prescribed curriculum objectives, and then a tentative timeline was established. Because many of the concepts and skills were at the introductory level, at least four weeks seemed necessary.

The Ginn *Journeys in Mathematics* 7 textbook was used as a resource, but the focus was on personally developed cooperative group activities that required concrete materials. Appropriate materials were selected to facilitate learning each objective. An inventory of on-hand manipulatives was taken, then a decision was made as to what other materials were to be borrowed or purchased. The daily activities were developed as the unit progressed.

Time had to be allowed for orienting the students to group work. As cooperative learning in mathematics was a new experience for most students, they needed to appreciate that the occasion was to be used for more than socializing. Routines had to be established. Rearranging the desks at the beginning and end of the class became progressively more efficient. Math classes were held four days per week. Each of the four students in a group was assigned responsibility for obtaining the folders, activity sheets, and trays of materials on a particular day.

The cooperative learning process required considerable time and patience for both the students and us. To encourage the students to work together, only one activity sheet was given to each group initially. However, many of the students were not sure how to share the sheet; some groups circulated it for each one to read separately and did the activities as individuals. Other groups attempted to read in unison but found it awkward to reposition themselves. After discussing the problem, Bryan and Marie compromised by giving each group two activity sheets. Even then, continual prompting was necessary to discourage them from working as separate pairs within a group. Frequent reminders were given to the students so that group members would monitor and help each other to understand and finish the activities.

Some groups worked more efficiently than others. For accountability, all students were expected to record their activities (for example, draw the result). Yet, some used too much time needlessly documenting (using pictures and writings) each step in detail. Pacing the activities was difficult for some students, as they were not used to making such decisions. As time went on, Bryan took a greater role in monitoring the amount of time they spent on each activity. The small-group approach with teacher as facilitator was a new teaching style for him, and the urge to go back to a more teacher-controlled situation was strong.

According to the curriculum, much of the fraction work in grade 7 should be at the concrete level. Not only did the exploratory hands-on nature of the activities place demands on time but, because few students had experience with concrete materials, more time had to be allowed for them to become familiar with the manipulatives. Journal writing was often short-changed. As this was deemed an important aspect of the experience, special effort was required to ensure that enough time was allowed. One solution was to start the class with journal writing on occasion.

Coping with time constraints was an uphill battle. Class time passed quickly; 40-minute classes were too short for an activity-based program. The teachers' and students' inexperience with cooperative learning and extensive use of concrete materials put extra demands on the time required. Further flexibility was required to accommodate school timetable changes and classes missed due to holidays and the teachers' convention.

LOOKING FOR SUPPORT

People feel stress when *someone else* (someone in authority) changes the boundaries, but the degree of stress is greater when the change is self-initiated, because the decision itself is questioned both internally and externally, and there is self-doubt in addition to criticism by others. For this reason, ongoing support is essential. Bryan and Marie continually looked for support from within, from each other and from the students. Unsolicited support from colleagues was relished.

Support from school administration is essential when a teacher departs from traditional methodology. Bryan's principal took an active interest, endorsing the project and visiting the classroom to observe the students as they worked in groups. Affirmative collegial support is also significant. Such encouragement came from a language arts teacher who, on several occasions, looked in on the class and each time expressed her pleasure to Bryan at seeing group work and journal writing being done in math.

The value of motivation cannot be overestimated.

Support from the students was manifested daily through their eagerness to enter class, set things up and use the materials enthusiastically. Even though they tended to react to the immediacy of a particular situation, their generally positive attitude was encouraging.

The availability and the practicality of using and storing the concrete materials was not a major problem: most of the students handled the materials reasonably and were accountable for their use. But, in view of the investment of money and teacher effort, Bryan found occasional lack of appreciation for the materials and subsequent misuse (at times from students in his other classes) frustrating.

High marks are often perceived as *concrete proof* of the success of a program. Because of this natural tendency to value test marks, Bryan and Marie were initially disappointed with the results of the first exam. Although the students' results were lower than hoped, they correlated with their marks from earlier in the year. Perhaps with a more traditional teaching and testing approach, the students may have been able to get more correct *answers* to routine questions, but Bryan and Marie doubt that the students would have had greater *understanding*.

Bryan admitted that even though long-term goals were critical and essential to him, a devil's advocate in the back of his mind kept telling him that he could get certain results faster with traditional methods. In fact, he pointed out that "good" students sometimes want the traditional approach because they have found it easy to be successful with it.

Bryan and Marie both considered the project a success. What they learned will guide their review and re-teaching of the unit. Some of the benefits had immediate impact on Bryan's teaching style. Following this unit, he implemented cooperative learning in a grade 9 unit on surface areas of prisms and cylinders, where he had previously used concrete materials. Not only has he shown more willingness to try innovative approaches with other classes, but also he feels an increased responsibility to continue and broaden his teaching techniques. All teachers who experience similar anxiety should have confidence knowing that moving out of comfort zones is an indication of professional growth.

This chapter was originally published as:

Findlay, C. M. (1993). Learning about computers and mathematics: A student perspective. *delta-K: Journal of the Mathematics Council of the Alberta Teachers' Association, 31*(3), 8–11.

CHAPTER 33

LEARNING ABOUT COMPUTERS AND MATHEMATICS

A Student Perspective

Craig M. Findlay

We are living in an "information age," and it seems that the technology of computers is here to stay. For many, including me, the computer has become an essential tool for information storage, manipulation, and output. In the last couple of years, following the purchase of my own computer, I too have been caught up in our technological era. As an educator of the future, the question then becomes, How can I integrate this technology into my profession so that it can aid in my teaching endeavors? In my research, I quickly discovered that many people before me have pondered this question. Winzer (1990, p. 112) says, "Computers cannot replace classroom teachers, but they are patient, consistent and accurate teaching tools that possess unlimited appeal and motivational value for students." These ideas first surfaced over 20 years ago when the computer was being billed as the educational utensil of tomorrow. In this light, the technology was proposed largely in the field of mathematics.

Selected Writings from the Journal of the Mathematics Council of the Alberta Teachers' Association, pages 255–261
Copyright © 2014 by Information Age Publishing
All rights of reproduction in any form reserved.

In our modern society, literacy refers to language, as well as mathematics (Mendoza, 1989 in Winzer, 1989). Bangs (1982, in Winzer, 1989) reveals very specifically that mathematics is indeed itself a language. In comparison with instruction in language, mathematics has received little attention when it comes to diagnosis, instruction and remediation (Winzer, 1989). Mathematics has been a domain in which I have not had a great deal of success. Therefore, when I saw a chance to learn about new ways to teach and explore mathematics in an area where I do have a great interest, computers, I obviously accepted the challenge. This becomes increasingly important when one realizes that in our rapidly changing technocratic society, people will use their arithmetic skills more than ever (Winzer , 1989). Those of us lacking in this burgeoning domain will find survival even more difficult than it already is and inevitably will be left behind.

Inherent in my discussion will be the use of computers in mathematics instruction for "students" in general, although I will point out where the computer can enhance the learning of exceptional students. Lerner (1981, in Winzer 1989, p. 319) describes certain principles that are applicable to all forms of mathematical learning and are key ingredients to effective teaching:

> Students understand concepts best when they move from the concrete to the abstract. They need plenty of drill and practice to develop automaticity about facts and operations. Finally, they need the opportunity to see mathematics as part of the real world.

Mendoza (1989, in Winzer, 1989) describes mathematics as hierarchical in nature. Thus, gaps in learners' backgrounds will go on to hinder their future successes. Drill and practice therefore become important components to mathematics teaching to promote the acquisition of fundamental facts and concepts. Many contemporaries in mathematics education would no doubt debate this, but this point still holds true for some, especially when dealing with learning disabled students. In attempting to gain a certain functional level of mathematics savvy, this traditional approach seems essential for special learners (Winzer, 1989). One benefit of computers emerges with their ability to perform repetitive tasks with immediate user feedback. This can help students, especially in mathematics, who require repetition of facts and concepts. Modern computers also allow for other minor alterations that will assist special learners. For example, font sizes can be enlarged, and Braille printouts can be made for visually impaired students. The speed of the presentation of material can also be altered to meet learners' needs. Some computer programs are based on "real-life" situations, making the content more functional than otherwise possible. Making education functional is vitally important to effective teaching. The graphics that modern computers offer enable users to manipulate seemingly concrete objects, making learning more genuine. On the other side of the spectrum, computer technology can help gifted students who wish to pursue more complex learning. Gifted students need expanded and enriched curricula that will stimulate higher-level thinking and will allow them to apply their skills in a variety of contexts (National Council, 1986, in Winzer, 1989). With this in mind, modern software is moving toward allowing the user to simulate certain ideas and concepts; opening new avenues of trial and error, exploration and higher-level learning.

In trying to understand computers and their use as an educational tool, I wanted to obtain a certain breadth of research. I chose to look at 11 different journal articles. In doing so, I obtained work from a variety of publications and from different time periods to represent as many perspectives as possible. The first article goes back to when computers were just being explored and their potential was only beginning to be forecast. The rest of the articles reflect more modern ideas and represent a transition from the computer "boom" of the 1980s to the present. The articles reflect several standpoints and highlight the computer as an increasingly important, if not controversial, instructional tool in education.

My bibliography also includes research that I have done outside of the 11 chosen articles. The concluding portion of the paper discusses the pieces in a more comparative light, recognizing that each article represents a certain aspect of computers in mathematics instruction. Finally, and in a much broader context, I have addressed whether or not the computer has lived up to mathematics teachers' expectations and to the expectations of educators as a whole.

DISCUSSION

The research that I have read constitutes somewhat of a "jarring" experience to my previous conceptions. My appreciation of the computer had been pedestaled largely because of my own perceptions of the technology. Despite the area of mathematics benefiting most from the advent of the computer, it too has not lived up to the early expectations beset on it in the field of education.

The computer was first conceived in terms of its value to educators in the late 1960s (Zion, 1969). From that time, the technology has advanced and experienced a large amount of growth through the 1980s to the present. The focus on computer education has itself seen a shift, one I have experienced. When I was in junior high school in the early 1980s, the emphasis in computing science, as the subject was called, was on programming. We focused on learning how to program the computer to meet our problem solving needs. Today, the computer is used as a practical tool, where large innovative software designers provide us with the programs. In these modern software packages, for the most part, we are limited within the confines of the program. Demario (1991) also sees this transgression, but from a feminist perspective. She argues that the software is somewhat limiting and suggests how software designed from the feminist standpoint, based on certain "feminine" characteristics, could eliminate many problems associated with present-day computer software.

Contradictory to the previous paragraph, proponents to certain software packages are out there. Within certain software applications, users can manipulate programs in a variety of ways and, unknowingly or not, immerse themselves in the traditional parameters of academia, including mathematics. Burnett (1987, 1988), Hoyles and Noss (1987), and Parker and Widmer (1989) have found computer applications to meet their own and, more important, their students' educational needs. These programs are the most useful and yet the most simple. Seymour Papert's Logo language as described in Burnett (1987), Land and Turner (1988), and Hayles and Noss (1987) and the development and use of the spreadsheet as highlighted in Burnett (1987, 1988) and in Parker and Widmer (1989) are two such programs. Logo is said to be an environment that promotes "mathematizing," while the spreadsheet is billed as a notational system for exploring ideas. These authors are perhaps more optimistic about the technology than the other researchers and have worked to find feasible uses for what is available.

Johnson (1988) claims that the research is too general and that it does not reflect the problems that students encounter in their work with computers. Other research proposes remedies for the situation. Zehavi (1988) argues that we need to design software for our students' specific needs. Backing this point up, MacGregor and Shapiro (1988) reveal that we must concentrate on individual learning and cognitive styles. This is something that most computer and software technology has failed to do. Land and Turner (1988) conclude that using certain programs only reveals that they help students with higher cognitive levels. In other words, students who do well in most areas are also going to succeed in the computer environment. Researchers also discovered that low-achieving students eventually reach a certain plateau in understanding mathematical concepts with a computer program helping them. This research would support the evidence that technology can help some students, but certain people need more than just "fancy" technology. The computer can be an effective tool for some students in specific situations, but to tap into its true effectiveness, more emphasis needs to be placed on computer use with "individual" needs in mind. Computers and computer programs cannot be seen as generic, no more so than can individual students in any given classroom. Parker and Widmer (1989) stress the importance of the teacher in the computer equation. They argue that the teacher must be responsible to students by identifying and selecting appropriate applications to be used in the classroom. Johnson (1988) outlines an additional concern about the use of computers by pointing out a situation where he saw the computer become an educational crutch to a student. Some educators are really concerned that students might become dependent on the technology, robbing them of their own intuitions and problem solving abilities (Demana & Waits, 1992; Zehavi, 1988).

Using computers does not come without costs. Zinn (1969) forecasted problems surrounding the cost of computer technology. Demana and Waits (1992) highlight similar modern-day concerns. They argue that there is too much pressure on students and educators alike to purchase and implement expensive computer systems. They go on to suggest that other forms of technology are much cheaper while still meeting the same instructional needs. For example, graphing calculators can aid secondary students with more complex mathematical concepts and related exercises. Today's students live in a society filled with innovation and technical gadgetry. Most students are engulfed in worlds of multimedia presentation (for example, television and videos) and video games. I fear that the novelty of the computer and computer software will eventually fade in the eyes of students. Many students gain motivation from using technology, and it is therefore up to the teacher, not the computer, to keep student interest and involvement (Demario, 1991; Johnson, 1988; Zehavi, 1988).

Duguet (1989) discusses the problem that the education field has faced with computer applicability: An obvious gap has existed between the hardware and the software. The main argument is that educators do not know enough about how students learn or exactly what they learn when they interact with computer-based materials. A review board or an organization needs to be established to study and screen software. The market is flooded with computer technology, and teachers cannot be expected to keep on top of it all. The international Organization for Cooperation and Economic Development (OCED) has started to set up such educational review centres. Statistics presented by OCED reveal that in mathematics only 49 percent of the software was recommended for use by teachers. Of the 457 software packages reviewed, only 223 were recommended (Duguet, 1989). This presents an obvious problem for teachers *and* their students.

Two other articles of interest relate directly to the use of computers and computer software for special learners. Biser (1986) discovered that few, if any, software titles are labeled as special education. This does not mean that the technology cannot be used for this portion of the population, but rather, modifications need to be made. Special educators need to look for two things in computer software. First, the programs need to be flexible and modifiable, and second, the software needs to have a record keeping option so that teachers can monitor student progress. These software attributes are a good indicator of software effectiveness in all realms, not just for special learners. Perhaps the most encouraging research that I discovered, in terms of special education, came from Divoky (1987). The Apple Computer Company announced the establishment of a National Special Education Alliance (NSEA). This organization provides resources and information about computers and other technology to the disabled population. Apple has also established an awareness program in its development of hardware and software. Serious efforts are being made to eliminate any obstacles to special learners—little things like making the repeat key optional with an on-off switch, which will help students with motor skill disabilities. Divoky (1987) lists the standard and special features offered to computer buyers.

Three main points contribute to the apparent dilemma that educators face regarding the use of computers in education:

1. Computers are a rapidly changing area of technology. Today's hardware and software will almost inevitably be obsolete in five years. This begs the question, Why get involved in an obviously unstable situation?
2. The expense of computer technology is staggering, especially in light of the rapidly changing nature of the industry. Personal and/ or school involvement demands a great deal of time and money, in terms of training and in hardware and software purchases.
3. Computers pose that threat of the unknown and symbolize "change," which many veteran professionals and laymen alike are weary of. Not understanding something can make people avoid and ignore it, creating "computer anxiety." The computer is another stepping stone we have yet to conquer in everyday life, as well as in education.

Computers are indeed going to be part of my educational career. Too much valuable technology exists out there that has yet to reach its full potential. There are of course concerns as with anything innovative, especially in such an important facet of society. We must remember though that education is the pathway to our future. Technology has begun to take over and navigate our journey. In 15 or 20 years, I will look back and laugh at the archaism of the instrument on which I composed this article. Change is inevitable; the real choice is whether or not you decide to jump on and enjoy the ride.

REFERENCES

Burnett, J. D. (1987, October). *Mathematical modeling using spreadsheets.* Paper presented at the Mathematics Council of The Alberta Teachers' Association Annual Conference, Calgary, Alberta.

Burnett, J. D. (1988, March). *Spreadsheets across the curriculum.* Paper presented at the Computer Council of The Alberta Teachers' Association Annual Conference, Calgary, Alberta.

Demario, S. K. (1991). Rethinking science and mathematics curriculum and instruction: Feminist perspectives in the computer era. *Journal of Education, 173*(1), 107–121.

Demana, F., & Waits, B. K. (1992). A case against computer symbolic manipulation in school mathematics today. *The Mathematics Teacher,* 5(3), 180–183.

Divoky, D. (1987). Apple sponsors a new alliance for disabled computer users. *Classroom Computer Learning, 8,* 46–49.

Duguet, P. (1989). Teaching: Software, hard choices. *The OCED Observer, 157,* 5–8.

Eiser, L. (1986). 'Regular' software for special education kids? *Classroom Computer Learning, 7,* 26–28.

Hoyles, C., & Noss, R. (1987). Synthesizing mathematical conceptions and their formalization through the construction of a Logo-based school mathematics curriculum. *International Journal of Mathematics Education in Science and Technology, 18,* 581–193.

Johnson, I. (1988). Computers in the math classroom: Computers, problem solving, and a belief. *The Computing Teacher, 16*(4), 24–25.

Land, M. L., & Turner, S. V. (1988). Cognitive effects of a Logo-enriched mathematics program for middle school students. *Journal of Educational Computing Research, 4,* 443–451.

MacGregor, S. K., & Shapiro, J. Z. (1988). Effects of a computer-augmented learning environment on math achievement for students with differing cognitive style.' *Journal of Educational Computing Research, 4,* 453–464.

Parker, J., & Widmer, C. C. (1989). Using spreadsheets to encourage critical thinking. *The Computing Teacher, 16*(6), 27–29.

Winzer, M. (1989).*Closing the gap: Special learners in regular classrooms.* Mississauga, ON: Copp Clark Pitman.

Winzer, M. (1990). *Children with exceptionalities: A Canadian perspective.* Scarborough, ON: Prentice-Hall.

Zehavi, N. (1988). Evaluation of the effectiveness of mathematics software in shaping students' intuitions. *Journal of Educational Computing Research, 4,* 391–401.

Zinn, K. L. (1969). *Implications of programming languages for mathematics instruction using computers.* Reston, VA: The National Council of Teachers of Mathematics.

CHAPTER 34

BUILDING A PROFESSIONAL MEMORY

Articulating Knowledge About Teaching Mathematics

Barry Onslow and Art Geddis

In many ways, teaching is a profession without a memory. Unlike architecture and engineering, few detailed records are kept of what teachers do and how they do it. Architects and engineers leave behind drawings, specifications, models, contracts, and buildings. Such artifacts provide a record of the problems faced, solutions tried, and products produced. Teachers, however, leave few descriptions that record their experiences introducing a new topic or their struggles with particularly challenging curriculum. Many records that are left do little to capture the complexity of teachers' pedagogy.

An integral part of the education of doctors, lawyers, and business people is the study of "cases" that record their profession's history. At present, there is no comparable body of literature to which beginning and practising teachers can turn to discover the wisdom of their predecessors. The development of a case literature of mathematics teaching should be a professional priority.

A useful focus for articulating the complexities of subject matter pedagogy is Shulman's (1987) view that effective teachers transform knowledge of subject matter into forms accessible to their pupils, rather than transmitting knowledge or pouring ideas into someone else's head. Shulman calls this amalgam of subject matter and pedagogy *pedagogical content knowledge.* It arises from deliberations about how to teach particular content to particular pupils in particular contexts and consists (among other things) of misconceptions pupils typically bring to instruction, alternative ways of representing subject matter and effective teaching strategies for changing misconceptions. To a significant degree, the acquisition of relevant pedagogical content knowledge—a way of thinking that helps the teacher understand the learner's difficulty and subsequently transform the content so the learner can understand—distinguishes effective mathematics teachers from those who are less effective (Figure 34.1).

When learning primarily involves acquiring information, instruction can proceed in a transmission mode. This typically involves motivating pupils, delivering content, providing opportunities for practice and evaluating learning. In these situations, pupils employ familiar ways of thinking to assimilate new information presented by teachers. Teachers have little need to use pedagogical content knowledge because they can transmit, relatively intact, their knowledge of the subject matter to their pupils. A good deal of mathematics, however, incorporates ways of thinking that are not intuitively obvious. Effective mathematics teaching demands that subject matter be transformed to allow it to be learned meaningfully by novices. Consequently, mathematics teachers find themselves in need of extensive repertoires of pedagogical content knowledge. This knowledge needs to be captured in case studies.

Having discussed mathematics teaching with many prospective teachers over the last few years, we have found that few are able to provide suitable representations for many rudimentary mathematical abstractions. This lack of understanding often stems from their own experiences of school mathematics and the resulting perception of mathematics as a collection of isolated rules to be memorized. We will use an elementary concept, division of fractions, to illustrate the importance of appropriate representations in

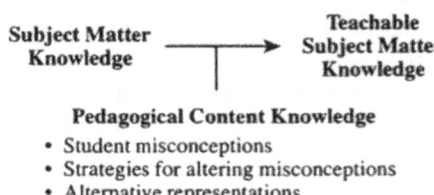

Figure 34.1 Transforming subject matter knowledge.

mathematics and the necessity for teachers to acquire pedagogical content knowledge regardless of the grade they teach.

TRANSFORMING KNOWLEDGE: DIVISION OF FRACTIONS

A question that often provides some indication of how a person has learned mathematics is 10 divided by ½. When answering this question, a few prospective teachers contribute the answer 5 (the answer being sensible for the representation they are using—10 shared in half), but most give the correct answer 20. However, when asked to explain why the answer is 20, a majority are unable to provide a representation or model. They resort to rules memorized during their own schooling. As one student teacher wryly explained, "Ours was not to reason why, just invert and multiply."

An understanding of pupils' misconceptions is important pedagogical content knowledge. Without this knowledge, teachers are not in a position to help pupils clarify their understanding. For example, in the situation described above, teachers need to know that many pupils will be comfortable with the smaller answer 5 when they divide 10 by ½ due to their previous experiences with natural numbers. Many pupils have developed the idea that "multiplication makes bigger" and "division makes smaller," which is true for natural numbers. Pupils will also find comfort in the partition or sharing model for division and happily share 10 in half, again obtaining 5 for the answer. Unless teachers are aware that such errors are extremely common among 12- and 13-year-olds, they are unlikely to spend the necessary time introducing the concept.

Being aware of possible misconceptions, however, is insufficient to enable teachers to advance their pupils' more complete understanding of division. To assist their pupils, teachers need a second model for division, sometimes referred to as the *quotition model*, and have available suitable representations for it. One representation of the quotition model occurs when one asks, "How many people can attend a party if there are 10 pizzas and everyone receives half a pizza?" The correct answer, 20, is usually quickly contributed, but often the connection between the verbal question and its equivalent mathematical expression, 10 divided by ½, is not made. When answering a question of the form 10 divided by ½, pupils have to ask themselves, "How many halves are there in 10?" if they are to make sense of the symbolism. This notion becomes especially important when dividing a fraction by another fraction, for example, ½ divided by 1/6. Many pupils want their answers to be a fraction and, what is more, a rather messy fraction. They are often surprised when they obtain the whole number 3. However, when pupils are provided with a pictorial representation (Figure 34.2) and are asked, "How many sixths in one half?", most grade 5 or 6 pupils understand why the answer is 3. Pupils

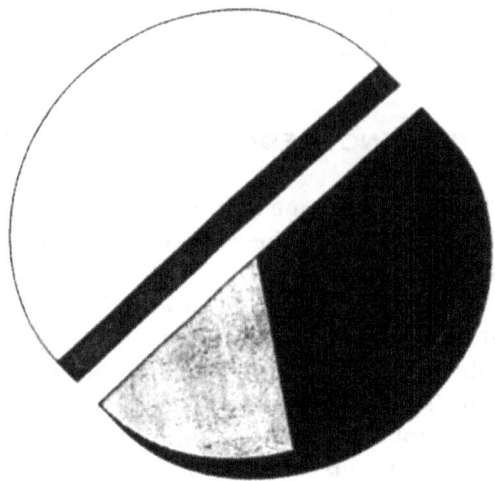

Figure 34.2   Pictorial representation to illustrate ½ ÷ ⅙.

are comfortable that there are six sixths in a whole pie, and therefore there should be three sixths in half the pie.

Unfortunately, children, and sometimes teachers, are concerned only with the final correct answer to mathematical questions, especially arithmetical questions. This preoccupation often leads to understanding becoming of secondary importance. It is even thought of as inconsequential by many pupils. Justifying the rationale underlying a procedure helps demonstrate its importance. Only when pupils travel comfortably between real entities and symbolic representations can the term *mathematically literate* be appropriately applied to them.

CONCLUSIONS

Having suitable representations for oneself and having the knowledge to help pupils develop their own models, stories, and analogies for mathematical symbolism comes not only with teaching experience but also with the philosophical outlook that such ideas are important. If mathematics is seen simply as a set of rules and procedures to be reproduced on a test, pedagogical content knowledge is unnecessary. However, if the teacher's role is to assist pupils to understand why their answer is correct or incorrect, the teacher must have more than a knowledge of general pedagogy and mathematics.

To transform subject matter content knowledge into a form accessible to pupils, teachers need to know particulars about the content relevant to its teachability, particulars that probably would not have been revealed until the task of teaching had been assumed. This pedagogical content knowledge is in some sense a result of the interaction of content and pedagogy. It is knowledge about the content derived from consideration of how best to teach it.

Certainly, teachers who are aware of a misconception, are cognizant of its origin, and who possess multiple representations to correct it have the pedagogical content knowledge to make the subject matter accessible to pupils. Seldom, however, is this expertise shared systematically among colleagues, and the wisdom of practice is lost to the teaching profession (Shulman 1986). Without records of experience, pedagogical content knowledge has to be continually reinvented by each new generation of teachers— consuming time and energy that should be used constructing new understanding.

Teaching is one of the few professions where most people expect the novice to perform in a similar fashion to the veteran. If teaching is to progress, we have to recognize the intricacies associated with it and understand that teachers are also learners. Teacher educators are gradually recognizing the importance and relevance of using case studies that capture the complexities of teaching. Consequently, there is an increased awareness that the teaching profession needs cases written by practising teachers (Cochran-Smith & Lytle, 1990; Shulman & Mesa-Bains, 1990). Through such cases, teachers hear their peers' voices. When we have built a shared professional memory that carefully articulates knowledge about teaching mathematics, present and future teachers will have at their command examples of exceptional teaching with which to face teaching's challenges and complexities. Only then will teachers really be able to benefit from peers' wisdom and practice.

REFERENCES

Cochran-Smith, M., & Lytle, S. L. (1990). Research on teaching and teacher research: Issues that divide. *Educational Researcher, 19*(2), 2–11.

Shulman, J. H., & Mesa-Bains, A. (Eds.). (1990). *Teaching diverse students: Cases and commentaries.* San Francisco, CA: Far West Laboratory for Educational Research and Development.

Shulman, L. S. (1986). Those who understand: Knowledge growth in teaching. *Educational Researcher, 15*, 4–14.

Shulman, L. S. (1987). Knowledge and teaching: Foundations of the new reform. *Harvard Educational Review, 57*(1), 1–22.

This chapter was originally published as:

Puhlmann, K. (1995). Enhancing mathematics teaching in the context of the curriculum and professional standards of the NCTM. *delta-K: Journal of the Mathematics Council of the Alberta Teachers' Association, 33*(1), 21–27.

CHAPTER 35

ENHANCING MATHEMATICS TEACHING IN THE CONTEXT OF THE CURRICULUM AND PROFESSIONAL STANDARDS OF THE NATIONAL COUNCIL OF TEACHERS OF MATHEMATICS

Klaus Puhlmann

This article briefly reviews the curriculum and professional standards of the National Council of Teachers of Mathematics (NCTM). These standards have driven and directed curriculum development, teaching, evaluation and professional development in mathematics since their publication in 1989. This article provides readers with an overview of the curriculum and professional standards for mathematics.

Inherent in the standards is a consensus that all students need to learn more, and often different, mathematics and that instruction in mathematics

Selected Writings from the Journal of the Mathematics Council of the Alberta Teachers' Association, pages 269–281
Copyright © 2014 by Information Age Publishing

must be significantly revised. The need for standards for school mathematics is clearly evident in that they ensure quality, indicate goals, and promote change. The NCTM considers all three reasons equally important. Schools, and in particular school mathematics, must reflect the important consequences of the current reform movement in mathematics if students are to be adequately prepared to live in the 21st century. Today's society expects schools to ensure all students have an opportunity to become mathematically literate, are capable of extending their learning, have an equal opportunity to learn, and become informed citizens capable of understanding issues in a technological society.

Educational goals for students must reflect the importance of mathematical literacy. Toward this end, the K–12 standards articulate five general goals for all students:

- That they learn to value mathematics
- That they become confident in their ability to do mathematics
- That they become mathematical problem solvers
- That they learn to communicate mathematically
- That they learn to reason mathematically

These goals imply that students should be exposed to numerous and varied related experiences that encourage them to value the mathematical enterprise, to develop mathematical habits of mind and to understand and appreciate the role of mathematics in human affairs; that they should be encouraged to explore, to guess, and even to make and correct errors so that they gain confidence in their ability to solve complex problems; that they should read, write, and discuss mathematics; and that they should conjecture, test, and build arguments about a conjecture's validity.

The mathematics classroom must be permeated with these goals and experiences so that they become commonplace in students' lives. Exposing students to the experiences outlined in the standards will ensure that students gain mathematical power. This term denotes an individual's abilities to explore, conjecture, and reason logically, as well as his or her ability to use various mathematical methods effectively to solve nonroutine problems. This notion is based on the recognition of mathematics as more than a collection of concepts and skills to be mastered; it includes methods of investigating and reasoning, means of communication and notions of context. In addition, for each individual, mathematical power involves development of self-confidence.

The NCTM (1989, 1991) has presented 78 standards divided among eight categories: grades K–4 curriculum, grades 5–8 curriculum, grades 9–12 curriculum, evaluation of students, teaching mathematics, evaluation of the teaching of mathematics, professional development of teachers, and support and development of mathematics teachers and teaching.

CURRICULUM STANDARDS

Curriculum standards for school mathematics are value judgments based on a broad, coherent vision of schooling derived from several factors: societal goals, student goals, research on teaching and learning, and professional experience. Each standard starts with a statement of what mathematics the curriculum should include, followed by a description of the student activities associated with that mathematics and a discussion that includes instructional examples. Three features of mathematics are embedded in the standards. First, "knowing" mathematics is "doing" mathematics. Doing mathematics is different from mastering concepts and procedures. That is not to say that informational knowledge has no value, only that its value lies in the extent to which it is useful in the course of some purposeful activity. Students clearly must know the fundamental concepts and procedures from some branches of mathematics; established concepts and procedures must be relied on as fixed variables in a setting in which other variables may be unknown. However, instruction should persistently emphasize "doing" rather than "knowing that."

Second, some aspects of doing mathematics have changed in the last decade. The computer's ability to process large sets of information has made quantification and the logical analysis of information possible in such areas as business, economics, linguistics, biology, medicine, and sociology. Because mathematics is a foundation discipline for other disciplines and grows in direct proportion to its utility, the mathematics community believes that the curriculum for all students must provide opportunities to develop an understanding of mathematical models, structures and simulations applicable to many disciplines.

Third, changes in technology and the broadening of the areas in which mathematics is applied have resulted in growth and changes in the discipline of mathematics. More than half of all mathematics has been invented since World War II. The new technology not only has made calculations and graphing easier but also has changed the very nature of the problems important to mathematics and the methods mathematicians use to investigate them.

Because technology is changing mathematics and its uses, the NCTM believes

- appropriate calculators should be available to all students at all times
- a computer should be available in every classroom for demonstration purposes
- every student should have access to a computer for individual and group work

- students should learn to use the computer as a tool for processing information and performing calculations to investigate and solve problems

Access to this technology is no guarantee that any student will become mathematically literate. Calculators and computers for users of mathematics are tools that simplify, but do not accomplish, the work at hand. Similarly, the availability of calculators does not eliminate the need for students to learn algorithms. Students should be aware of the choices of methods when calculating an answer to a problem. When an approximate answer is adequate, students should estimate. If a precise answer is required, students must be capable of choosing an appropriate procedure. Many problems should require students to conduct mental calculations or use paper and pencil. For more complex calculations (for example, long division or column addition), students should be able to use calculators. Finally, if many iterative calculations are needed, a computer program should be written or used to find answers (for example, finding a sum of squares).

With respect to mathematical content, the standards represent the minimum that all students will need to be productive citizens. The standards do not specify alternative instructional patterns prior to grade 9. For grades 9–12, the standards have been prepared in light of a core program for all students, with explicit differentiation in terms of depth and breadth of treatment and the nature of applications for college-bound students. There is an implied expectation that all students have an opportunity to encounter typical problem situations related to important mathematical topics.

Student activities are the second aspect of each standard. Two general principles have guided the description of these activities: first, activities should grow out of problem situations; and second, learning occurs through active as well as passive involvement with mathematics. Traditional teaching emphases on practice in manipulating expressions and practising algorithms as a precursor to solving problems ignore the fact that knowledge often emerges from the problems. Thus present strategies for teaching may need to be reversed: Knowledge often should emerge from experience with problems. Furthermore, students need to experience genuine problems regularly. A genuine problem is a situation in which, for the individual or group concerned, one or more appropriate solutions have yet to be developed. However, instruction should vary and include opportunities for

- appropriate project work
- group and individual assignments
- discussion between teacher and students and among students
- practice on mathematical methods
- exposition by the teacher

Another premise of the standards is that problem situations must keep pace with the mathematical and cultural maturity and experience of the students. For example, the primary grades should emphasize the empirical language of the mathematics and whole numbers, common fractions, and descriptive geometry. In the middle grades, empirical mathematics should be extended to other numbers and the emphasis should shift to building the abstract language needed for algebra and other aspects of mathematics. High school mathematics should emphasize functions, their representations and uses, modeling, and deductive proofs.

The standards specify that instruction should be developed from problem situations. Situations should be sufficiently simple to be manageable but sufficiently complex as to provide for diversity in approach. They should be amenable to individual, smallgroup, or large-group instruction; involve a variety of mathematical domains; and be open and flexible as to the methods to be used.

The first three standards for each grade level are problem solving, communication, and reasoning, although these vary between the levels on what is expected of students and of instruction. The fourth curriculum standard at each level is mathematical connections. This label emphasizes the belief that, although it is often necessary to teach specific concepts and procedures, mathematics must be approached as a whole. Concepts, procedures, and intellectual processes are connected. Thus the curriculum should include deliberate attempts, through specific instructional activities, to connect ideas and procedures among different mathematical topics and with other content areas. Following the connections standard, nine or ten specific content standards are stated and discussed. Some have similar titles, which reflects that a content area needs emphasis across the curriculum; however, others emphasize specific content that needs to be developed at that level.

STUDENT EVALUATION STANDARDS

These standards are viewed and presented in three categories. The first set consists of three evaluation standards and discusses general assessment strategies related to the curriculum standards. These standards present principles for judging assessment instruments, including alignment, multiple sources of information, and appropriate assessment methods and uses.

The second set contains seven standards under student assessment and focuses on providing information to teachers for instructional purposes. They closely parallel the curriculum standards of problem solving, communication, reasoning, mathematical concepts, and mathematical procedures, in addition to two separate standards on mathematical disposition and mathematical power. These seven standards are to be used by teachers to judge students and their mathematical progress.

The final set of four standards falls under program evaluation and addresses the gathering of evidence with respect to the quality of the mathematical program. The standards are indicators of program evaluation, curriculum and instructional resources, instruction, and evaluation team. These standards are to be used by teachers, administrators, and policymakers to judge the quality of the mathematics program and the effectiveness of instruction.

STANDARDS FOR TEACHING MATHEMATICS

Central to the curriculum and evaluation standards is the development of mathematical power for all students. Mathematical power includes the ability to explore, conjecture, reason logically, solve nonroutine problems, communicate about and through mathematics, and connect ideas within mathematics and between mathematics and other intellectual activities. Mathematical power also involves the development of personal self-confidence and a disposition to seek, evaluate, and use quantitative and spatial information in solving problems and in making decisions. Students' flexibility, perseverance, interest, curiosity, and inventiveness also affect the realization of mathematical power.

To reach the goal of developing mathematical power for all students requires the creation of a curriculum and an environment, in which teaching and learning are to occur that are very different from much of our current practice. The image of mathematics teaching needed includes elementary and secondary teachers who are more proficient in

- selecting mathematical tasks to engage students' interest and intellect (that is, worthwhile mathematical tasks)
- providing opportunities to deepen their understanding of the mathematics being studied and its applications
- orchestrating classroom discourse in ways that promote the investigation and growth of mathematical ideas
- using, and helping students use, technology and other tools to pursue mathematical investigations
- seeking, and helping students seek, connections to previous and developing knowledge
- guiding individual, small-group, and whole-class work

The professional standards for teaching mathematics are a major shift away from the current practice to mathematics teaching for student empowerment. Current practice is characterized by a predictable sequence of activities. First, answers are given for the previous day's assignment. Difficult problems are worked on by the teacher or students at the chalkboard. This is followed by the introduction of a new concept and the assignment of homework. Students work on the homework for the remainder of the class period, with the teacher moving around the room to answer questions.

The professional standards for teaching mathematics call for a change in the environment of mathematics classrooms. We need to shift toward

- classrooms as mathematical communities and away from classrooms as simply a collection of individuals
- logic and mathematical evidence as verification and away from the teacher as the sole authority for right answers
- mathematical reasoning and away from merely memorizing procedures
- conjecturing, inventing, and problem solving and away from an emphasis on mechanistic answer finding
- connecting mathematics, its idea, and its applications and away from teaching mathematics as a body of isolated concepts and procedures

The phrase "all students" is used throughout the standards. This means that schools and communities must accept the goal of mathematical education for every child from kindergarten to grade 12. This does not mean that every child will have the same interests or capabilities in mathematics. It does mean that we will have to examine our fundamental expectations about what children can learn and can do and that we will have to strive to create learning environments in which raised expectations for children can be met.

STANDARDS FOR THE EVALUATIONS OF THE TEACHING OF MATHEMATICS

This section presents eight standards for evaluating the teaching of mathematics organized under two categories. The first category describes the process of evaluation and includes standards dealing with the evaluation cycle, teachers as participants in evaluation, and sources of information.

The evaluation process should reflect that the overall intent is to improve instruction, that it should be a dynamic and continual process, that teachers should be an integral part of that process, and that because of the complexity of teaching, it should involve a variety of sources of information gathered in various ways. The standards emphasize that teachers should be encouraged and supported to engage in self-analysis and to work with colleagues in improving their teaching. When evaluation involves supervisors or administrators, their relationship with teachers should be collegial with the intent to improve instruction.

The second category of standards in this section describes the foci of evaluation and includes five standards dealing with mathematical concepts, procedures, and connections; mathematics as problem solving, reasoning, and communication; mathematical disposition; assessing students' mathematical understanding; and learning environment.

These standards, particularly the standard dealing with mathematical disposition, emphasize the importance of significant mathematics when reevaluating mathematics teaching. Through encounters with significant mathematics, students develop mathematical power. But attaining mathematical power requires more. It requires a disposition to do mathematics and an environment in which the processes of doing mathematics are continually emphasized.

This can occur only when teachers present stimulating tasks and create an environment in which problem solving, reasoning and communication are valued and promoted. Further, the message teachers send students should not be limited to instruction alone; it must also include what and how mathematical learning is assessed. Through assessment, we communicate to our students what mathematical outcomes are valued.

A consistent message throughout the standards for the evaluation of teaching is the importance of teachers reflecting on their teaching and working with colleagues and supervisors to improve their teaching. While the standards provide a focus for improvement, such improvements will occur only when teachers consciously decide to engage in ongoing professional development. This, in turn, requires support and encouragement at all levels.

STANDARDS FOR THE PROFESSIONAL DEVELOPMENT OF TEACHERS OF MATHEMATICS

Five standards are presented in this section: experiencing good mathematics teaching, knowing students as learners of mathematics, knowing mathematical pedagogy, developing as a teacher of mathematics, and the teacher's role in professional development.

Mathematics teachers must have good role models during their preservice and continuing inservice training. Teachers often teach the way they have been taught. Therefore, preservice instructors need to address the major components of teaching: tasks, discourse, environment, and analysis of teaching.

The education of mathematics teachers should develop their knowledge of the content and discourse of mathematics. Teachers' comfort with and confidence in their own knowledge of mathematics affects what they teach and how they teach it. Their conceptions of mathematics shape their choice of worthwhile mathematical tasks, the learning environments they create, and the discourse in their classrooms. Knowing mathematics includes understanding specific concepts and procedures as well as the process of doing mathematics. Mathematics involves the study of concepts and properties of numbers, geometric objects, functions, and their uses—identifying, counting, measuring, comparing, locating, describing, constructing, transforming, and modeling. The relationships and recurring patterns among these objects and the operations on these objects lead to building such mathematical structures as number systems, groups, or vector spaces and to studying the similarities and differences among these structures.

Such knowledge ought not to be developed in isolation. The ability to identify, define, and discuss concepts and procedures; to develop an understanding of the connections among them; and to appreciate the relationship of mathematics to other fields is critically important.

The standards clearly state that, to sufficiently understand the mathematical topics specified for each level, teachers teaching mathematics should have not less than nine semester hours of coursework in content mathematics at the K–4 level, fifteen semester hours of coursework at the 5–8 level, and the equivalent of a major in mathematics at the 9–12 level.

The preservice and continuing inservice training of teachers of mathematics should provide multiple perspectives on students as learners of mathematics by developing teachers' knowledge of research on how students learn mathematics; the effects of students' age, abilities, interests, and experience on learning mathematics; the influence of students' linguistic, ethnic, racial, and socioeconomic backgrounds and gender on learning mathematics; and ways to affirm and support full participation and continual study of mathematics by all students.

Knowing mathematical pedagogy is an important standard that can be achieved through preservice and continuing inservice training of teachers of mathematics by developing

- teachers' knowledge of and ability to use and evaluate instructional materials and resources, including technology
- ways to represent mathematics concepts and procedures
- instructional strategies and classroom organizational models
- ways to promote discourse and foster a sense of mathematical community
- means for assessing student understanding of mathematics

Developing as a teacher of mathematics is really at the heart of teaching. Teachers of mathematics must have ongoing opportunities to examine and revise their assumptions about the nature of mathematics, how it should be taught, and how students learn mathematics; observe and analyze a range of approaches to mathematics teaching and learning, focusing on the tasks, discourse, environment, and assessment; work with a diverse range of students individually, in small groups, and in large-class settings with guidance from and in collaboration with mathematics education professionals; analyze and evaluate the appropriateness and effectiveness of their teaching; and develop dispositions toward teaching mathematics.

Essentially, being a teacher of mathematics means developing a sense of self as such a teacher. Such an identity grows over time and is built from many different experiences with teaching and learning. Further, it is reinforced by feedback from students that indicates they are learning mathematics, from colleagues who demonstrate professional respect and acceptance, and from a variety of external sources that demonstrate recognition of teaching as a valued profession.

Teachers develop as professionals on an ongoing basis. The standard regarding the teacher's role in professional development suggests that teachers of mathematics should take an active role in their own professional development by accepting responsibility for experimenting thoughtfully with alternative approaches and strategies in the classroom; reflecting on learning and teaching individually and with colleagues; participating in workshops, courses, and other educational opportunities specific to mathematics; participating actively in the professional community of mathematics educators; reading and discussing ideas presented in professional publications; discussing with colleagues issues in mathematics and mathematics teaching and learning; participating in proposing, designing, and evaluating programs for professional development specific to mathematics; and participating in school, community, and political efforts to effect positive change in mathematics education. Schools and school systems must support and encourage teachers in accepting these responsibilities. What is essential is that teachers of mathematics view themselves as agents of change, responsible for improving mathematics education at many levels: the classroom, the school, the district, the region, and even the nation.

STANDARDS FOR SUPPORTING AND DEVELOPING MATHEMATICS TEACHERS AND TEACHING

Professional Standards for Teaching Mathematics (NCTM, 1991) presents a vision of teaching that calls for a teacher who is educated, supported, and evaluated in ways quite different from current practice. To create a teaching environment as described in the standards, teachers must have access to educational opportunities over their entire professional lives that focus on developing a deep knowledge of subject matter, pedagogy, and students.

Teachers can, and do, implement successful mathematics programs with little help or encouragement. However, sustaining them or expecting that they flourish without adequate support is not reasonable. The changes called for by the curriculum and evaluation and the professional teaching standards need the support of policymakers in government, business, and industry; school administrators, school board members, and parents; college and university faculty and administrators; and leaders of professional organizations.

Policymakers in government, business, and industry should take an active role in supporting mathematics education by accepting responsibility for

- participating in partnerships at the national, provincial and local levels to improve the teaching and learning of mathematics
- supporting decisions made by the mathematics education professional community that set directions for mathematics curriculum, instruction, evaluation and school practices
- providing resources and funding for, and assistance in, developing and implementing high quality school mathematics programs that reach all students, as envisioned in *Curriculum and Evaluation Standards for School Mathematics* (NCTM, 1989) and *Professional Standards for Teaching Mathematics* (NCTM, 1991)

School administrators and school board members should take an active role in supporting teachers of mathematics by accepting responsibility for

- understanding the goals for the mathematics education of all students set forth in the NCTM standards and the needs of teachers of mathematics in realizing these goals in their classrooms
- recruiting qualified teachers of mathematics, with particular focus on the need for the teaching staff to be diverse
- providing a support system for beginning and experienced teachers of mathematics to ensure that they grow professionally and are encouraged to remain in teaching
- making teaching assignments based on the qualifications of teachers
- involving teachers centrally in designing and evaluating programs for professional development specific to mathematics
- supporting teachers in self-evaluation and in analyzing, evaluating and improving their teaching with colleagues and supervisors
- providing adequate resources, equipment, time and funding to support the teaching and learning of mathematics as envisioned in the standards
- establishing outreach activities with parents, guardians, leaders in business and industry, and others in the community to build support for quality mathematics programs

- promoting excellence in teaching mathematics by establishing an adequate reward system including salary, promotion, and conditions of work

College and university administrators need to actively support mathematics and mathematics education faculty by accepting responsibility for establishing an adequate reward system, including salary, promotion and tenure, and conditions of work, so that faculty can and are encouraged to

- spend time in schools working with teachers and students
- collaborate with schools and teachers in the design of preservice and continuing education programs
- offer appropriate graduate courses and programs for experienced mathematics teachers
- provide leadership in conducting and interpreting mathematics education research, particularly school-based research
- cooperate with pre-college educators to articulate the K–16 mathematics program
- make concerted efforts to recruit and retain teacher candidates of quality and diversity

The leaders of professional organizations need to take an active role in supporting teachers of mathematics by accepting responsibility for

- promoting and providing professional growth opportunities for those involved in mathematics education
- focusing attention of the membership and the broader community on contemporary issues dealing with the teaching and learning of mathematics
- promoting activities that recognize the achievements and contributions of exemplary mathematics teachers and programs
- initiating political efforts that effect positive change in mathematics education

SUMMARY

In *Curriculum and Evaluation Standards for School Mathematics* (NCTM, 1989) and *Professional Standards for Teaching Mathematics* (NCTM, 1991), the argument is made that what is needed is a design change strategy. This means that new ways of doing things within the system—new roles for teachers and students, new goals, new structures—must be explored to find solutions to persistent problems that result in students failing to become mathematically powerful.

While the standards have directed us to some degree to the issues that need to be addressed, they are not a prescription for what must be done at each grade level. However, if we make a long-term commitment to the standards set forth, if we endeavor to persevere, and if we continue to modify our course as new knowledge comes to the fore, we will make progress toward the goal of developing mathematical power for all students. Such massive change as is proposed in the standards will take time and much work and dedication from teachers and many others.

First, we must challenge all those charged with responsibility to teach mathematics in our schools to work collaboratively in using the curriculum and evaluation and professional standards as the basis for change so that the teaching and learning of mathematics in our schools is improved.

REFERENCES

National Council of Teachers of Mathematics (NCTM). (1989). *Curriculum and evaluation standards for school mathematics.* Reston, VA: Author.

National Council of Teachers of Mathematics (NCTM). (1991). *Professional standards/or teaching mathematics.* Reston, VA: Author.

Stenmark, J. K. (Ed.). (1991). *Mathematics assessment: Myths, models, good questions, and practical suggestions.* Reston, VA: National Council of Teachers of Mathematics.

This chapter was originally published as:

Dickie, A. (1996). Computers in classrooms: Essential learning tool ... or program for disaster? *delta-K: Journal of the Mathematics Council of the Alberta Teachers' Association, 33*(2), 18–20.

CHAPTER 36

COMPUTERS IN CLASSROOMS

Essential Learning Tool ...
or Program for Disaster?

Alison Dickie

Editor's Note: Reprinted with permission from *Home & School,* Volume 2, Number 5, May 1995, pp. 30–33. This article also appeared in *One World,* Volume XXXJJ1, Number 2, 1995.

Jennifer was only seven when she started pestering her parents for a computer of her own. Her grade 2 teacher had told her to ask her parents for one because all but three of the other children in her class were using home computers to complete their school projects. Jennifer's teacher believed that it was important for students to become comfortable with this technology and the sooner the better.

Teachers like Jennifer's are one reason sales of computers and software are booming. Between 1991 and 1992, sales of educational software jumped 50 percent. In the past three years, British Columbia schools spent $35 million on computers, and Ontario schools spent $140 million.

Ottawa and the other provinces have invested hundreds of millions more, and former Ontario Premier Bob Rae pledged $500 million for computers in schools over the next five years. And then there's the $475 million Canadians spent last year on video games—many people are under the impression the games have educational merit.

But not everyone is convinced that computers are effective learning tools. A growing number of parents and teachers are questioning the value of devoting so much scarce money and teacher time to a technology that is largely experimental. Some educators argue that most so-called educational software isn't good enough to introduce into the classroom.

"There is this idea that decent educational software is out there," says American learning theorist Roger Schank. "It isn't. The software you are seeing is a bad imitation of books. It's a good imitation of quizzes, but who wants quizzes as a form of instruction?"

Schank says the quality of most programs is poor because they have been designed by computer scientists who know nothing about how children really learn. As a result, they make programs Schank characterizes as "shoot the verb when it goes by."

Susan Kiil, a Toronto ecology consultant and author of several books on ecology for children, agrees. "Children are limited, not only by the parameters of the software but also by the creativity of the person who designed it," Kiil says.

BAD IMITATION OF BOOKS

Because much educational software involves fill-in-the-blank or drill-and-rote work, Kiil believes we may be creating a generation of vertical, not lateral, thinkers. At the same time, she is concerned that using this technology in the primary years subtly devalues a child's own handwriting and drawing. Not only does the computer create a homogenizing effect because every piece of work looks the same, but also children are left with a feeling that work produced by hand is less worthy than work that comes from a machine. In the long run, she says, we could be creating kids with self-esteem problems.

"Children are first and foremost creatures of sense," says Roma Lupenec, a special education teacher and parent of a four-year-old.

Children are attracted to computers because of the strong visuals and the mechanical aspect of playing with the keyboard. But ultimately the computer is a one-sense experience. Real learning, she says, nourishes the child's imagination by engaging all the senses.

Lupenec is particularly skeptical of software programs that claim to teach eye–hand coordination. One example she cites is a game that simulates a handball court. But because young children don't play handball and are unlikely to have visited a court, the game has little meaning for them.

Artistic activities, like painting, yarn work, cutting, or sewing, are far more effective in teaching eye–hand coordination than any computer game.

"It is much healthier for a child to be standing using an art easel," Lupenec says, "than to be sitting almost motionless at a monitor."

Most progressive educators agree that early learning experiences should be child-centred and concrete, using examples drawn from a child's experience and appropriate to his or her level of development. To develop skill with numbers, for instance, children should first be given concrete objects like blocks or pebbles to play with.

Learning should always proceed from the concrete to the theoretical, says Valdemar Setzer, a math teacher and computer science lecturer. Computers, says Setzer in his book *Computers in Education*, force a child to work backward, moving from the abstract to the concrete.

For this reason, he is also highly critical of computer geography programs. The best way to teach geography, he believes, is by personal recognition—explore and describe the neighborhood surrounding the school before moving on to the complexities of map reading. Setzer believes that the best time to introduce computers into classrooms is when students are in high school and more capable of abstract thinking.

But if, as some proponents argue, computers can speed up the learning process by teaching children to read, write, or do basic arithmetic more quickly, shouldn't they have access to them as early as possible?

Not necessarily. Pushing children into an activity too soon may force them to use a part of their brain that was meant for another function and can actually interfere with the learning process, says American psychologist Jane Healy.

Children's minds are already bombarded with too much fast-paced sensory stimulation from electronic sources, Healy argues. In her book *Endangered Minds: Why Children Don't Think, and What We Can Do About It*, Healy says that too much electronic stimulation may actually be changing the structure of children's brains.

BOMBARD A CHILD

Television, computers, and video games bombard a child with visual information, and do not leave enough time for quiet reflection, concentration, or conversation. These quiet moments are essential, she says, if children are to develop into thoughtful people with the inner control necessary to manage their own lives.

Healy believes that electronic overstimulation, especially from computers and televisions, may be contributing to the rise in the incidence of learning disabilities, such as auditory-processing problems and attention deficit disorder. Children who are not taught to listen can easily develop habits that let them avoid exercising—and thus building—important auditory-processing connections in the brain.

"This very act of remembering lays down physical tracks in the brain, but children can quite easily avoid having to build these systems," she says.

Because some children now get more information from pictures than from talking, Healy argues that their brains are simply not trained to understand and retain language. She points out that teachers believe that the listening skills of children in schools today are much worse than those of previous generations.

Like Healy, Susan Kiil believes that computers spell trouble for children who are not visual learners. Furthermore, the pressure to embrace computer technology in the school system is siphoning funds from other areas, such as the creative arts, which are critical to the development of the whole child. Music, for instance, is known to have beneficial effects on math as well as literacy skills. Not only does the ability to play an instrument bring children joy and a sense of accomplishment, but also it allows them to develop focus and discipline. Yet many school boards have cut back or eliminated music programs, only to spend money on computers. Kiil also fears that widespread use of technology is devaluing the creative teacher as well as the child.

Proponents of computer technology cite studies that seem to show that it can improve writing and arithmetic skills, sharpen critical thinking, and motivate children who want to learn. However, critics have taken issue with these studies, which are often initiated by the companies that market the software.

When Henry Jay Becker of Johns Hopkins University analyzed the results of computer evaluation reports from elementary and middle grades, he concluded that some studies substantially over-reported the effectiveness of computers.

"The poor quality of most evaluations, and the likely bias in what does get reported, all provide too weak a platform for district purchasing decisions," he concluded in an article in the 1992 *Journal of Educational Computing Research*.

More recently, a Japanese study explained the effectiveness of using microcomputers in teaching on a sample of 803 primary school children. The researchers concluded that computer use neither improved intellectual activities, such as creativity, nor motivated the grades 1 and 2 children to study more. A 1994 American study designed to test the effectiveness of computer-based geography programs showed that students who used computers learned no more than their counterparts who used maps and atlases.

Susan Kiil worries that children who are given computers too early will grow up without an essential critical perspective on the very technologies they're using. If we want children to have an appreciation of the natural world and their place in it, we should give them more opportunities to experience this world, through regular field trips that take them out of the classroom. If the educational system persists in leaning so heavily on technology, she thinks we are in danger of losing the knowledge of how to use other, more imaginative technologies that may be more beneficial in teaching and learning.

Oh, and what about seven-year-old Jennifer, whose parents were told she needed a computer because all the other students had one?

Jennifer's mother went for advice to her sister, Charlene Watson, who is an office manager for a software developer. Watson knows enough about computers not to be impressed and advised her sister not to buy a computer the family couldn't afford and didn't need.

"Grade 2 is just too young to be using this technology," she said.

This chapter was originally published as:

Sawada, D. (1996). Mathematics as problem solving—A Japanese way.
*delta-K: Journal of the Mathematics Council of the Alberta Teachers'
Association, 33*(2), 44–47.

CHAPTER 37

MATHEMATICS AS PROBLEM SOLVING—A JAPANESE WAY

Daiyo Sawada

We know from many international studies that Japanese students (and Asian students in general) do quite well in mathematics on knowledge-level questions and on deeper conceptual and problem-solving items (Stevenson & Stigler, 1992; Stedman, 1994; and many others). A question arises: How are mathematical concepts and problem solving approached in Japanese elementary schools?

My response to this question was largely formed during a research visit to the University of Michigan in Ann Arbor at the invitation of Harold W. Stevenson during spring/summer 1993.[1] His project team was in the midst of analyzing voluminous data gathered from classrooms in Taipei, Beijing, Sendai, and Chicago. Dr. Stevenson kindly made it possible for me to immerse myself in a set of classroom observations of 160 lessons taught by 40 teachers of grades 1 and 5 in Sendai, Japan.

To share with you what I found, I do not want to provide a "once over lightly" treatment of a few general characteristics of problem solving in Japanese elementary schools. Rather I have selected a typical grade 5 lesson taught by a typical grade 5 teacher with typically 40 students in her

Selected Writings from the Journal of the Mathematics Council of the Alberta Teachers' Association, pages 289–296
Copyright © 2014 by Information Age Publishing
All rights of reproduction in any form reserved.

classroom. This teacher does not have many striking characteristics that would mark her as exceptional among the set of 40 teachers included in the data set. I would describe her as competent, careful, and caring. That description covers most of the other teachers as well. I present the lesson in some detail to provide readers with a feel for the problem. I then identify and comment on some aspects that were particularly striking to me. I close the article with a question and a partial answer to the main question. [Note: Parenthetical material is enclosed in square brackets.]

LESSON DESCRIPTION (GRADE 5)

The teacher, Ms. Sato (a pseudonym), introduces the lesson with a discussion of "crowdedness" (a notion quite meaningful for people of Japan). After pointing out that one cannot say that a place is crowded simply because it has a lot of people, she focuses discussion on a particular comparison: "Which is more crowded, Hokkaido or Okinawa?"

She reminds the class of problems they have already worked on in previous classes and raises one of them again: "Which tatami room is more crowded, the one that measures 10 mats and has 1 person, or the room that measures 10 mats and has 10 persons?" [Note: The area of a tatami room is measured in terms of the number of tatami mats it takes to cover the floor. The floor space of a tatami room is always designed as a particular tessellation of tatami mats. A tatami mat itself is 1 m by 2m.] A student volunteers a response: "The second room." Ms. Sato agrees and explains that this problem is easy because the rooms are the same size: "But in our case, Hokkaido and Okinawa have different areas." She then asks if the two areas can be compared. A student suggests figuring out how many times bigger Hokkaido is than Okinawa. The teacher explains that this method is similar to the one used in the "quantity of salt experiment" they worked on earlier, and that this too can be a good method.

Another student suggests that one can "turn over a certain amount of area from Hokkaido to Okinawa to make them the same size." Other students object, suggesting that this method is too cumbersome to use here. Another student volunteers, "Like what we did with the tatami room problem, we should try to figure out crowdedness for the same amount of area first" (instead of messing around with exchanging area, as in the previous suggestion). Ms. Sato takes up this suggestion asking, "What then shall be done with the tatami room case?" A student responds, "Compare number of persons per mat." The teacher writes this suggestion on the board while repeating it orally.

Ms. Sato recalls the figures from the tatami mat problem from a previous lesson (Room A—9 persons in a 6-mat room; Room B—15 persons in a 15-mat room) and calculates the number of persons per mat for each tatami room. She then asks, "Which room is more crowded?" She confirms the answer by drawing two diagrams on the board showing one and one-half bodies per mat for Room A and one body per mat for Room B. She summarizes saying that we calculated "quantity per unit."

Turning to the Hokkaido–Okinawa problem, she asks what unit to use. A student responds with "one square kilometer." Accepting this, the teacher points out again that they can base the calculation on how many persons per square kilometer. She then summarizes with "population divided by area = population per square kilometer" and writes this on the board. Students copy this statement into their notebooks. The teacher explains further that by dividing population by area, one figures out how many live in the same amount of space.

In preparation for the calculations, the teacher reminds students about the number of significant digits to keep when rounding. A student responds, "Round off to significant digits when doing calculations" and explains that the numbers given in the problem are already rounded off to two significant digits, and the teacher confirms. She tells students to solve the problem starting with Hokkaido and helps by having students identify the rounded numbers they should use and writes them on the board. Students work on the calculations at their desks. Ms. Sato circulates. She then asks for an answer and gets "70.8" from a student. She tells him to round it off and then states that Hokkaido has about 71 people per square kilometer.

Students then calculate using the numbers for Okinawa, and the teacher again reminds them to use two significant digits when calculating and not to round off until after they find the answer. A student volunteers "478," and Ms. Sato asks, "Rounding this number to two significant digits we have?" A student responds with "480." The teacher then asks which island is more crowded, receives the answer that Okinawa is more crowded and confirms this by reiterating that Hokkaido has 71 people per square kilometer, whereas Okinawa has 480 per square kilometer.

Ms. Sato directs students to look at page 79 in the textbook and asks, "What do we call this 'crowdedness' that we are trying to figure out?" In a choral manner, students respond, "Population density." The teacher confirms, saying, "We have just learned to figure out population density today."

At this point, 35 minutes have elapsed. During the remaining 7 minutes, students do two more problems involving population density. [Two days later, students were still working on density problems but with some variation: the opening problem involved iron and silver—250 cc of iron weighing 1,975 g; 350 cc of silver weighing 3,675 g. This problem took 25 minutes to solve and discuss. The next day, the concept of average was developed as a particular case of density. The problem consisted of comparing the output per factory of two kinds of production where the procedures for handling density generated a quantity which could be called the "average production" (per factory).]

LOOKING BACK

On reflection, these points seem worthy of notice:

- A deliberate effort was made to connect the present problem to previous problems and solutions.
- Simpler problems were used.
- Teacher functioned as a guide with a definite agenda.
- Crowdedness provided a meaningful context for embedding the concept of population density.
- In turn, the concept of density provided a context for embedding the concept of average.
- More than 35 minutes were spent working on one problem.
- Student contributions were used to determine the content of the lesson as well as its flow.
- The notion of rounding and significant digits was used consistently when doing the computations.
- The problem was rich in social studies content.
- Formulaic statements did not enter the lesson until they could be used as summary confirmations.
- The lesson was not followed up by assigning several application exercises. Instead, two related problems were discussed and solved, each widening the interpretation of the idea of density.

COMMENTARY

The comments to follow also reflect my study of the 159 other lessons taught by Sendai teachers in the data set.

Embedding

Problem solving occurs pervasively in this lesson. Problem-solving techniques are used not only in solving the problem about density but also in the approach to the lesson structure itself. I have used the term *embedded* to describe this more pervasive use of problem solving in teaching mathematics in which problem solving is the medium for mathematics teaching and learning. In the lesson described, there is as much emphasis on the medium (problem solving process) as there is on the message (the particular mathematical content to be learned). This emphasis on the medium in many grade 5 lessons in Sendai often takes precedence over the message so that instructional decisions (how to respond to mathematical "errors," for example) are made with the health of the medium as much in mind as the need to assess the correctness of the message. (For example, a so-called "error" is often discussed at length and thereby contributes to the diversity of the "solution space" and therefore to the health of the medium, and thus is valued even if not particularly "correct" as a message.) Indeed, such contributions, because they do enrich the solution space are actively sought by teachers through such simple leads as, "Any other methods?" None of the responses to such questions are rejected at the outset, and this decision not to reject indicates that the health of the medium is at least as important as the correctness of the message. This priority on the medium occurred in the lesson in many particular as well as general ways.

From a mathematical viewpoint,

- crowdedness was used as a medium for (to embed) density
- density was used as a medium for rate/ratio
- in its turn, rate/ratio was used as a medium for average

From a pedagogical viewpoint,

- the general problem situation was a medium for the structure of the lesson
- the particular problem was a medium for the mathematical concept
- the multiple solutions generated by students were a medium for the "correct" solution

From a learning viewpoint,

- the process of solving the problem was a medium for learning of mathematics
- right or wrong, student answers de facto were the medium for exploring the solution space
- generating and discussing solutions was the medium for arriving at and critiquing answers

Spending 35 Minutes or More on One Problem

Spending nearly the whole period on one problem might seem an inefficient and drawn-out way to teach a mathematical concept. On the other hand, if only four or five minutes were spent on a nonroutine problem, I would have grave doubts that a problemsolving approach was being used at all. For a rich problem-solving process to occur, there needs to be time for investigating the nature of the problem, generating and proposing multiple solutions, discussing and critiquing various types of solutions, assessing and comparing the relative merits of each solution within the set of solutions, reflecting on what one has learned and what else could be learned. This sounds like a lot of rhetoric, but it happens regularly in most grade 5 classes in Sendai (statistically speaking, 86 percent of Sendai grade 5 teachers use problem solving to embed mathematical concepts, while the comparable figure for Chicago elementary schools is 14 percent), and it means that the whole period will often be spent on only one problem.

Acting Out the Problem

Although the students in the lesson described did not act out the solution process, in other Sendai classrooms this happened quite often. For example, the teacher would bring cushions to simulate tatami mats, children would select which one(s) they would prefer to sit on, and a problem of density would be enacted. After this, rooms with so many mats would be drawn on the board with students selecting which room they preferred, and the density problem thus enacted would be ready to be solved.

Multiple Solutions

Multiple solutions are the "rice and fish" of the problem-solving approach used in Sendai. If only one solution exists, it would have to be the correct one so nothing would function as, nor need to function as a medium for anything else. The correct solution would be demonstrated or illustrated for all to learn. Multiple solutions and establishing the conditions for generating, articulating, understanding, comparing, and critiquing multiple solutions are required components to the way problem solving happens in Sendai classrooms.

Interesting Problems to Solve

While the problem of the relative population density of Hokkaido and Okinawa may be of interest to Japanese students, as a generalization, most problems are rather mundane. Many of them remind me of typical word problems occurring by the hundreds in Canadian texts. For the classrooms in Sendai, the problems per se are not different; the way these problems become the medium of instruction and of learning is contrastive.

Use of Manipulative Aids

The NCTM publication *Making the Grade in Mathematics: Elementary School Mathematics in the United States, Taiwan, and Japan* (Stevenson et al., 1990) reports a surprising finding: When manipulatives are introduced into American lessons, the amount of talk decreases, while in Japanese elementary classrooms it increases. This finding may have been surprising, but when problem solving is the pervasive mode of teaching and learning, introducing manipulative aids contributes immensely to the variety of multiple solutions to be generated, assessed, compared, and so on, thus providing many more opportunities for talk.

Interpretation Rather Than Application

Currently in North America, we talk about developing problem-solving strategies and skills and then applying them. In this sense, problem solving is split into two rather distinct parts:

1. The learning of the concepts and skills (perhaps through problem solving)
2. The use or application of these concepts and skills in similar situations

This second part is often taken to be the full extent of problem solving. It is a matter of applying concepts learned, not for learning concepts and skills. Our textbooks are organized this way: concepts are taught (often by demonstration or explanation, as well as problem solving), and then students are given a collection of similar "problems" to do. Because of this practice, lessons in North America are likely to end with students working at their desks. The contrast in Sendai is striking. Classes often end with discussion. And when children are working at their desks at the end of the class, they are not only applying the concepts just learned but also interpreting problem situations that extend the ideas beyond the initial circumstances. In the lesson presented, the problem solving approach is not two parts but just one. Problem solving becomes two parts when the concepts learned (the messages) become so important that they need to be separated and dealt with differently (as applications). On the other hand, if we keep the problem-solving process intact and pervasive, the messages learned will never dominate the medium that created them.

CONCLUSION

Several other aspects about problem solving in Sendai deserve mention, such as the deliberate making of errors (by the teacher) or the occurrence of memorization, but I think the important points have been made. I close this article with a question and a partial answer. Why has the problem-solving approach as medium taken hold so pervasively in Japanese elementary schools but remains largely problematic in North American schools despite a decade in which nothing in mathematics education received more attention than problem solving? While not wishing to oversimplify, I should like to suggest that, in Japan, accepting ways of doing things as being as important as the things themselves is a familiar and comfortable stance. We see it in martial arts, flower arranging, the tea ceremony, pottery making, and in teaching, too. Problem solving as a medium for teaching mathematics is another of these important ways.

NOTE

1. My research with Dr. Stevenson was in part supported by grants from the Japanese Redress Foundation, and the Social Sciences and Humanities Research Council of Canada.

REFERENCES

Becker, J., et al. (1990, October). Some observations of mathematics teaching in Japanese elementary and junior high schools. *Arithmetic Teacher, 38,* 12–21.

Lee, S. Y., Graham, T., & Stevenson, H. W. (in press). Teachers and teaching: Elementary schools in Japan and the United States. In T. Rohlen (Ed.), *Teaching in Japan.* Berkeley, CA: University of California Press.

National Council of Teachers of Mathematics. (1989).*Curriculum and evaluation standards for school mathematics.* Reston, VA: Author.

Stedman, L. C. (1994). Incomplete explanations: The case of U.S. performance in the international assessment of education. *Educational Researcher, 23*(7), 24–32.

Stevenson, H. W., et al. (1990). *Making the grade in mathematics: Elementary school mathematics in the United States, Taiwan, and Japan.* Reston, VA: National Council Teachers of Mathematics.

Stevenson, H. W., & Stigler, J. W. (1992). *The learning gap: Why our schools are failing and what we can learn from Japanese and Chinese education.* New York, NY: Summit.

This chapter was originally published as:

Sanders, H., & Vivone-Vernon, G. (1998). Western Canadian Protocol: The common curriculum framework (K–12 mathematics). *delta-K: Journal of the Mathematics Council of the Alberta Teachers' Association, 35*(1), 25–28.

CHAPTER 38

WESTERN CANADIAN PROTOCOL

The Common Curriculum Framework (K–12 Mathematics)

Hugh Sanders and Gina Vivone-Vernon

Editor's Note: This article was originally published in *Early Childhood Education,* Volume 29, Number 2, Fall/Winter 1996, pages 31–34, the journal of The Alberta Teachers' Association's Early Childhood Education Council.

WHAT IS PROTOCOL?

Curriculum development in western Canada has been done in the context of two protocol agreements that have been signed by the ministers of education in Canada. These agreements provide the authority and direction for the various projects to be undertaken.

WESTERN CANADIAN PROTOCOL

In December 1993, ministers of education from the four western provinces and two territories (Manitoba, Saskatchewan, Alberta, British Columbia, the Yukon Territory, and the Northwest Territories) signed the Western Canadian Protocol for Collaboration in Basic Education, Kindergarten to Grade 12 (WCP). This agreement encourages and enables the participating jurisdictions to work more closely together because of the importance they place on

- common education goals
- high standards in education
- removing obstacles for student access to educational opportunities, which includes improving the ease of transfer from jurisdiction to jurisdiction
- optimum use of educational resources

The WCP covers six major areas: curriculum, distance learning and technology, special education, student assessment and standards of student performance, aboriginal education, and teacher preparation and certification. There is also provision in the agreement for launching other types of cooperative projects in the future.

The WCP group has agreed to develop common curriculum frameworks of general and specific outcomes—what students will be expected to know and be able to do—for mathematics and English language arts. They will also be working together on the review of learning resources.

Alberta is playing a lead role in three of the projects: mathematics curriculum for kindergarten to grade 12, curriculum in languages other than English and French, and distance learning and technology. We are also participating in the language arts project for kindergarten to grade 12, which is being coordinated by Manitoba.

PAN-CANADIAN PROTOCOL

After discussions in 1994, a general framework for collaboration, the Pan-Canadian Protocol for Collaboration on School Curriculum, was adopted in February 1995 by the Council of Ministers of Education, Canada.

Recently, ministers of education have initiated several joint projects to develop common frameworks for school programs. Work has begun on the development of the pan-Canadian science project, which will potentially involve all Canadian provinces and territories, except Quebec.

FRAMEWORK DEVELOPMENT FOR K–12 MATHEMATICS

The development of the common curriculum framework tor mathematics commenced with the first interprovincial writing in August 1994. Since then, 64 educators from six participating jurisdictions have developed the framework. It is important to note that the participants represented practising teachers (K–12, as well as some postsecondary instructors) and curriculum specialists from the various departments of education.

In Alberta, The Alberta Teachers' Association nominated teachers to the Alberta team, as did Alberta Education.

The development of the K–12 framework took place in two phases. The first phase ended in June 1995 with the distribution of the K–9 component. The second ended in June 1996 with the distribution of the 10–12 component.

THE FRAMEWORK

The Common Curriculum Framework for mathematics does several things:

- It has grade-level student outcomes organized as
  - general outcomes
  - specific outcomes
  - illustrative examples
- It identifies four strands within which all student outcomes are organized.
- It has a K–12 focus.
- It identifies key mathematical processes that are critical elements affecting student learning.
- It discusses the nature of mathematics.

Student Expectation

The content of the common curriculum framework is stated in terms of outcomes. These outcomes are measurable and identify what students are required to know and do.

The outcomes are developed and based on the expectation that they are appropriate to a large majority of the students. Outcomes are stated at the level where they are expected to be "mastered." There may be some time delays between where students first encounter the learning and where they are expected to demonstrate knowledge of, or mastery in, that learning.

General outcomes are general statements that identify what students are expected to know and be able to do on completion of a grade.

Specific outcomes are statements identifying the component knowledge, skills, and attitudes of a general outcome.

Illustrative examples are sample tasks that demonstrate and elaborate on the general and specific outcomes. They are important in conveying the richness, breadth, and depth intended in outcomes.

Four Strands

All student outcomes are organized into four strands. Regardless of grade level, there are student outcomes in each of the following strands:

- Number
- Patterns and Relations
- Shape and Space
- Statistics and Probability

K–12 FOCUS

The design of the common curriculum framework focuses on mathematics as a K–12 program rather than as a subset of this, which has been the practice historically.

There are two major sections of the document that show this focus. The general outcomes are presented on four two-page spreads, one for each strand. The specific outcomes in the K–9 component are also presented so that the reader can easily see the student outcomes across multiple grades.

Mathematical Processes

The seven mathematical processes that are identified are critical elements that students must encounter in a mathematics program to achieve the goals of mathematics. Students are expected to

- communicate mathematically
- connect mathematical ideas to other concepts in mathematics, to everyday experiences and to other disciplines
- use estimation and mental mathematics where appropriate
- relate and apply new mathematical knowledge through problem solving

- reason and justify their thinking
- select and use appropriate technologies as tools to solve problems
- use visualization to assist in processing information, making connections and solving problems

The common curriculum framework incorporates these seven related mathematical processes that are intended to permeate teaching and learning.

NATURE OF MATHEMATICS

By enriching our view of mathematics and the learning environment, the outcomes of the common curriculum framework can be accomplished.

The brain is constantly looking for and making connections. "Because the learner is constantly searching for connections on many levels, educators need to orchestrate the experiences from when learners extract understanding.... Brain research establishes and confirms that multiple complex and concrete experiences are essential for meaningful learning and teaching" (Caine & Caine, 1991, p. 5).

There are additional critical components that must be addressed in a mathematical program beyond those listed as mathematical processes. The components discussed are Pattern, Number, Shape, Change, Constancy, Dimension (size and scale), Relationships, Quantity, and Uncertainty. They are used to describe mathematics in a broad way to establish the variety of connections that can be made among the various strands.

FRAMEWORK SUMMARY

The components of the conceptual framework for K–12 mathematics, as described, dictate what should be happening in mathematics education. The components are not meant to stand alone but are to be related to enhance one another. Activities that take place in the classroom should stem from a problemsolving approach built on the *mathematical processes* and lead students to an understanding of the *nature of mathematics* through specific knowledge, skills, and attitudes related to each strand.

1996 ANNOTATED BIBLIOGRAPHY OF K–9 MATHEMATICS RESOURCES

The annotated bibliography of the K–9 mathematics resources identifies the English language resources endorsed by and common to all WCP jurisdictions implementing the common curriculum framework. The bibliography also identifies authorized Alberta resources that are not listed in earlier annotated bibliographies.

The WCP resources in the annotated bibliography were selected through a collaborative review process, based on their high level of fidelity with the rationale, philosophy, mathematical processes, and outcomes of the common curriculum framework for K–9 mathematics. These resources have undergone an intensive review and were found to be the most suitable of those submitted.

In addition, the annotated bibliography includes authorized K–9 resources in Alberta. These resources have been evaluated in Alberta and not by WCP evaluation process.

Learning resources for kindergarten to grade 6 that were listed in the 1995 elementary mathematics authorized resources annotated list have not been repeated in the 1996 bibliography. Listings of elementary learning resources authorized since publication of the 1995 list and all resources authorized for grades 7–9 are included in the 1996 bibliography. Both the 1995 list and the 1996 bibliography are required for a complete listing of authorized resources for kindergarten to grade 9.

A complete list all authorized learning resources is also available on the Education in Alberta Web site at http://ednet.edc.gov.ab.ca in the Students and Learning section.

ALBERTA MATHEMATICS K–9 PROGRAM OF STUDIES

Alberta Education has incorporated the common curriculum framework for K–9 mathematics into the program of studies for K–9 mathematics. The relationship between the common curriculum framework and the program of studies can best be described as "copy and paste." A grey background has been inserted to illustrate the required components. This amendment to the program of studies was sent to schools in June 1996.

1995 KINDERGARTEN PROGRAM STATEMENT

The 1995 Kindergarten Program Statement was revised in June 1997 to include the Kindergarten components of the common curriculum framework for mathematics and language arts.

IMPLEMENTATION TIMELINE FOR K–12 MATHEMATICS

In September 1996, the mathematics programs for grades 7 and 9 were implemented, while the programs for Kindergarten to grade 6 were available for optional implementation.

In September 1997, the kindergarten through grade 6 and grade 8 programs were implemented.

DIFFERENCES BETWEEN ALBERTA'S 1994 INTERIM K–6 AND THE 1996 K–9 PROGRAM OF STUDIES

For the purposes of this article in *Early Childhood Education*, the focus will be on the differences in the K–3 portions of 1994 Interim K–6 and the 1996 K–9 programs of studies.

The organization and presentation of the documents vary. The common curriculum framework presents a multiple-grade focus. This type of presentation enhances the horizontal version of the 1994 interim program of studies. There are fewer outcomes that indicate teaching strategies; however, the illustrative examples are used to get at teaching methodology. There is more reference to technology in the 1996 program of studies as well.

In the Number strand, competency with higher numerical values is expected, and there are links between the process of estimation and mental mathematics.

In the Patterns and Relations strand, the emphasis on patterns is maintained.

In the Shape and Space strand, there is an increased emphasis on the use of language related to the concepts of *parallel, intersection, perpendicular,* and *congruent,* which are introduced in grade 3. Position language to document motion is also expected. The concept of area has its beginning in kindergarten. There is an increased focus on standard units of measure.

DOCUMENT ACQUISITION

The following documents can be obtained by contacting the Learning Resources Distributing Center at 427-2767, fax 422-9750:

- *The Common Curriculum Framework for K–12 Mathematics: Western Canadian Protocol for Collaboration in Basic Education,* Alberta Education, 1995; product code 302183-01, $11.60
- *Alberta Program of Studies for K–12 Mathematics,* Alberta Education, 1996; product code 317653-01, $6.90
- *Kindergarten to Grade 9 Mathematics Resources: Annotated Bibliography,* Alberta Education, 1996; product code 319154-01, $8.40

BIBLIOGRAPHY

Alberta Education. (1995). *The Common Curriculum Framework for K–12 Mathematics: Western Canadian Protocol for Collaboration in Basic Education.* Edmonton, AB: Alberta Education.

American Association for the Advancement of Science (AAAS). (1989). *Science for all Americans.* New York, NY: Oxford University Press.

American Association for the Advancement of Science (AAAS). (1993). *Benchmarks for science library.* New York, NY: Oxford University Press.

Armstrong, T. (1993). *Seven kinds of smart: Identifying and developing your many intelligences.* New York, NY: NAL-Dutton.

Caine, R. N., & Caine, G. (1991). *Making connections: Teaching and the human brain.* Menlo Park, CA: Addison-Wesley.

Hope, J. (1990). *Charting the course: A guide for revising the mathematics program in the Province of Saskatchewan.* Regina: Saskatoon Instructional Development and Research Unit (SIDRU), Faculty of Education, University of Regina.

National Council of Teachers of Mathematics (NCTM). (1989). *Curriculum and Evaluation Standards for School Mathematics.* Reston, VA: Author.

Steen, L. A. (Ed.). (1990). *On the shoulders of giants: New approaches to numeracy.* Washington, DC: National Academy Press.

This chapter was originally published as:

Liedtke, W. W. (1999). Multiculturalism and equities issues: Selected experiences and reflections. *delta-K: Journal of the Mathematics Council of the Alberta Teachers' Association, 36*(1), 63–66.

CHAPTER 39

MULTICULTURALISM AND EQUITIES ISSUES

Selected Experiences and Reflections

Werner W. Liedtke

At first I did not react when the College of Teachers announced that all teachers-to-be would receive training on dealing with multiculturalism and equity issues. After all, I had taken anthropology courses that included information about minorities and taught me how to include various groups of First Nations' students in a classroom. I had been part of a writing as well as an editorial team that was made aware of appropriate ways of dealing with these types of issues. And I had helped to raise three daughters. Frankly, I felt a little smug since I believed that I had dealt quite adequately with these issues all along. All that was left to do was to elaborate on some of the statements on my course outlines.

My complacency was short lived. My level of excitement rose when during a meeting, a few colleagues, including some in my subject area, suggested that multiculturalism and equity issues were really only relevant for areas other than mathematics teaching/learning. Many of the statements made during the discussion bothered me and caused me to reflect. That was not the end of it. The 1997 NCTM yearbook, *Multicultural and Gender Equity in the Mathematics Classroom: The Gift of Diversity*, that arrived the following summer provided me with a lot of additional food for thought. One sentence had a special impact on my reflections. In this chapter Koontz (1997, p. 186) states, "Gender bias is so subtle that we barely recognize it. It is so ingrained in how we define ourselves that it is essentially invisible. If we are to create gender-inclusive learning environments, we must first recognize our own biases." The question that came to mind immediately after I read the quote was, could it be that this statement is applicable to all biases that might be part of one's existence?

The paragraphs that follow include selected experiences and reflections about some possible biases, subtle or otherwise. These thoughts illustrate to me that we have come a long way but that there is always more that can and needs to be done. For anyone who has not had an opportunity to examine the 1997 NCTM Yearbook, the ideas that are referred to from this reference provide a hint about the intriguing information it includes. If, by chance, a few of the recorded examples cause a reader to smile, the effort of recording the ideas was worthwhile. If, at the same time, information causes reflection or a few new insights to be made, this effort has been worthwhile.

Anyone who teaches or has contact with schools in large urban centres knows that the makeup of classrooms has and is changing. Recently, I found this graphically illustrated in a local paper waiting for a connecting flight at the Calgary International Airport. One picture showed an elementary teacher and her students today. Another picture showed the same teacher and her students of several years ago. What a difference in ethnics! When students in one local grade 1 class were asked to name the country where their parents came from, their answers included India, China, and Vietnam. (One young boy, who had listened to these responses, raised his hand and said, "Squamish." No doubt he remembered his parents talking about "living in the country," and living "in the" or "in a" country may not be all that different for most children at this age level.)

Contacts with different schools and classrooms allow me to teach different multicultural groups of students. Because of this, I am concerned about questions such as, are my planned requests and comments appropriate? how are suggestions and questions interpreted by those whose language is not English? what possibilities for misinterpretations might or do exist? As a result of personal experience, I am aware of some possible misinterpretations, subtle or otherwise. Some examples in the 1997 NCTM Yearbook refreshed my memory. Beyond that, many issues discussed in this book generated a new awareness that I think cannot help but benefit all my future students.

One of my anthropology instructors at university had been a missionary as well as a principal of a school on a reservation. As students, we were continually reminded of the importance of knowing about First Nations' customs and taboos. Daily assignments included writing a letter to a colleague, real or imaginary, and elaborating on what should be kept in mind while trying to accommodate First Nations' students in one's classroom.

During one lecture the instructor mentioned that, as principal of a school on a reservation, he had overheard a grade 5 teacher asking one boy in his class, "What's the matter, did the cat get your tongue?"

Another time, he found out that a boy had been asked, while all his coed classmates were listening, to assist with cleaning the chalkboard. Each of these incidents illustrates cultural misunderstandings.

The first boy had learned how to be successful while hunting gophers, rabbits, coyotes, and foxes. He was moving on to bigger game. To suggest to him that a lowly animal like a cat was not only able to sneak up to him but also on top of that, to bite his tongue was inappropriate and degrading. He lost face in front of all the other boys. As far as the second episode is concerned, it would have been appropriate to make that request in a one-to-one setting or in front of an all-male audience. However, having girls listen to that request, and having to do cleaning chores in front of the girls, is degrading and results in having a young "hunter" lose face.

After this course, I was determined to do my best to do things right. My grade 5 class in Edmonton consisted of 37 students with only one First Nations' student. I attributed his reluctance to initiate tasks and his tendency to look at what others were doing after assignments were given to cultural differences. One day as he was walking out of the classroom, I reminded him that a soccer practice was scheduled. When he did not react, I ran after him only to discover that he was hard of hearing. When I called his home to discuss the incident, his foster parents asked, "What? Was he not wearing his hearing aid?" The first thing he had done every morning after stepping out of the house was to put the aid into his pocket.

Thomas (1997), in a chapter about teaching mathematics in a multicultural classroom, uses a story about a student in a Catholic school who read *angle* as *angel* to illustrate how the child's view of the world can impinge on comprehension. (Every time I see the word angle, I have to stop and think. How many ESL students are in the same predicament? Is there an easy way to make the distinction for these students?) Thomas raises an important point concerning comprehension when mathematics is embedded in text with the reference to results of a study showing that when Caucasian children were given reading passages depicting characters from either Caucasian or African American background, 55 of 57 scored higher when a Caucasian was the protagonist. Thomas includes a fascinating example to reinforce the fact that numerous possibilities exist for interpreting the context in which mathematics is presented. This example refers to the different interpretations of a test item that involved the movement of cattle from one part of a farm to another. Children from wheat farming areas were satisfied with an answer that was less than the number of cattle they started with. That was not the case, as one might expect, for children from dairy farming areas. This outcome reminded me of an incident reported by Lesh and Zawojewski (1988, p. 60). The author of an area problem involving goats decided to test the item by taking it to students who knew about these animals. The students were asked to calculate the grazing area for a goat tied by the neck with a 50-unit rope to the corner of a rectangular barn with dimensions 20 by 40 units. According to these children, a goat tied with a rope by the neck would chew through it, thus making the answer and resulting discussion much different from what the author had envisioned.

Although Lubienski (1997, p. 56) warns us that categorizing students by a single variable may be useful for some purposes, danger exists whenever there is stereotyping. The author reports results that show that beliefs about the nature of research differ according to socioeconomic status and that "these differences seemed to influence the ways in which they interpreted and said they would use mathematical claims in the media." After each visit to a rural school, I am more than a little despondent. I believe the difference between students in the "have" urban schools and some of the small "have-not" rural schools is too great. What chance do many young children have when they lack proper nourishment and care? The one thing that is encouraging is that these children are able to think mathematically, and I have found that they can be enticed to do and talk mathematics for long periods. Will we ever be able to solve the complex problem of keeping these students in school?

Thomas (1997) reminds us that context and linguistic difficulty make culturally unbiased testing extremely difficult. She uses the apparently simple question, "If a man takes three minutes to run 500 metres, how long will he take to run a kilometre?" to point out that even for young English-speaking students this task is difficult because it requires several assumptions and connections. Some ESL students will focus on *how long* and on *kilometre*. As a result they do not come up with the expected answer. If it is the long-term goal to have ESL students read questions that increase in linguistic complexity, Thomas suggests that during the early stages of coping with English, it is advantageous to present questions in the simplest form possible. Even if this is done, many possibilities for subtle misinterpretations may exist as the examples that follow illustrate.

A lady from Germany successfully appealed a parking ticket—she parked under a sign that clearly stated, "Fine for Parking." Prior to their first Christmas in Canada, two relatives who were recent arrivals (both ESL) decided to go shopping. They came away disappointed, empty-handed, and a little puzzled after having followed the promising sign that declared, "Lots for Sale."

There are many opportunities for errors, mispronunciations, and a little embarrassment for ESL students. My daughter, who teaches ESL, shared the following experience recounted by one of her students. While strolling on a city sidewalk shortly after arriving in Canada, he was approached by a young lady who inquired, "Do you have the time?" Thinking that here existed an excellent opportunity of getting to know a young Canadian lady, he responded with, "Yes." A finger pointing at her wrist and the expression on her face told him that he had misinterpreted the question.

During the summer, a full-page article about school in the local newspaper said, "It's a Girls' World." Statistics were cited that pointed to female dominance in honor roll and secondary scholarships. Yet, as far as mathematics learning is concerned, there are girls who will buy into the myth that boys are smarter than girls in math. Many girls tend to underestimate their mathematical ability and knowledge. Part of attempting to create a gender-fair program means that we need to search for and recognize many of the subtle biases that, according to Koontz (1997, p. 186), can and do exist.

On more than one occasion, I have found that tutoring sessions for high school girls become more effective after the girls had been convinced that they may know much more than they think they do and that it is appropriate, advantageous, and acceptable to talk while trying to solve a problem or to develop or apply an algorithmic procedure. It is very rewarding to observe when something is revealed and discussed by a girl who had previously protested, "I don't know how to do this." The reaction to the observations "You know a lot" or "You know almost all of it" is frequently greeted with a satisfying smile. Getting students to talk and increasing wait-time are effective as well as rewarding strategies that I had too often overlooked during my early stages of teaching/tutoring.

My observations have led me to conclude that unfair problems exist because they do not include sufficient background information. During one tutoring session, a grade 8 girl produced problems from a text that asked her to consider different distances between threads to calculate the efficiency of jacks, costs of renting different types of cars, and the cost of flying airplanes at different altitudes. During the discussion, it became obvious that she did not know what a jack was nor what the threads were, why someone would want to rent a car or why different cars should vary in rental costs. After she "plugged" values into the formula for flying airplanes at different heights, she concluded that there was an error and something seemed wrong. According to her, flying at a lower altitude should be less expensive or more efficient since the cost of climbing is avoided. Her answers did not confirm this belief.

It could be that boys are more familiar with some of the ideas in these types of problems. However, that is not my reason for including this incident as part of the discussion. One could assume that the girl's responses are not related to gender equity. If examples of this nature were part of an assessment, they would discriminate against anyone lacking the general background knowledge to make sense of the variables presented. If that is the case, test results would indicate weaknesses as far as mathematical thinking is concerned, even though other factors are in fact responsible.

The ideas included in this article are indicative of some of the many issues discussed in the 1997 NCTM Yearbook. Should the reader, by chance, still feel a little complacent about issues related to multiculturalism and equity, the following quotation from Wick and Kenschaft (1997, p. 209) provides sufficient and somewhat sobering ammunition for further reflection:

> Small incidents that do not in themselves have much significance, called "microinequities," constitute a significant portion of discrimination. Noticing them can seem petty, and even the victims can find them humorous in retrospect. But cumulatively, they render us invisible, require us to work harder, and steal time and energy as we contemplate whether to fight back, laugh, or pretend not to notice. They generally divert us from functioning as competent professionals. In classrooms, inequities in the interactions between teacher and students or among students themselves send subtle messages about who can do mathematics. Microinequities wear away at our strength, bit by bit, like drops of water on a rock. Moreover, they create a climate in which outrageous acts are accepted.

REFERENCES

Koontz, T. (1997). Know thyself: The evolution of an intervention gender-equity program. In J. Tentacosta (Ed.), *Multicultural and gender equity in the mathematics classroom: The gift of diversity* (1997 Yearbook of the National Council of Teachers of Mathematics, pp. 186–194). Reston, VA: NCTM.

Lesh, R., & Zawojewski, J. S. (1988). Problem solving. In T. R. Post (Ed.), *Teaching mathematics in grades K–8: Research based methods* (pp. 40–77). Toronto, ON: Allyn & Bacon.

Lubienski, S. T. (1997). Class matters: A preliminary excursion. In J. Tentacosta (Ed.), *Multicultural and gender equity in the mathematics classroom: The gift of diversity* (1997 Yearbook of the National Council of Teachers of Mathematics, pp. 46–59). Reston, VA: NCTM.

Thomas, J. (1997). Teaching mathematics in a multicultural classroom: Lessons from Australia. In J. Tentacosta (Ed.), *Multicultural and gender equity in the mathematics classroom: The gift of diversity* (1997 Yearbook of the National Council of Teachers of Mathematics, pp. 34–45). Reston, VA: NCTM.

Wick, C. A., & Kenschaft, P. C. (1997). Microinequity skits: Conversation about gender issues. In J. Tentacosta (Ed.), *Multicultural and gender equity in the mathematics classroom: The gift of diversity* (1997 Yearbook of the National Council of Teachers of Mathematics, pp. 209–213). Reston, VA: NCTM.

This chapter was originally published as:

Lowen, A. C. (1990). Implementing manipulatives in mathematics teaching. *delta-K: Journal of the Mathematics Council of the Alberta Teachers' Association, 28*(1), 4–11.

CHAPTER 40

IMPLEMENTING MANIPULATIVES IN MATHEMATICS TEACHING

A. Craig Loewen

Editor's Note: A. Craig Loewen is an assistant professor in the Department of Education at the University of Lethbridge, Lethbridge. Alberta.

Manipulative materials have emerged in mathematics instruction as more than just a means to add variety to lessons; they are an essential element for effective mathematics instruction. However, for manipulatives to be used successfully in the classroom, a great deal of thought must precede their implementation. What role do manipulatives play in mathematics instruction? What factors influence the effective implementation of manipulatives in the instructional process? This paper presents alternative answers to these questions.

A DEFINITION OF MANIPULATIVES

The purpose of manipulatives is to make mathematics more concrete. Manipulatives enable students to play with, experience, and develop for themselves mathematical principles, relationships, and ideas. For manipulatives to have any place in the mathematics classroom, they must embody or physically represent specific mathematical concepts (Wiebe, 1983). Consider two examples and one counterexample.

A concept that many elementary mathematics students struggle with is why the remainder after division can never exceed the divisor. This concept may be illustrated when teaching division using a balance beam (Knifong & Burton, 1985). To model the equation 7 ÷ 2, the student would place I weight on the balance a distance of 7 units to the left of the fulcrum (see Diagram 40.1).

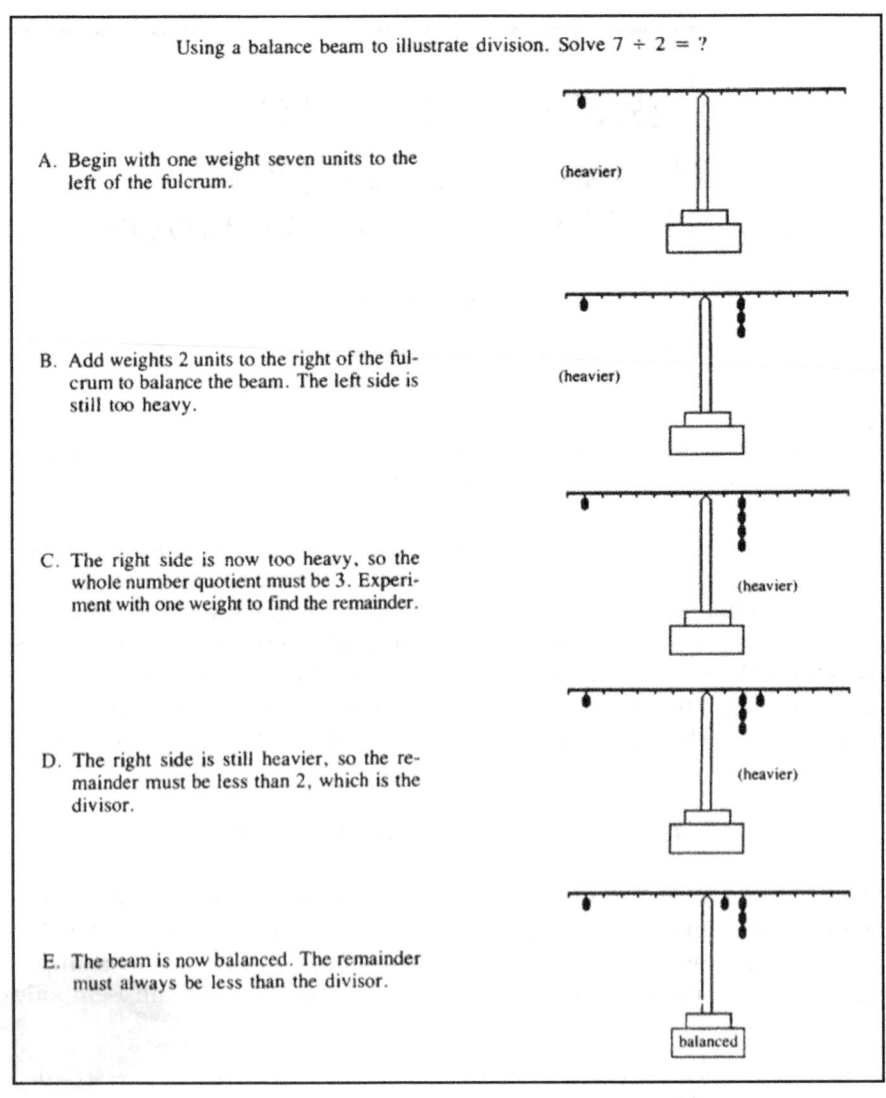

Using a balance beam to illustrate division. Solve 7 ÷ 2 = ?

A. Begin with one weight seven units to the left of the fulcrum.   (heavier)

B. Add weights 2 units to the right of the fulcrum to balance the beam. The left side is still too heavy.   (heavier)

C. The right side is now too heavy, so the whole number quotient must be 3. Experiment with one weight to find the remainder.   (heavier)

D. The right side is still heavier, so the remainder must be less than 2, which is the divisor.   (heavier)

E. The beam is now balanced. The remainder must always be less than the divisor.   balanced

Diagram 40.1

Because the divisor is 2, weights are hung 2 units from the fulcrum on the right side until the beam is balanced. The situation quickly arises that when 3 weights are hung on the right side, the beam tips to the left, but when another weight is added, the beam tips to the right. Where then should the final weight be hung in order to balance the beam? Through experimentation, it is obvious that hanging the weight any further to the right (e.g., a value greater than the divisor) is counterproductive, and thus a position closer to the fulcrum (e.g., a value less than the divisor) must be selected. In this case, the weight must be hung 1 unit to the right of the fulcrum to completely balance the beam. The remainder must always be less than the divisor; this mathematical concept is actually embodied within the manipulative materials.

Poker chips may be manipulated to model the subtraction of negative integers. Assume that a blue poker chip represents +1 and a red poker chip represents −1. Thus the subtraction of −4 from 3 in the equation $3 - -4 = ?$ may be modeled as follows. Set out 3 blue poker chips (+3) and then remove 4 red ones (−4). It is obvious that no red chips may be removed because there are only blue chips available. However, note that a blue and a red chip together total zero (e.g., $-1 + 1 = 0$). Thus, any number of pairs may be added without changing the value of the expression. Pairs are added until there are enough red chips such that 4 may be removed (add 4 pairs). Now, when 4 red chips are removed, 7 blue chips remain. This process illustrates that $3 - -4 = 7$. This model makes it clear why the difference is greater than the minuend when subtracting a negative subtrahend.

As a counterexample, consider the common exercise in which students pair numbered cards with corresponding word cards (see Diagram 40.2).

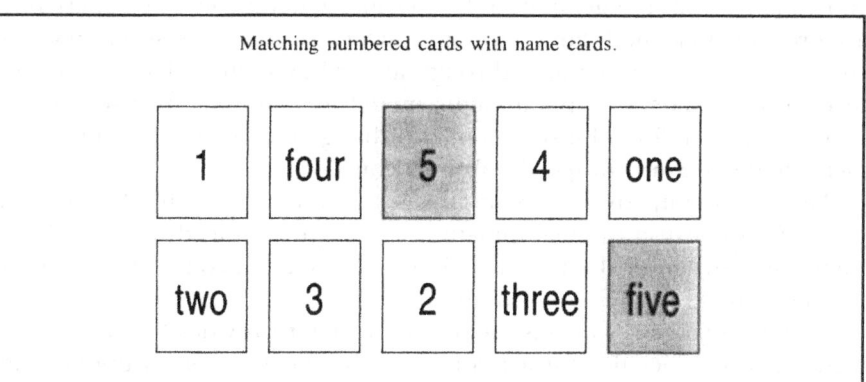

Matching numbered cards with name cards.

The shaded pair is a match and may be removed. This activity is sometimes played as a game which begins with all cards face down. Two cards, one at a time, are turned over by a player. If a match is found then those cards are removed from the game. If no match is found, the cards are turned back face down and the other player takes a turn. The winning player is the one having made the greatest number of matches once all the cards have been used.

Diagram 40.2

This activity, and others like it, may be called manipulative only in that students are given some object (cards) that they may touch and move. The cards do not embody any mathematical concept, however, and this exercise only serves to help students develop correspondence between names and symbols (a vocabulary exercise). For the student, no greater understanding of "oneness," "twoness," or "fiveness" is developed simply by matching symbols to words; memory skills are drilled.

THREE IMPLEMENTATION MODELS

What role do manipulatives play in mathematics instruction? Where do manipulatives fit into the typical instructional sequence: introduce, develop, review, and evaluate? The following three general models for implementing manipulatives offer some alternative answers to these questions. These models may be applied to either individual lessons or to complete units.

The first implementation model is called the introductory model (see Diagram 40.3) because the manipulatives are used only in the beginning stages of instruction. The purpose of the manipulative in this model is to introduce the mathematics concept to be learned and to provide a body of concrete experiences that can be drawn upon or synthesized during later formal instruction. The manipulative also fulfills the purpose of increasing student interest and motivation. In some cases, the manipulative also provides a sense of relevance to the later formal instruction delivered by the teacher. The learning sequence flows from the concrete to the abstract. In this model, knowledge is organized from the general to the specific; general concrete experiences are provided first and followed by highly structured formal sessions in which specific concepts are revealed through an oral exposition delivered by the teacher. The major assumptions of this model are that students require a context for effective formal instruction, and that general concrete experiences facilitate the learning of specific abstract concepts.

The second implementation model is called the tertiary model (see Diagram 40.3) because the manipulatives are not introduced until the latter stages of instruction. Early instruction is teacher-controlled, but later experimentation is less closely monitored.

In the initial stages of this model, the teacher provides formal focused instruction on specific abstract concepts; the focus is not on understanding but on the awareness of principles. These specific principles are later linked to create more general knowledge through informal experimentation with manipulatives; knowledge is organized from the specific to the general, while experiences are organized from the abstract to the concrete. The manipulative serves a synthesis role and functions as a context in which learned concepts may be applied. In this model, the manipulative may also

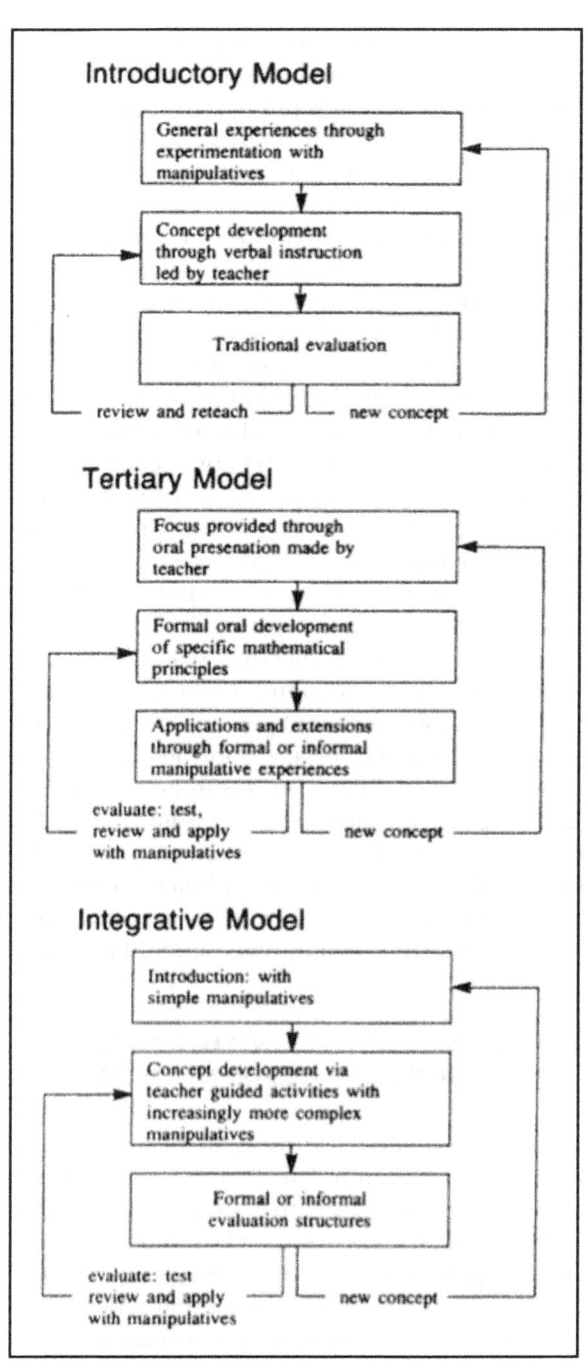

Diagram 40.3   Implementing models.

serve as a means for the teacher to evaluate student progress and understanding, as well as a means to undertake review of specified concepts. The second model is built upon the assumption that students require basic skills and knowledge before they can fully benefit (e.g., draw conclusions and formalize mental structures) from the experiences and environment manipulatives provide.

The final model is the integrative model (see Diagram 40.3). In this model, manipulatives are used continually throughout instruction; knowledge and skills are introduced, developed, reviewed, and evaluated through concrete experiences with physical representations of mathematical concepts. By using manipulatives at all points during instruction, it is hoped that high motivation and interest levels will be maintained throughout the entire instructional cycle. Using a manipulative for all phases of instruction eliminates the need to introduce more than one set of materials. The major assumptions of this model are that students learn better and retain longer what is learned in a familiar context, and that all phases of instructional cycle may be delivered easily using manipulatives.

No one model is correct or better than another. Instead, the teacher should use the model that best suits the material to be taught, the needs of the students and his or her own instructional style. The teacher may wish to consider the mathematics skills and motivation levels of the students, the students' learning styles, the synthesis and generalization skills of the students, the ease with which students master and apply learned concepts, as well as the complexity of the mathematics concepts to be taught. Each model possesses its own advantages, disadvantages, and assumptions. The teacher must select the model in which the disadvantages are minimized, the assumptions appear realistic, and the advantages are exploited. When these conditions exist, the purpose of the manipulative is maximized and effective implementation is achieved and measured by improved student learning.

SOME FACTORS INFLUENCING EFFECTIVE IMPLEMENTATION

Manipulatives may be evaluated according to a variety of criterion. Hynes (1986) has suggested that manipulatives may be evaluated according to both their pedagogical and physical attributes. With respect to pedagogical attributes, manipulatives must provide a clear representation of a mathematical concept, be appropriate for the student level, interest the students, be versatile, contribute to the building of a mathematical concept, assist in developing vocabulary, improve spatial visualization, promote problem solving, provide a sense of proof, and promote creativity. With respect to physical attributes, the manipulative must be durable, simple, attractive, manageable, cost-effective, and reasonable in terms of the quantity required. Not all manipulatives exemplify all of these attributes, but generally, the better the manipulative, the more conditions it will satisfy.

The attributes that Hynes describes are valuable when discussing the relative differences between manipulatives, but the true value of manipulatives lies in how effectively they may be employed in teaching and learning situations. The most versatile, motivating, and attractive manipulative will not be effective unless properly employed. Therefore, the manner in which the activity is conducted may be just as important as the materials themselves.

The first implementational consideration is the degree to which the student has control over concept development. If given time to simply experiment and play with the objects, will students develop the desired concept on their own? To what extent must the students' interaction with the manipulatives be guided by the teacher? To allow students to discover and develop concepts independently is often too time consuming, and there is no guarantee that the concept will ever be clearly or correctly formalized; however, concepts that are developed independently are more likely to be retained and treasured. The teacher must decide which form of manipulative is preferable based upon such considerations as students' past experiences with discovery learning, students' learning styles, the time available for the development of a concept, the motivation level of the students, and the ease with which the concept may be summarized from the play experience.

The second implementational consideration pertains to the degree to which the student may control the manipulative. Is it desirable that each student has his or her own set of manipulatives, or is it sufficient that the teacher manipulate one set for the benefit of all? When the teacher manipulates the materials for the students, then the visual experience is substituted for the tactile. When working with individuals or small groups, the discovery learning approach is possible, and this approach necessitates a tactile experience. When working with a large group, a visual learning experience is more practical. In essence, the teacher defines his or her own role as that of consultant or group leader. The choice of role dictates the degree to which the teacher intervenes in and controls the concept development process.

The third implementational consideration is the degree to which the mathematical concept embodied within the manipulative is obvious to the students. When the concept is obvious, then the materials are appropriate for developing mathematical relationships or facts. When the concept is less obvious, then the manipulative serves as a data-keeping tool. When the manipulative is used as a data-keeping tool, the student is relieved of data-keeping functions and may concentrate more fully on process skills. The following examples clarify the data-keeping and concept development natures of manipulatives.

The first example is illustrated in Diagram 40.4. In this example, paper is folded and cut (Bober & Percevault, 1987) in such a way as to illustrate why $a^2 - b^2 = (a - b)(a + b)$. Once the activity is completed, an inherent sense of proof or obviousness makes it difficult to contest that the relationship is true.

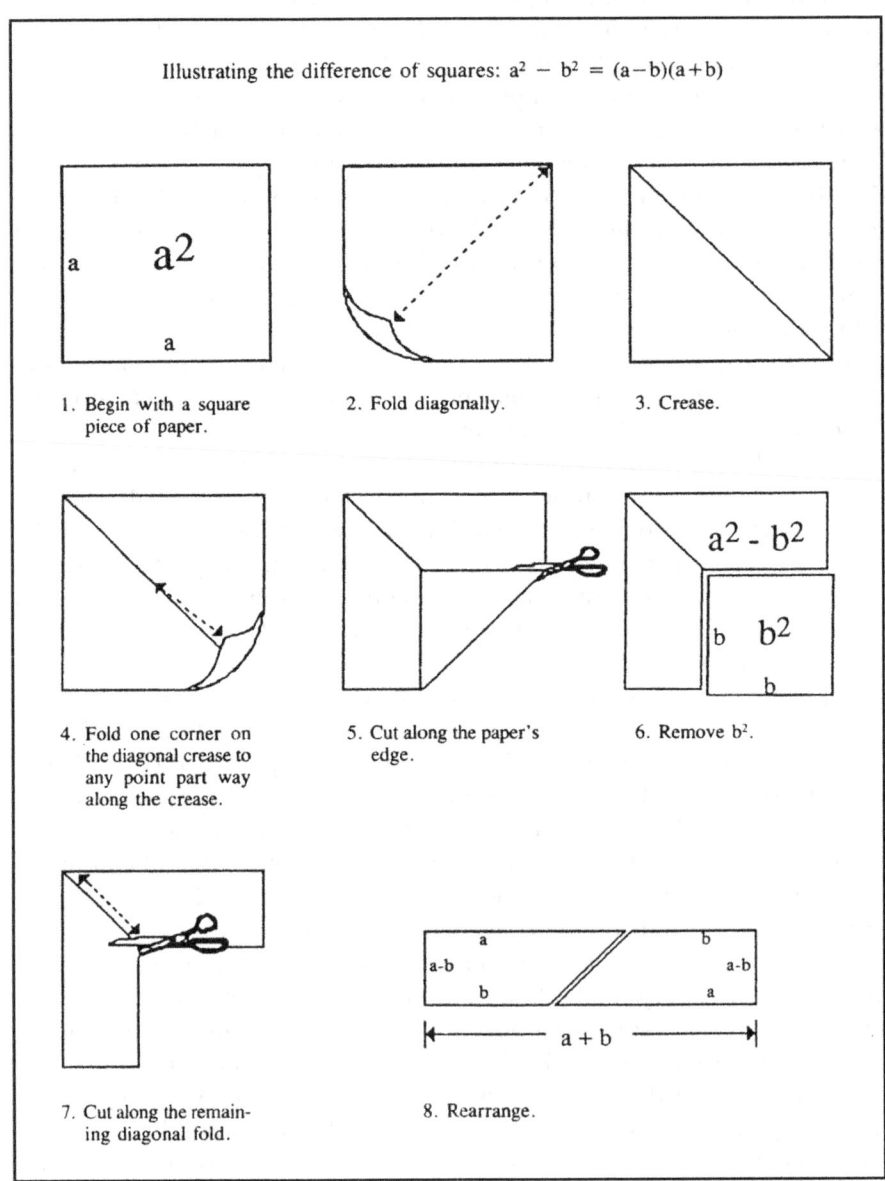

Diagram 40.4

The second example is illustrated in Diagram 40.5. In this example, the students model the process of solving simple algebraic equations through the manipulation of ordinary objects such as paper cups and circles cut from colored construction paper. In this exercise, the process of solving equations is emphasized, and the materials serve the purpose of keeping track of the various symbols and quantities found on each side of an algebraic

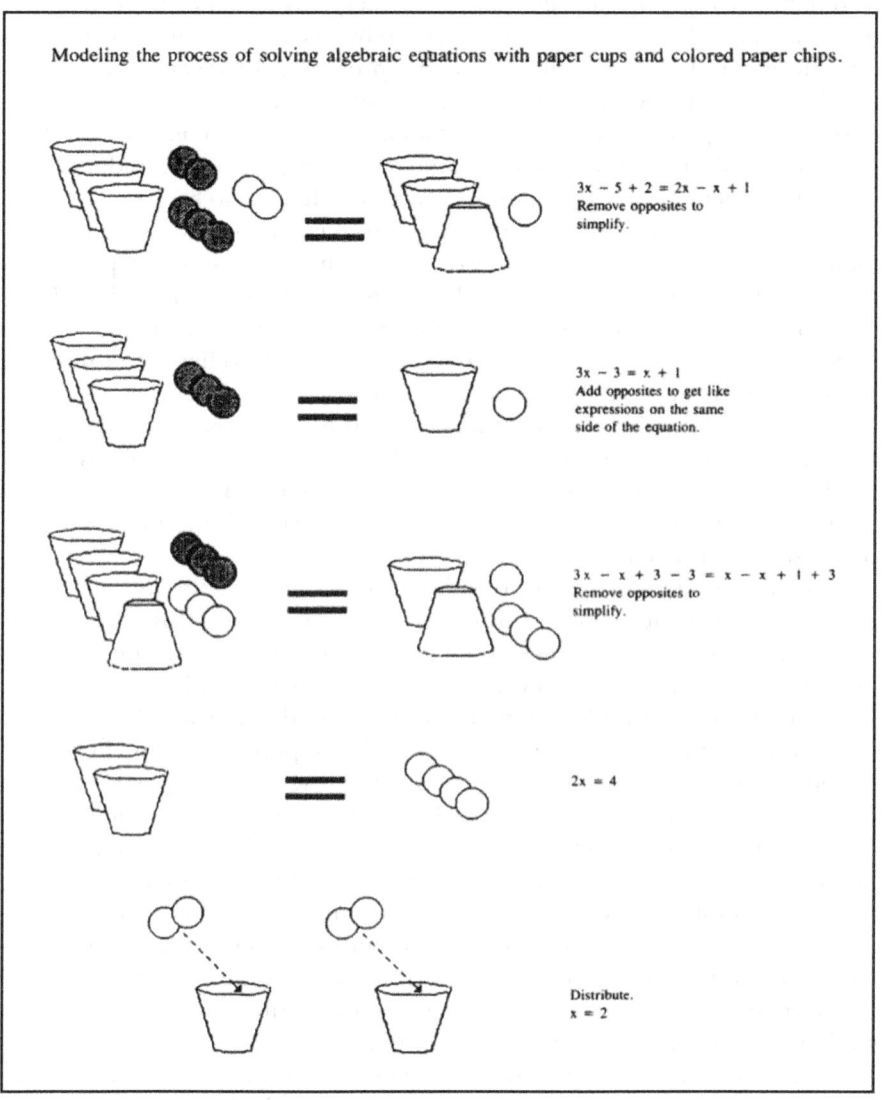

Modeling the process of solving algebraic equations with paper cups and colored paper chips.

$3x - 5 + 2 = 2x - x + 1$
Remove opposites to simplify.

$3x - 3 = x + 1$
Add opposites to get like expressions on the same side of the equation.

$3x - x + 3 - 3 = x - x + 1 + 3$
Remove opposites to simplify.

$2x = 4$

Distribute.
$x = 2$

Diagram 40.5

equation. Certain key relationships, such as $-x + x = 0$ and $-1 + 1 = 0$, are not made more obvious through this exercise.

Experience with these manipulatives simply provides the students with an alternate way of conceptualizing and remembering a process; it does not necessarily impart a greater understanding as to why the inherent relationships within the process are true.

The manipulatives are useful in learning the process because it is easier to remember how to distribute paper chips equally to paper cups than it is to correctly divide both sides of an equation by a constant value. The teacher must decide under what circumstances it is preferable to select manipulatives that facilitate a deeper or more complete understanding of a concept as opposed to manipulatives that simply promote an alternate conceptualization of a process. Both alternatives have value depending upon the students with whom the teacher is working and the objectives that are to be met.

The fourth implementational consideration is the number of concepts a manipulative supports within an instructional unit. If many concepts within an instructional unit are to be taught using manipulatives, then it is desirable to use similar materials for each topic within the unit. Using similar materials helps students link and relate these topics, relieves the need to constantly introduce and familiarize students with new materials, and provides a sense of continuity and coherence to the unit. However, a manipulative can be effective even if it supports only one concept, especially when used to review a concept or provide a brief extension to a previously developed concept. The manipulative must fit the instructional purposes and processes that the teacher has designed.

The fifth implementational consideration is the degree of familiarity students need with the materials in order to use them properly. How much time must be spent introducing the materials to the students and developing necessary vocabulary? If students are not properly familiarized with the materials, they will spend less time focusing on mathematical principles and more time trying to remember the manipulative procedure. Furthermore, if students are not familiar with the materials, they will not possess the vocabulary or language necessary to ask questions of themselves and others or to summarize their new knowledge. Certain materials require a longer introduction time, and generally, materials that require more introduction are less desirable. In order to justify a longer introduction time, the teacher must consider how well the manipulative embodies the mathematical concept, the number of concepts that may be taught using the materials, the required degree and extent of teacher–student interaction and whether students will work with their own sets of materials.

Well-constructed manipulative materials do not guarantee effective instruction. Even good manipulative materials will only be as effective as the process through which they are employed, and this process requires careful thought and reflection by those who understand the mathematics curriculum as well as children's thinking processes, capabilities, needs, and interests.

REFERENCES

Bober, W., & Percevault, J. (Eds.). (1987). *Make it—take it, Math Monograph No. 9.* Edmonton, AB: Mathematics Council of the Alberta Teachers' Association.

Hynes, M.C. (1986). Selection criteria. *Arithmetic Teacher, 33*(6), 11–13.

Knifong, J. D., & Burton, G. M. (1985). Understanding word problems. *Arithmetic Teacher, 32*(5), 13–17.

Wiebe, J. H. (1983). Physical models for symbolic representations in arithmetic. *School Science and Mathematics, 83,* 492–502.

COMMENTARY

DELTA-K IN THE 1990s
Learning Mathematics with Meaning

Elaine Simmt

The 1990s in Alberta have been described as "the restructuring nineties" because of the significant changes to education that were brought about by the Klein government over that period: New school ward boundaries, changes in the levying of school taxes and distribution of the funds, and the introduction of charter schools are some of the most significant changes to education since Alberta became a province (Ell, 2002). In North American research, constructivism emerged as the most common theory of learning in mathematics education (see *Journal for Research in Mathematics Education*, V. 25), and around the world educators called for mathematics education for all (e.g., UNESCO, 1984). Throughout the decade, *delta-K* offered readers a range of topics discussed by various people involved in mathematics education, from classroom and university teachers to educational researchers and policymakers. Article themes ranged from mathematics problems and solution strategies to teaching methods, learner characteristics, curriculum reform, and the use of technology.

Although in this paper I integrate ten papers selected from the 1990s *delta-K* issues, I spent some time leafing through all of the issues published

Selected Writings from the Journal of the Mathematics Council of the Alberta Teachers' Association, pages 323–334
Copyright © 2014 by Information Age Publishing
323

in that decade. I found that scan uplifting because it suggested that the journal's editors were less interested in what mathematics wasn't doing or unable to do and more interested in possibilities for mathematics teaching and learning. For instance, we do not find reports of students' performance on provincial achievement tests, diploma examinations, or international comparisons of students, although such reports were beginning to attract more media attention. The collection of papers selected for this issue reflect a journal that in the 1990s was focused on enhancing the teaching and learning of mathematics. Those papers were written for teachers and discussed teaching mathematics, learning mathematics, teaching teachers of mathematics, curriculum reform, and innovations brought about by advances in research into learning and the rapid growth of information communication technology (ICT). These themes reflect the spirit of a decade when constructivism, the National Council of Teachers of Mathematics *Curriculum and Evaluation Standards for School Mathematics,* international comparisons of student achievement, and rapid advances in ICT were the dominant conversations in the mathematics education community in North America. Another theme was teaching all learners meaningful and relevant mathematics.

In this chapter I will discuss the papers selected for this monograph based on five themes: (1) attending to the learner, (2) mathematics lessons for diverse learners, (3) integrating ITC into mathematics education (4) the teacher as a learner, and (5) curriculum reform.

ATTENDING TO THE LEARNER

The 1990s refocused educators' attention on the needs of the learner in the mathematics classroom. The reasons for this are multifaceted and complex but three in particular stand out: (1) the neo-Piagetian movement and constructivist learning theory (see Steffe & Kieren, 1994), (2) international comparisons of learners' achievement in mathematics (e.g., Institute of Education Sciences National Centre for Education Statistics, TIMSS, 2012), and (3) a movement that called for mathematics for all (UNESCO, 1984). As I reflect on the articles, I see them in relation to themes that were present in the international discourse in mathematics education: learner diversity; reshaping mathematics lessons framed with a theory of constructivist learners; using technology to teach and learn mathematics; teachers teaching mathematics; and reforming mathematics curriculum for all.

In the late1980s and early 1990s, the neo-Piagetian movement, known as constructivism, had grown to dominate as a theoretical frame for mathematics education researchers in the United States and Canada. Constructivism provided a much more powerful theory of mathematics learning than behaviourism, and it focused researchers' attention on children's mathematics rather than on modes of transmitting and transforming pre-given (adult) mathematics into bite-sized pieces for children to learn:

> It is perhaps this [the constructivist researcher's] phenomenological consideration of children's mathematics arising in inter-action with a teacher in very particular spaces of mathematical possibilities that has led to whatever influence constructivist research has had on reformers in mathematics curriculum and teaching. Observing and listening to the mathematical activities of students is a powerful source and guide for teaching, for curriculum, and for ways in which growth in student understanding could be evaluated. (Steffe & Kieren, 1994, p. 723)

Constructivism had become so dominant in research that in the 1990s it began to enter teacher education (both preservice and inservice) and subsequently influenced teaching in K–12 schools. Curriculum developers and mathematics teachers rethought mathematics education from the perspective of the learner as a meaning-maker who constructs mathematical understanding on the basis of his or her history, experiences, and interaction in the world.

It is not surprising, then, that we find authors in the 1990s *delta-K* papers who wrote about the nature of the learner that teachers encounter in the classroom. Hubber (1990), in "It's All Greek to Me: Math Anxiety," writes about a phenomenon that Sheila Tobias (1995) identified in some of her university students. Tobias wondered how learners who were very competent in some areas could encounter such difficulty with mathematics. In Hubber's reflections on math anxiety, she attributes much of it to the teaching methods used in school, methods that suppress meaning for rote learning. She contends that such methods result in students struggling to make meaning of solutions to mathematics problems, which in turns leads to a perception that mathematics is an "incomprehensible mystery." Hubber also wonders about the impact on students when they are unable to make sense of mathematics in their day-to-day living or in communications that they encounter that are composed of difficult mathematical language and symbolism. In her paper she offers some advice to teachers for creating better environments for their learners: incorporate mathematics into the life of learners, reduce competition, encourage creative problem solving, pay attention to reading demands, systematically instruct problem solving, and most important, dispel the myth that mathematics involves a secret code that only the elite can know.

Mathematics classes have a great deal of diversity, some of which may be attributed to schooling itself (say, confidence and anxiety in mathematics or background knowledge of the content) but diversity comes in many other forms that a learner brings to the classroom: gender, ethnicity, cultural practices, language, and activity preferences are just a few examples. In the article by Liedtke (1999), "Multiculturalism and Equities Issues: Selected

Experiences and Reflections," we read a thoughtful reflection by a pedagogue who wonders about the impact his biases have on students in his mathematics classes. Although this paper was written in 1996, it points to issues that continue to be relevant today. Indeed, one of the great challenges of preservice teacher education in 2013 is to educate teachers and prepare them for the diversity their learners will bring to the classroom.

The Hubber and Leidtke articles reflect a concern for inclusive mathematics education. In 1984, UNESCO published the reports of the Mathematics for All working group of ICME 5. Calling for "mathematics for all" has meant different things in different parts of the world. In developing nations, the call was to provide mathematics as part of universal basic education, which, at that time, was not a basic right of every child in a number of countries. In nations like Canada, and in Alberta where basic education is a right, it was a call for mathematics education to be appropriate for meeting the needs of children and youth—gifted learners and students with cognitive disabilities; learners who favour abstraction, formalism and symbolism and learners who favour the concrete, informal and everyday language; city dwellers and country dwellers; First Nations and newcomers to Canada; English speakers and English learners. Leidtke points out that teachers need to be aware not only of the diversity of the learners in classrooms but also of strategies and approaches for mathematics lessons that cause no harm to students and that make learning mathematics meaningful and relevant. For such learners, Leidtke notes how teachers must be attentive to their turns of phrase, the examples they use, technical language, and classroom activities.

As our understanding of diversity grows, we start to see that the great potential of diversity: One student's way of seeing something is not a drawback so much as an opportunity for the student and others to make meaning of mathematics. There is a message in Hubber's and Leidtke's papers for today's mathematics teachers; that is, by reflecting on our own understandings, biases, and patterns of behaviour we can create opportunities for more inclusive lessons so that our classrooms are sites of mathematics for all.

ATTENTION ON MATHEMATICS LESSONS FOR DIVERSE LEARNERS

With constructivism and the calls for mathematics for all learners in the 1990s, mathematics teachers were encouraged to think differently about their classroom instruction. Creating meaningful mathematics lessons was a challenge for teachers who had previously placed a great deal of emphasis on direct instruction that involved definitions, worked examples, and guided practice followed by independent practice. Kieren (1995) labels such teacher-centric instruction as TIRE (Tell, Interrogate, Respond, and Evaluate) and suggests that such modes of instruction do not help learners make meaning. To make meaning, rather than preparing lecture notes and practice questions, teachers should prepare lessons in which learners experience mathematics in the form of patterns and relations, change and constancy, shape and space, number and measure (see Sanders & Vivone-Vernon, this volume). Loewen (1990), in "Implementing Manipulatives in Mathematics Teaching," shows how to create alternative lessons: that is, how teachers could move from telling to selecting and setting up experiences in which learners manipulate materials from which they are expected to construct their mathematics. Though the notion of using manipulatives to teach mathematics has become part of the teacher's repertoire today, in 1990 it was quite novel, so much so that Loewen begins his article by defining a manipulative. A manipulative "must embody or physically represent specific mathematical concepts" (1990, p. 4). He suggests that there are three purposes for manipulatives: (1) introducing concepts through experimentation with manipulatives, (2) applying mathematical principles by using manipulatives, and (3) integrating manipulatives of different complexity throughout a lesson to help students achieve multiple layers of understanding. A preservice teacher reading Loewen's paper today will likely find it just as informative as teachers did in the 1990s.

Working with manipulatives is one form of problem solving. Problem solving continued to be stressed in mathematics education into the 1990s but came to include a discussion of what constituted a problem and why we solve problems. Preceding this discussion was a common view in the United States and Canada that word problems and applications were content for problem solving after students developed concepts and skills. But closer examination of problem solving and international comparisons of student achievement triggered comparative classroom studies that illustrated how differently problem solving could be used in instruction. Stevenson and Stigler (1992) and Sawada and Stevenson studied Japanese mathematics lessons. In Sawada's (1996) article, "Mathematics as Problem Solving—a Japanese Way," we see a classroom-based example of a typical Japanese lesson in which a class of learners work with a teacher on a problem set in a realistic context. With the introduction to Japanese lessons, a model of extended class time spent on a single problem to develop mathematics concepts became another possibility for Alberta teachers to modify their instruction to create more learner-centred lessons that encourage meaning making. The Japanese case illustrated how a teacher could work with learners on a complex problem, interpret it as a set of smaller problems, develop multiple solution strategies with the students, use manipulative aids, and focus on interpreting problems, not simply practising problems similar to ones previously demonstrated by the teacher. Mathematics teaching in Japanese schools continues to be of interest to Canadian teachers, and recent interest in lesson study has become a form of professional development.

INTEGRATING ICT INTO MATHEMATICS EDUCATION

One of the most striking possibilities for mathematics classrooms in the 1990s was the introduction of information and communication technology (ICT) in the form of personal computers (microcomputers) and graphing calculators (in essence, a handheld computer decades before the iPad). Interestingly, of the 17 issues of *delta-K* from 1990–1999 that I scanned, 11 had articles about using computers and (after 1995) graphing calculators for doing mathematics. Most of the articles provided activities that used computers or graphing calculators and provided illustrations of how computers could be used for mathematical problem solving in a number of different areas. The decade was an interesting time with respect to ICT since there were calls for integrating computers and calculators into the teaching and learning of mathematics (see, for example, the *NCTM Standards* and the *Common Curriculum Framework for K–12 Mathematics: Western Canada Protocol for Collaboration in Basic Education (WCP)*) and educators were providing illustrations of how computers could be integrated in mathematics. However, there was a stumbling block—computers were often housed in business labs, so mathematics students did not have a great deal of access to them.

Given that the 1990s was a beginning point in the integration of ICT in school mathematics, it is not surprising that the technologically based mathematics lessons presented in *delta-K* from that time do not reflect contemporary conversations as much as the articles on the use of manipulatives or the nature of the learner. The two articles selected for this volume provide reflections on the anticipated use of computers in mathematics, but neither mentions the role of the graphing calculator in high school mathematics. Instead, authors discuss the use of computers in the mathematics classroom. In "Learning about Computers and Mathematics: A Student Perspective," Findlay asks himself, "How can I integrate this technology into my profession so that it can aid in my teaching endeavours?" (1993, p. 8). He begins with a thought experiment, asking who among his students might benefit and what kind of resources would be of value. He saw an opportunity for special-needs learners, both those who are challenged by mathematics as it is currently taught and those who need to be challenged. But his review of the research uncovered impediments to using technology in the ways he anticipated, some of which continue to be expressed today; for example, computers rapidly change and obsolescence is an ongoing concern; computer technology is expensive; and computers pose a threat of the unknown (pp. 10–11). Dickie (1996), in "Computers in

Classrooms, Essential Learning Too or Program Disaster?," highlights the threat of the unknown. The article, a reprint from *Home & School* magazine, expresses the author's deep concern with the integration of computers in the mathematics classroom. She worries that fill-in-the-blank and drill-and-rote programs are far too common and they stifle creative learning, that the ability to work from the concrete to the abstract will be lost, that children are already bombarded with visual information from television, and that children who receive computers too early will grow up without a critical perspective on them.

Even though many *delta-K* articles in the 1990s focused on computers in mathematics classes (though only Dickie mentions the concerns just listed), it was calculators and graphing calculators that made their way onto students' desks. Indeed, Smith (in this volume) points to the graphing calculator as one of the most significant changes in high school mathematics.

THE TEACHER AS A LEARNER

The introduction of constructivism as a theory of learning, new manipulative materials, teaching aids, computers, and calculators affected teachers' professional development because they had to rethink their instructional practices. Given that constructing one's own knowledge is not restricted to students of mathematics and that the same learning theory applies to all of us in all domains of knowing, teachers had some learning to do. It is not surprising that in the articles selected for this decade we find one that illustrates teacher meaning making and another that points to the need for teachers to transform their mathematics knowing into pedagogical content knowledge so that they can create educative experiences for their learners. In Hauk and Quinn's (1992) article, "Moving Out of the Comfort Zone," we read about a teacher's experience of trying to transform his mathematics instruction. We learn how he modified his instruction to include manipulatives, small-group discussions, and student journalling to create a classroom where students would have "opportunities to be involved responsibly and actively in their own learning" and where they could have "concrete experiences in personally meaningful problem solving contexts" (p. 4). This modification of practice was principled and deliberate, and based on educational research. But it was not easy. In this case study, teachers can learn from an innovator brave enough to talk about the difficulties and the rewards he encountered when making change. On one hand, teachers reading the article today may be surprised to learn how novel working in small groups was for the learners and how exciting they found it; on the other, teachers will surely empathise with the challenge Quinn faced trying to fit the manipulative-based activities, small-group work, and journalling into the time allocated for math class. Another interesting aspect about this article is its contemporary tone. It is much more common today than it was when this article was published to use student voices (qualitative research) to provide insights into what students value, what they believed impeded their learning, and what meaning they made of the mathematics they were learning.

The Hauk and Quinn paper is the kind of work called for by Onslow and Geddis (1995) in "Building a Professional Memory: Articulating Knowledge about Teaching Mathematics." They argue that "effective mathematics teaching demands that subject matter be transformed to allow it to be learned meaningfully by novices" (p. 21) and that case studies are needed for that learning. They point to the need for resources for teaching pedagogical content knowledge (Shulman, 1986) to new teachers. Since the 1990s, a large community of scholars interested in what mathematics teachers need to know and how can that be taught (e.g., Lowenberg Ball, Thames, & Phelps, 2008) has arisen. Today, it is much more common to see strategies described by Hauk and Quinn used in the mathematics classroom, though there continues to be a need for case studies that illustrate pedagogical content knowledge for teaching mathematics, especially given the demand for mathematics teaching that has emerged out of curriculum reform over the last twenty years.

CURRICULUM REFORM

Whereas constructivism and ICT informed the shape of mathematics lessons in the 1990s, the National Council of Teacher of Mathematics' *Curriculum and Evaluation Standards for School Mathematics* in 1989 (commonly known as the *NCTM Standards*)[1] affected mathematics education more generally and informed curriculum reform in the United States and Canada. The *NCTM Standards* offered a vision of mathematics education based on research and expert opinion of the content, methods, and nature of mathematics, and argued that children and youth needed an appreciation of mathematics as well as mathematical skills and content for full participation in contemporary society. In "Enhancing Mathematics Teaching in the Context of the Curriculum and Professional Standards of the NCTM," Puhlmann (1995) introduces Alberta teachers to the *NCTM Standards*. The "Curriculum *Standards* for school mathematics are value judgements based on a broad coherent vision of schooling derived from several factors: societal goals, student goals, research on teaching and learning, and professional experience" (p. 21). With this statement we learn that the *NCTM Standards* are as much a political document as they are an educational resource. The document is divided into sections based on grade levels. Each section delineates the mathematics content that students should learn and a set of core standards across all grade levels: problem solving, communication, reasoning, and mathematical connections. Further, in the *NCTM Standards*, it is asserted that technology should be integrated throughout all school mathematics. A close read of Puhlmann's article against current curriculum and instruction locally will lead the reader of this volume to see how the *NCTM Standards* have permeated the consciousness of the mathematics education community in Alberta and Canada in the decades that followed their publication. Although the *NCTM Standards* documents have been revised over the past twenty years and are now known as the *Principals and Standards for School Mathematics*, they continue to influence mathematics education across the world.

The influence of the NCTM Standards is most evident in the curriculum reform of the 1990s in Alberta in mathematics education. In "The Western Canadian Protocol (WCP): The Common Curriculum Framework (K–12 Mathematics) (CCF)," Sanders and Vivone-Vernon (1998) elaborate on the WCP and the process by which the new program of studies for Alberta came about. That curriculum led to one of the most significant reforms of high school mathematics education in Alberta in decades. The authors discuss how the ministers of education from the four western Canadian provinces and two territories agreed to develop a common curriculum of general and specific outcomes for mathematics. Although the intention was for the six signatories to the protocol to implement the new framework for mathematics education K–12 in their jurisdictions, only the K–9 curriculum was implemented by all. In the end, only a couple of jurisdictions developed high school programs of studies from the CCF. Alberta, as the lead province of this initiative, did use the framework in developing pure and applied math courses, which replaced Math 10-20-30 and Math 13-23-33.

The CCF included four strands within which all content outcomes were organized: number, shape and space, patterns and relations, and statistics and probability. It also referenced seven mathematical processes outcomes that were intended to permeate all strands: (1) communicate mathematically, (2) connect mathematical ideas, (3) use estimation and mental mathematics, (4) relate and apply new knowledge through problem solving, (5) reason and justify thinking, (6) select and use appropriate technology, and (7) use visualization in mathematics (p. 26). Not surprisingly, the WCP framework, a product work of the 1990s, was compatible with the other movements we have read about in this paper. For example, the use of manipulatives and ICT was promoted as valuable for problem solving, reasoning, visualization and making connections. Even today teachers will point to the *CCF for K–12 Mathematics* as a significant curriculum reform that has affected mathematics in Alberta over the past two decades.

THE RESTRUCTURING 1990s

Although today's mathematics classrooms may appear on the surface similar to those of the 1990s, there are contexts and practices we take for granted today that were just being introduced twenty years ago. Constructivism as a theory of learning continues to inform mathematics teaching; however, the constructivism of today has been elaborated to include the role of the social. Manipulatives, multiple representations, problem posing, and problem solving all continue to be used in Alberta classrooms as teachers create opportunities for students to engage in meaningful and relevant mathematics. Technology introduced to the 1990s mathematics classrooms, especially graphing calculators, has changed mathematics in high schools. Children and youth of the 2010s don't need to purchase expensive software packages to access mathematics lessons, practice exercises, problem-solving activities, simulations, or games. Such things are all available on the Internet and come in a variety of forms that the teachers of the 1990s couldn't even imagine. In large cities and smaller towns throughout the province, a booming economy has continued to make Alberta a province of newcomers, and the ethnocultural makeup of our classrooms continues to diversify. As well, policies to integrate all students in the classroom have resulted in classes in which the mathematics instruction is differentiated through the use of small groups, manipulatives, and variable-entry problem solving. *Delta-K* in the 1990s published plenty of articles that are as relevant today as they were 20 years ago. The 1990s was a pivotal decade for mathematics education. Of all of the innovations and trends of the 1990s, there is no doubt that constructivism, the call for mathematics for all, the *NCTM Standards*, and graphing calculators were the most significant factors in the restructuring of mathematics education in Alberta.

NOTE

1. Dr. Thomas Kieren, currently Professor Emeritus of the University of Alberta was the only Canadian consultant on the working group that created the document.

Elaine Simmt is a professor of secondary mathematics education at the University of Alberta. She completed her doctoral studies in the 1990s under the direction of Tom Kieren. A former mathematics and physical sciences teacher, Elaine has embraced the call for meaningful mathematics for all in the preservice and inservice teacher education in which she is involved.

REFERENCES

Damerow, P., Dunkley, M., Nebres, F., & Werry, W. (Eds.). (1985). *Mathematics for All: Problems of cultural selectivity and unequal distribution of mathematical education and future perspectives on mathematics teaching for the majority.* Report and papers presented in theme group 1, at the 5th International Congress on Mathematical Education. Paris, France: Division of Science Technical and Environmental Education UNESCO.

Dickie, A. (1996). Computers in classrooms: essential learning tool . . . or program for disaster? *Delta-K 33*(2), 18–20.

Ell, J. (2002). History of education. Alberta Teachers Association: Edmonton, Alberta. Retrieved from www.teachers.ab.ca/Teaching%20in%20Alberta/History%20of%20Public%20Education/Pages/Index.aspx.

Findlay, C. (1993). Learning about computers and mathematics: a student perspective. *Delta-K, 31*(3), 8–11.

Hauk, M., & Quinn, B. (1992). Moving out of the comfort zone. *Delta-K, 30*(2), 4–9.

Hubber, D. (1990). It's all Greek to me: math anxiety. *Delta-K, 28*(2), 13–15.

Institute of Education Sciences National Center for Education Statistics. (n.d). *Trends in international mathematics and science study (TIMSS): Overview.* Electronic Retrieval September 23, 2012 from http://nces.ed.gov/pubsearch/pubsinfo.asp?pubid=2013009rev

Kieren, T. (1995). *Teaching in the middle.* Paper presented at the Gage Educational Conference, Queens University, Kingston, ON.

Liedtke, W. (1999). Multiculturalism and equities issues: selected experiences and reflections. *Delta-K, 36*(1), 63–66.

Loewen, C. (1990). Implementing manipulatives in mathematics teaching. *Delta-K, 28*(1), 4–11.

Lowenberg Ball, D., Thames, M., & Phelps, G. (2008). Content knowledge for teaching: What makes it so special? *Journal of Teacher Education 59*(5), 389–407.

National Council of Teachers of Mathematics. (1989). *Curriculum and evaluation standards for school mathematics.* Reston, VA: Author.

National Council of Teachers of Mathematics. (1994). *Journal for Research in Mathematics Education,* vol. 25. Reston, VA: Author.

Onslow, B., & Geddis, A. (1995). Building a professional memory: articulating knowledge about teaching mathematics. *Delta-K, 32*(3), 21–23.

Puhlmann, K. (1995). 33(1) Enhancing mathematics teaching in the context of the curriculum and professional standards of the NCTM. *Delta-K, 33*(1), 21–27.

Sanders, H., & Vivone-Vernon, G. (1998). Western Canadian Protocol: The common curriculum framework (K–12 mathematics). *Delta-K, 35*(1), 25–28.

Sawada, D. (1996). Mathematics as problem solving—A Japanese way. *Delta-K, 33*(2), 44–47.

Shulman, L. (1986). Those who understand: Knowledge Growth in Teaching. *Educational Researcher, 15*(2), 4–14.

Steffe, L. P., & Kieren, T. (1994). Radical constructivism and mathematics education. *Journal for Research in Mathematics Education, 25,* 711–733.

Stevenson, H., & Stigler, J. (1992). *The learning gap: why our schools are failing and what we can learn from Japanese and Chinese education.* New York, NY: Summit.

Tobias, S. (1995). *Overcoming math anxiety.* New York, NY: W.W. Norton.

Von Glasersfeld, E. (1983). Learning as a constructive activity. In J. C. Bergeron & N. Herscovics (Eds.), *Proceedings of the Fifth Annual Meeting of the North American Chapter of the International Group for the Psychology of Mathematics Education* (pp. 41–69). Montreal, QC: University of Montreal.

2000s INTRODUCTION

SOME MENTIONS
IN THE AUGHTS

Mark Mercer

The aughts seem distinct as a decade by a changed emphasis: algorithms. Algorithms can be thought of as a sequence of steps, highly routine, that produce a desired result. For example, as a math student and teacher, I study and teach math with and without calculators and emphasize their use in various ways. I study mathematics and math education with a sense of speed through an algorithm and what is available for thoughtful reflection. This blend seems unique. Algorithms can also be thought of as a play between a teacher's craft and the effects of technology on a purposeful direction, our young people as students and learners. We also choose the value we bring to our classrooms and this also might be thought of as algorithmic. Indeed, this value forms our views of assessment that bridges what we have done and what we will do: the reasons for our craft.

The aughts uniquely position students within a changed view of algorithms. High school students in the early aughts most likely studied elementary numeracy with calculators—unseen algorithms. In my elementary years, to substitute calculators for mental arithmetic meant not learning. An appreciation for the beauty of number patterns and changing magnitude

Selected Writings from the Journal of the Mathematics Council of the Alberta Teachers' Association, pages 335–339

with ordered operations had limited development. In contrast, students entering secondary school in the aughts have a variety of experiences. And elementary students in the aughts learn multiple algorithms based on place value with manipulatives exposing certain number patterns, like friendly facts, through a variety of methods, tasks, and even games. A catalyst for this sort of emphasis seems to be technology. For example, during the same year early in the aughts I taught with a chalk board, complete with chalk holder, and with a dedicated math computer lab using Microsoft Excel's conditional statements to provide feedback about experiments, tables, and formulae students entered answering a sequence of questions, all automatically graded: a paperless finance unit. During those few weeks I asked the same questions differently. Students still experienced curricular algorithms in paced ways, but outside my direct and carefully controlled instruction. I programmed how their experience started and they experienced mathematics with unseen algorithms to expose and interpret other algorithms through an interpretation on Excel. Due to this interpretation, a selected algorithm became the point of questions in more ways than could be given on a chalk board.

Not surprisingly, ways of asking questions also changed. A previous pedagogic notion of developing a simple and elegant algorithm to learn and apply to various problems and then use in carefully constructed ways to explore its boundaries became an issue of repeating a teacher. This notion decreased in popularity, especially evident in the front matter of curriculum documents. Students, and their questions, became a focal point, and a craft emerged where prompts and starters initiated students' questions about portions of this simple and elegant algorithm. Initial student skills focused on ways of thinking, such as metaphors, with an emphasis on vernacular and reading comprehension through the use of short stories and longer word problems. More traditional arithmetic and algebraic practice for these simple and elegant algorithms became a topic of debate as technology revalued what was worth remembering. Indeed, formula sheets and question stems changed through the aughts to match these trends.

Just a few years before the aughts, grade 12 graduation became one of the measures of success. Questions about whether a student had the choice to fail within the purposeful design of a teacher's classroom guided professional development and matched a change in society's construction. Trends with students not graduating illustrated a consistent pattern of behaviours and were becoming unsustainable. In my grade school years, not graduating usually meant entering a trade, and many dropouts experienced success. For example, a common profile of a millionaire in the aughts is a small business owner providing a trade with low overhead and high productivity able to withstand harsh economic cycles, often without grade twelve and sometimes only grade eight credentials (starting their career twenty and more years earlier). Yet, in the aughts, the trades required a different teenage experience. Profoundly, education became focused towards asking questions from personal experience. Previous experiences and how students co-create classroom environments became crucial parts of Alberta's curricular experience. The teacher became responsible for a transparent design of inclusion and leadership. Indeed, in 2011, Alberta Education held an information meeting for superintendants and other curriculum leaders focusing curriculum interpretation towards transformation, journey, and experience. The skills of what is learned became aimed at the unknown future, not simply repeating algorithmic work of a past classroom.

Curriculum topics changed in the aughts and these were mixed with more changes of classroom culture and a mandated secondary school calculator. Multiple revisions of Math 30 and 33 changed to Pure and Applied, and this redefined secondary streaming. Bluntly and simply written, what was formerly understood to group academic or matriculation students became the course for many. For these students, streaming became somewhat of a moot issue. The mandated calculators enabled the completion of a Pure-level Diploma examination, and curriculum topics seemed split between those taught with more formal arithmetic and algebraic notations and those with calculators. For the first time in high school, an electronic calculator could check what were presumed to be high-level and calculator-less math questions. Profoundly, the mathematical notations that previously guided problem solving's reading and writing became a point of question, or more accurately, how a student internalized arithmetic-based algorithms for modeling and adjusting for errors. An appreciation for the beauty of mathematical modeling and errors, or applications, became an expectation of the classroom culture. This fundamentally changed a value of assessment; all students were expected to learn and demonstrate reading, writing, and arithmetic with a collection of unseen algorithms. Indeed, by the end of the aughts, another curriculum change brought Math 10 Common, with streaming pushed another year closer to graduation; instructional resources for modelling were multimedia, interactive, and widely available through the Internet; and pedagogy focused what students were capable of and teaching focused mathematical processes while students were engaged creatively.

Evidence for these math classrooms being changed can be understood in how teachers supplement resources. More evidence is technology. For example, elementary education journals from the 1980s suggest a variety of strategies for teaching arithmetic algorithms. These thoughts are reflected in current approved resources. Some secondary resources that expound calculator how-to's mimic a simple and elegant sequence for chunking, facilitating memory. Sometimes what continued were disconnections between a learner's thoughts and the symbols written. Elementary expectations of purposefully disconnecting formal thought from writing and later reconnecting these intentionally became contrasted with preparing students' demonstration on high-stakes exams. These students were prepared for a known and very immediate future unlike society's expectation. By the end of the aughts, technology continued to change and focus an unknown future. Learners experienced dynamic graphing and image-building software, knowledge-based search engines, and handwriting-recognition calculators. Answers became available through images, by asking questions, and by arithmetic done by writing formally (not memorizing calculator buttons). These tools focus an education of proofs, asking questions, and writing formally in different ways.

The aughts brought classroom questions that took student beliefs, expectations, and prior knowledge and exposed foundational math concepts by creating situations that required questions. For example, to explore a one-dimensional line, a string might hold clothes pins attached to recipe cards with fraction representations. Mathematical beauty occurs just before the third recipe card is hung. Hanging the second fraction suddenly creates unit relationships and definitions of "one." To attach the third recipe card accurately, a learner might use referent measures, measuring devices, proportions, and various formulaic representations. For younger or more immature learners, teachers can always define endpoints and proportions, focusing more ordinal concepts. Another question takes paper folding and constructs metric and imperial rulers helping learners with referents. Then, these learners might represent a repeating and non-terminating decimal and compare it to an irrational number. A final question example is a variety of stories related to a sort of number line bent in a circle with 50 tic marks where every third tic is erased as these are counted in a circular way. Modeling and predicting this situation can utilize concepts like logarithms. Assessing this kind of work and learning reevaluated what it meant to have fun and persevere. A teacher created a classroom culture of learning and students transformed from a collaborative skill set to functioning individually with confidence. These situations are intended to present a cross-section of how the aughts redefined questions answered with questions (from the definition of "one," to recollecting a historical perspective of mathematics based in measurement and wealth, to an expanding sense of number systems, to modeling with more advanced algorithms).

The aughts also brought many methods to add variety to address reading comprehension methods for more traditional and formal notations. For example, three solutions for multistep determinations of quadratic functions from a parabolic curve (general, vertex, and factored forms) can be cut up, piled, and reordered. Compared to copying notes from a chalkboard, students are required to be more sensitive as learners, wondering about where to start with a pile of paper "shredded into strips."

Over the past few years, the Education Act, amongst other items, has been questioned in Alberta. Mobile technology has progressed into personal education and mobile learning. In fact, in 2012, Alberta Education held a Math Digital Diploma Item Writing session where answers for these items were almost impossible to do by paper and pencil. If this sort of assessment changes, then it will match a growing sense of what is required of numeracy with seen and unseen algorithms, how knowledge is valued, and how we ask questions. In many ways, that is still to be decided. The aughts brought us students uniquely positioned with unseen algorithms and various experiences between knowing simple and elegant algorithms, their applications, and developing math skills that form ways of thinking and using words. It also brought us a changed value in graduation and classroom culture. How students access and use available technology will influence a value of knowledge. The value of these experiences should only increase. So, the questions we always had become used in ways to underpin a fact that mathematics is a collection of definitions because we say so...it helps us make sense of our world.

Mark Mercer currently teaches with Edmonton Public Schools. He enjoys teaching math and math education, and he is keenly interested in the value of a-teacher-in-the-classroom amongst vast amounts of technology and choices. For him, using imagination is the most valuable result of math education whether it is through reading, writing, or calculations.

This chapter was originally published as:

Liedtke, W. W. (2000). Enhancing numeracy in the early years. *delta-K: Journal of the Mathematics Council of the Alberta Teachers' Association, 37*(1), 38–42.

CHAPTER 41

ENHANCING NUMERACY IN THE EARLY YEARS

Werner W. Liedtke

WHY IS NUMERACY IMPORTANT?

Willoughby (1990, p. 2) reminded us that the world is becoming more mathematical and "muddling through without the appropriate attitude and abilities in mathematics has become, and will continue to become, more difficult—for both individuals and society." Since everyone can learn mathematics (Romberg, 1990), why is it that, "innumeracy, an inability to deal comfortably with the fundamental notions of number and chance, plagues far too many otherwise knowledgeable citizens" (Paulos, 1990, p. 3)? Why are there so many people who are willing to admit that, "I am not good and never was good at mathematics?" Why is it that even those who have done well on written tests and have mastered school survival skills may actually know less than they think they do since many lack the ability to talk about the mathematics they have learned (Davis, 1986)? The answer to these questions must lie in the way mathematics has been taught.

Since a large part of our population lacks important elements of numeracy that are necessary for being wise consumers, we should be able to conclude that for the majority of students, "traditional teaching" (Battista,

1999) or "Parrot Math" (O'Brien, 1999) did little or nothing to foster numeracy. Without a focus on numeracy as part of all mathematics teaching, the most important outcome of mathematics learning will not be reached.

WHAT IS NUMERACY?

A pamphlet produced by the British Columbia Association of Mathematics Teachers (1998) entitled *Numeracy* answers the question with,

> Numeracy can be defined as the combination of mathematical knowledge, problem solving and communication skills required by all persons to function successfully within our technological world. It includes the ability to apply various aspects of mathematics to understand, predict, and control routine events in people's lives.

One important indicator of numeracy at every stage of learning mathematics is related to the outcome that everything should make sense to the learner. Students should be able to visualize, or as many of them state during interviews "see 'it' in my brain," the numbers, operations and relationships that are being examined and discussed. Students should be able to connect what they are learning not only to previous learning but also to their experiences outside the classroom. Since the reaching of these goals is dependent upon skillful teachers, a sense of efficacy is appropriate and well deserved. Without teachers who know how to select effective materials, create appropriate classroom settings, and orchestrate meaningful dialogues and discussions that involve high-order thinking questions, these goals would not be reached.

HOW CAN NUMERACY BE DEVELOPED
PRIOR TO SCHOOL?

The development of numeracy can and should begin at home. Parents exert a powerful effect on children"s attitude and achievement in mathematics (Leder, 1992). Whenever possible, parents need to be informed about the types of activities and settings that can contribute to fostering this development. For most parents this may not be an easy or natural task. It is more than likely that their mathematics learning resulted from experiences that can be classified as "parrot math" (O'Brien, 1999). Their learning could be described as closed, since mathematics was taught mainly by telling them what and how to think. For the vast majority, these experiences did not contribute to conditions that are favorable for the development and growth of numeracy, namely, risk taking, flexible thinking and visualization.

Sample activities can be used to illustrate to parents open-ended settings that tease children to think and can serve them well with future encounters of mathematics. For example, when parents use the familiar sorting tasks that involve finding the answers to the questions: *Which is not like the other? Which of these isn't the same?* (Which they might sing—as it is done on *Sesame Street*), adults should put the onus on the child by saying that they do not have a correct answer in mind and that more than one answer is possible. This can be done by asking, *Which of these do you think does not belong? Which of these do you think is different?* After a response is elicited the question, *Is another answer possible?* or *What is another possible answer?* is posed. These types of activities and questions have been shown to stimulate discussion, risk taking, and curiosity and can contribute to self-confidence (Liedtke, Kallio, & O'Brien, 1998; Spungin, 1996), which are important requisites for numeracy.

To illustrate to parents the difference between closed teaching, or teaching-by-telling, and teaching that encourages flexible thinking and visualization, use examples from Spungin (1996) (see Figure 41.1).

Figure 41.1

Parents are likely to be familiar with the question, What comes next? for settings like—■ ● ■ ●? or 1 2 3 ?

A closed setting is very likely to have a key that considers one answer to be correct. An open setting that assesses flexible thinking would accept any answer as correct as long as a reason is provided. For the sorting tasks in Figure 41.1, each member of the display could become the correct answer. The sequences can be extended in different ways. The five year old who extended the shape pattern with ▲, justified it with, "I want to see a triangle!" and then he continued by building a pattern around this triangle. When a 6 appeared for the second sequence, my initial guess about addition being part of the child's thinking was proven to be incorrect when the response to why was, "I am six years old." (Sequences need to accommodate important numbers!) These types of responses are indicative of flexible thinking and visualization that can transfer favorably to future experiences in numerical situations.

Willoughby (1990, p. 20) made the comment that, "If we teach children to use their fingers intelligently in the early grades, they should be able to get along without using them later." Parents can begin this intelligent use of fingers with children while talking about numbers before their children come to school. For example, after several fingers on one hand are briefly shown, ask the child to state the appropriate number name. Use counting to verify the response. This type of finger-flash naming and counting activity can develop the ability to visualize and match an appropriate number, or cardinal number, with the number name that is being used in conversations. The foundation for numeracy can be established before children come to school. If children did not have experiences prior to school that fostered self-confidence, risk taking, flexible thinking, and visualization, mathematics teaching will present an added challenge for the teachers of these children.

HOW CAN NUMERACY BE DEVELOPED IN THE EARLY GRADES?

What types of activities can foster numeracy and should be part of the ongoing learning of mathematics? What types of activities are unlikely to contribute to reaching this important goal?

The foundation of numeracy will have to be laid with each encounter of mathematics learning. Two general hints are provided by Howden (1989, p. 11) who expresses the view that number sense, the most important component of numeracy, develops gradually as a result of exploring numbers and visualizing them in a variety of contexts, and by Greeno (1991, p. 173), who suggests that "it may be more fruitful to view number sense as a by-product of other learning than as a goal of direct instruction." Even if that is the case, classroom settings and activities that are conducive to developing number sense or numerical power require careful planning and a skillful orchestration of discussion.

The growth of number sense in the early grades will be enhanced if, as part of their mathematics learning, students experience the type of open-ended settings that were described for parents. Such settings can foster or continue to foster self-confidence, risk taking, visualization as well as flexible thinking. It is beyond the scope of this paper to discuss all of the components of numeracy. The illustrative examples will be restricted to selected ideas of the most important component-number sense. These ideas will deal with number meanings, basic facts, estimation and mental mathematics.

Reys et al. (1999, p. 61) state that number sense "results in a view of numbers as meaningful entries and the expectation that mathematical manipulations and outcomes should make sense." Van de Walle (1990, p. 63) suggests that number sense contributes to problem-solving abilities and flexible thinking in numerical situations. He concludes that "without a major commitment by a curriculum to experiences that develop number sense, many children will never understand numbers in any way other than counting" (p. 64). How can the meaning of numbers be developed and flexible thinking be fostered?

VISUALIZING NUMBER

Before students examine numerals, they should be able to visualize the numbers that are labeled by these names. Without this visualization, mathematics learning can become a rote and meaningless experience. One aspect of this visualization consists of being able to match a group of 10 or fewer fingers that is briefly (too briefly to count them) shown with the appropriate number name. Another aspect would be for a student to answer not only, *How many fingers* (thumbs are defined to be fingers) *did you see?* but also, *How many did you not see?* (or *How many more to show 5?*, if one hand was considered, and *How many more to show 10?* if both hands were considered). The use of fingers to foster visualization and recognition of numbers without counting (subitizing) can be extended by asking students to show numbers in different ways.

The fingers on either one hand or both hands can be used as a referent, a group of objects briefly looked at to suggest whether there were more than, fewer than, or about the same number as the referent that was used. As is true for any estimation task, care must be taken not to judge student responses as right or wrong, but to give credit for willingness to take risks. This experience can be beneficial for fostering self-confidence.

The visualization of number is just as important as two-digit numerals and counting sequences are examined. When number names like 14, 32, or 58 are displayed, students should be able to think of these in terms of referents. For example, *What is the fewest number of students required to show each numeral?* (For 32, the answer is four. Three students show ten fingers each—3 tens, and one student shows two fingers—2 ones.) The important idea that numbers can be shown in different ways and can have different names can also be illustrated by having students display fingers. This experience can be transferred to representing an amount of money in various ways by using different combinations of different coins. It is very unlikely that chanting number names or rote counting while pointing to numerals on a number line, a calendar (Schwartz, 1994), or on different number charts fosters visualization or growth in numeracy. Having students as a group recite number names and clap at every fifth name, for example, does not help these students think of the numbers behind these names. If, on the other hand, for every fifth name another hand is shown, the numerousness behind each of these number names is illustrated, and visualization is enhanced.

BASIC FACTS

Students will know—and we want them to know—the basic facts of the four operations. More important than that, if an answer is forgotten, we want them to be able to have at least one way of re-inventing that forgotten answer. A sense of number of visualization and flexible thinking will enable students to do this. For example, consider someone who has forgotten the answer, or is asked to teach how to find the answer, for 8 and 7. The student could think,

> The seven could be thought of as two and five, or the eight as three and five. This allows me to think of going to ten first. Then five more will give the answer.
>
> I could think of eight as one and seven, then the familiar double seven plus seven is used. One more will give me the answer.
>
> I could think of seven as eight minus one and use the double eight plus eight and then subtract one to arrive at the answer.

This illustration of flexible thinking is not a result of lessons that focused on teaching students how to think. For most students, it will be a result of visualizations that were developed while illustrating number activities with fingers and/or objects. For example, the "going to 10" strategy described above is directly related to the finger-flash activities that involved the instruction and questions, *Show me a number in another way,* and *How many did you see? How many did you not see?* or *How many more to show 10?*

Strategies that foster recall of the answers to the doubles can also be based on visualizations formed while thinking of number arrangements with fingers. For example, for 8 + 8, two students are asked to show eights by using the familiar arrangements of 5 and 3 fingers, respectively. As the display of fingers is examined, the answer for 5 + 5 + 3 + 3 or 10 + 6 becomes visually obvious. The next level of visualization would involve showing one of a double on one's fingers, and finding the answer by considering each group of displayed fingers twice.

These types of settings illustrate that, in order to use fingers or materials intelligently, appropriate questions and discussions that focus on thinking are required. Without an appropriate orchestration, the desired connection between objects or fingers and the mind is unlikely to take place. Another determinant related to enhancement of number sense is the selection of effective activities and meaningful games. Appropriate selection can provide opportunities for furthering visualization and reinforcing flexible thinking. Activities and games that are closed and do not provide for decision making as moves are made, or focus on who is fastest, or who is first, are very unlikely to contribute to development of number sense.

Thinking strategies are important since they enable students to reinvent forgotten facts. These strategies can transfer to other mathematical settings. In the conclusion of their article entitled "Strategies for Basic Facts Instruction," Isaacs and Carroll (1999, p. 514) summarize existing research to identify several advantages and several possible disadvantages. The advantages follow:

- Children do learn their facts.
- Learning and retention arc facilitated.
- Facts arc organized into a meaningful network.
- Strategies supply links to other mathematical concepts.
- Students' understanding and confidence are enhanced.
- Use of common sense is encouraged, thus supporting concept development.
- Cost in instructional time is low: delayed practice often means less practice.
- Children's success at learning their facts reassures parents about their children's mathematics program.

Possible disadvantages follow:

- Children might learn strategies by rote so that mindless memorization is replaced by equally mindless "strategies."
- Class discussions might degenerate into the tedious recitation of every imaginable method, with little critical appraisal of approaches.

- Multiple ways may lead some students to conclude that memorizing the facts is not important.

The authors believe that, in most situations, thoughtful and sensitive teachers can avoid these pitfalls or hazards.

ESTIMATION AND MENTAL MATH

Number sense can be enhanced when the topics of estimation and mental math are dealt with. Once students know the difference between guessing and estimating, the essence of dealing with estimation should be on trying to have students develop strategies that are indicative of their personal development of number sense, rather than imposing strategies that attempt to teach how to think. To illustrate this important notion, consider the request to estimate the answers for 248 + 769 and 624 + 13.

For the first example, different students could think:

200 + 700; 300 + 700; 300 + 800; 200 + 800; 250 + 750;
24 tens + 76 tens and so on.

For the second example, different students could think:

600 + 20; 600 + 15; 600 + 12; 600 + 20; 600 +10 and so on.

If students have a notion about the relative magnitude of numbers, they are likely to know whether their estimates are greater than or less than the actual answers, an ability that should be part of their estimation strategies. The assumption should be made that students will use the numbers that are indicative of their sense of numbers. They should be encouraged to use numbers that they feel comfortable with rather than using numbers that someone else thinks are appropriate and that are associated with some sort of rote procedure. As far as assessing ability to estimate is concerned, the focus needs to be on examining the strategies students use rather than on numerical responses. The goal should be to encourage confidence and risk taking and thus foster growth in numeracy, which will allow for flexible use and refinement of estimation strategies.

The aforementioned notion about flexibility can also be part of activities that involve mental math. Rather than prescribing or suggesting a prescribed procedure, students should be given the opportunity to experiment or try different approaches. For example, in order to calculate the answer for 248 + 769, different starting points and sequences can be tried. After possible differences and/or similarities are discussed, students could identify their preferred strategies and state reasons for this being the case.

CONCLUSION

The major component of a mathematics curriculum that contributes to developing number sense and enhancing numeracy is a skillful teacher with a sense of efficacy. Neither worksheets, textbooks, nor computer programs can provide the input that is required to enhance this development. It takes a teacher to get students to think and to think about their thinking. Reys (1971, p. 558) concludes, "There is not now, never has been, and it is hoped, never will be a genuine substitute for a good teacher who knows how and what children need to learn and when they need to learn it!"

REFERENCES

Battista, M. (1999). The mathematical miseducation of America's youth. *Phi Delta Kappan, 80*(6), 425–433.

Davis, R. (1986). *Learning mathematics: the cognitive science approach to mathematics education.* Norwood, NJ: Ablex.

Greeno, J. (1991). Number sense as situated in the conceptual domain. *Journal for Research in Mathematics Education, 22*(3), 170–218.

Howden, M. (1989). Teaching number sense. *Arithmetic Teacher, 36*(6), 6–11.

Isaacs, A., & Carroll, W. (1999). Strategies for basic-facts instruction. *Teaching Children Mathematics, 5*(9), 508–515.

Leder, G. (1992). Mathematics and gender: Changing perspectives. In D. Grouws (Ed.), *Handbook of research on mathematics teaching and learning* (pp. 597–622). Toronto, ON: Maxwell MacMillan Canada.

Liedtke, W., Kallio, P., & O'Brien, M. (1998). Confidence and risk taking in the mathematics classroom (grade 1). *Primary Leadership, 1*(2), 64–66.

O'Brien, T. (1999). Parrot math. *Phi Delta Kappan, 80*(6), 434–438.

Paulos, J. (1990).*Innumeracy: Mathematical illiteracy and its consequences.* New York, NY: Vintage Books.

Reys, R. (1971). Consideration for teachers using manipulative materials. *Arithmetic Teacher, 18*(8), 551–558.

Reys, R., et al. (1999). Assessing number sense of students in Australia, Sweden, Taiwan, and the United States. *School Science and Mathematics, 99*(2), 61–70.

Romberg, T. (1990). Evidence which supports NCTM's curriculum and evaluation standards for school mathematics. *School Science and Mathematics, 90*(6), 446–481.

Schwartz, S. (1994). Calendar reading: A tradition that begs remodeling. *Teaching Children Mathematics, 1*(2), 104–109.

Spungin, R. (1996). First and second grade students communicate mathematics. *Teaching Children Mathematics, 3*(4), 174–179.

Van de Walle, J. (1990). Concepts of number. In J. Payne (Ed.), *Mathematics for the young child* (pp. 63–87). Reston, VA: The National Council of Teachers of Mathematics.

Willoughby, S. (1990). *Mathematics education for a changing world.* Alexandria, VA: Association for Supervision and Curriculum Development.

CHAPTER 42

TEACHING MATHEMATICS FOR PEACE

David Wagner

Although mathematics education is important to me, I consider the quest for peace to be my vocation. I can work at bringing peace to my world through mathematics education. I recognize that my goal of a peaceful world is likely unattainable in my lifetime, but perhaps it is the mathematician in me that is content to approach the inaccessible and the infinite.

The idea of relating the teaching of mathematics to an interest in peace is strange to many people. I can tell by the looks on the faces of those whom I share my passions with. To many people, mathematics is culturally and socially neutral—sterile. I am quite familiar with this view of mathematics because it was my own attitude as well when I started my mathematics teaching career.

In my first five years of teaching in Alberta, I strived to be a caring presence in my classroom. I listened to my students and tried to help them understand both their world and their mathematics, but separately. I saw myself as a peacemaker who happened to be teaching mathematics, which was, for me, a neutral subject that was unconnected to my students' experience outside the classroom. I disagreed with other mathematics teachers who

Selected Writings from the Journal of the Mathematics Council of the Alberta Teachers' Association, pages 351–356

constantly reminded their pupils that mathematics is the most useful of the subjects. Perhaps elementary and junior high mathematics is used in the real world, but high school mathematics seems quite foreign to the experience of most adults. For me, the value of the subject was only found in its parallels to outside experience, not in its connection to outside experience.

My interest in the connection between mathematics and society has been sparked by my recent experiences with the new Alberta mathematics curriculum in juxtaposition to my experiences teaching mathematics in rural Swaziland for two-and-a-half years.

I realized that methods and values in the mathematics classroom were connected to methods and values in the outside world. I also began to see that mathematics is often both implicated in violence and used in the name of peace.

D'Ambrosio (1994, p. 443) notes the importance of mathematics in this century's enormous technological advances.

> Humanity has seen the smallest reaches of imagination and talks about reaching the boundaries of the universe. And yet, this same century has shown us a despicable human behaviour.... Much of this paradox has to do with an absence of reflections and considerations of values in academics, particularly in the scientific disciplines, both in research and in education.

I agree that more reflection is needed in and about the classroom. The reason I am sharing my thoughts about peace and mathematics education is to promote this reflection. In this article, I will ask some simple questions that have no simple answers. However, I assert that considering these questions is a good beginning for thoughtful adjustments to teaching practice.

WHAT IS PEACE?

Before thinking about how we can structure mathematics classroom experiences to promote peace, it is important to ask what we mean by peace. The word peace can be used to describe vastly different things. After two airplanes were flown into New York's World Trade Center towers on September 11, 2001, I heard a radio reporter on the scene exclaim in panic, "The worst thing is that no one knows what will happen next." I wondered if anyone ever knows what will happen next. In the days that followed, I reflected on ways of living in an unpredictable world.

One approach to a complex existence in an unpredictable world is to try to control the environment and build a network of security. With this approach, the world is structured around a certain view of what a peaceful world should look like. This approach equates peace with security. Wars against terrorism and jihads against infidels are extreme examples of this approach.

An opposite approach is to become aware of the complexity and change-ability of the world and to find our place in it. This approach focuses on awareness and equates peace with harmony. The former approach tries to mould the world to match certain standards. The latter approach involves moulding ourselves to the world in order to find our place in it.

SECURING A RIGHT WORLD—MATH POWER

How would mathematics pedagogy look if we were trying to secure a peaceful world? The security approach embodies teleological ethics, in which the ends justify the means. Mathematics and other tools become important in proportion to their utility for pursuing an end. Although this view favors applied mathematics, pure mathematics is also valued because applications have been found as a result of its development.

If we think that the ends justify the means, then the answer to a question will be of more interest than the process of arriving at the answer. Multiple-choice and numeric response examinations, where students' correct answers are credited no matter how they are found, embody this view. The purpose of a mathematics class would be to train students to use technology and knowledge efficiently and powerfully. Words relating to effectiveness and power would appear in our scoring rubrics and in our classroom resources and decorations.

Mathematics does provide tools that can be used to solve problems in our world, and problem solving has long been an important part of mathematics pedagogy. What if we used our mathematical word problems to have students make calculations about fair wealth distribution or efficient structuring of peacekeeping forces? Gerofsky (1996, p. 41), in her exploration of mathematical word problems, questions this and other expectations of word problems:

> The claim that word problems are for practicing real life problem solving skills is a weak one, . . . unlike real life situational problems, no extraneous information may be introduced.

> Still, word problems that express an interest in a peaceful world are preferable to, for example, missile trajectory calculations.

We must keep our students aware that their word problem contexts are unrealistically simple. When they address real problems outside of the mathematics classroom, we would expect them to consider all the related information. If they select a few numbers that relate to the situation and simply manipulate them using mathematical algorithms, they will likely exacerbate the problem rather than solve it.

Mathematics may provide helpful tools for solving local and global problems, but there are other good tools as well. It is wise to have a diversity of tools available and to use them appropriately.

BUILDING AWARENESS—ETHNOMATHEMATICS AND CRITICAL MATHEMATICS

How would mathematics pedagogy look if we were interested in harmony and finding our place in a world that we view as an interconnected system? Mathematical tasks would be evaluated by their potential for building awareness. There are two recent movements in mathematics education scholarship that prescribe more awareness: ethnomathematics and critical mathematics education.

Ethnomathematics is interested in the mathematical activity of diverse cultural groups. It can be incorporated into the curriculum in a number of ways, including using mathematically rich artifacts from other cultures as starting points for mathematical discussions or tasks. This movement assumes that an awareness of other cultures' processes helps us see our place in the world. The International Study Group on Ethnomathematics can be found online at www.rpi.edu/-eglash/isgem.htm.

Critical mathematics education involves a wide array of activities and reflections that uncover mathematics education's assumptions and values. For example, Borba and Skovsmose (1997) criticize the typical fixation on single right answers in mathematics classrooms. They call it the ideology of certainty and worry about how it prepares students to think about problems outside the classroom. Borba and Skovsmose suggest that mathematical tasks be structured to make a variety of approaches viable and a diversity of answers correct.

With their approach, the teacher is involved in critical thinking. Their assumption is similar to mine—that students will benefit from having teachers who are more aware of the connections between mathematics and society. D'Ambrosio (1998, p. 72) suggests that teachers engage students in critical dialog about the connections between mathematics and their world.

I propose questions as "What do you think of a current event or a philosophical question?" let some discussion follow and then come with another question, "What does mathematics have to do with this?"

THOUGHTFUL ACTION—A MIDDLE WAY

The two views of peace that I have described are quite polar. I put myself somewhere in the middle of these two approaches to peace, closer to the harmony end of the spectrum. Although I consider it necessary to be aware of our place and the interconnectedness of things in a constantly changing, complex world, we still need to act. If we spend all our time merely noticing, we cannot participate. Somewhere in between the two extremes of thoughtless action and detached analysis is a place for thoughtful action. A degree of engagement with the world is necessary for more thorough understanding and higher awareness.

There are three planes of thoughtfulness. While engaged in mathematical activity, students can be mindful of the connection between different forms of mathematical thinking, the importance of their individual choices within their activity and the connection between generalizations and the particular cases they relate to.

Mathematics curriculum is typically fragmented. It divides mathematical activity into pure and applied streams, broad categories within these streams (such as geometry and algebra), smaller categories in each broad category (for example, in algebra, there are equations, factoring, and so on) and concepts in these narrower categories (for example, in factoring, there are differences of squares, second-degree trinomials, and so on).

When students are given a problem, they are expected to classify it according to known types. In doing this, however, they ignore the people and places involved. After classification, they follow a standardized procedure designed for problems of that particular category. Only one result is acceptable. By drawing boundaries between classes of problems and between useful and extraneous information, students are encultured to ignore the interconnectedness of mathematics and the connections between problems and their mathematical or real-life contexts.

Connection-making in mathematics can be encouraged by presenting students with open-ended questions and problems, and not giving any clue as to how they might be answered. These tasks could be given either as formal assignments or through informal class discussion. Either way, students would need to consider the breadth of their mathematical experiences to find some mathematical knowledge that might apply. They would then need to decide how it applies and how they can know if they are applying it correctly.

These tasks cannot only open up the possibility of seeing connections in mathematics, but help students develop ownership of their mathematical ideas. The mathematician is connected to the mathematics. As students listen to each other's mathematical ideas or look at each other's mathematical writing, they will see a diversity of viable approaches and forms of presentation.

With this realization, they may see that the form of mathematics is closely connected to the people who construct it in response to particular problems.

Although the connection between mathematicians and their mathematics grounds the discipline in particular contexts, generalization is at the heart of mathematical processes. Generalizations can, however, be grounded in context for students. As we direct students to make generalizations, they can become aware of the limitations of generalization as well as the power of generalization. When students are simply presented with a rule to use, they struggle with its application and wonder when and how to use the rule. By contrast, if they are engaged in the activity of constructing rules, they are more likely to see the connections between general rules and the contexts from which they were born or in which they may be applied.

Skovsmose (2000) describes how traditions of activity in mathematics classrooms format society by providing a framework for solving problems in the world outside the classroom. It is my hope that mathematics classrooms can format the world for peace by establishing traditions of classroom mathematics activity that encourage thoughtful action.

CONCLUSION—DOING GOOD

In my first year of teaching, a grade 11 student asked me, "Did I do good on yesterday's math test?" I responded playfully, "No. Doing good is feeding the hungry, clothing the naked, healing the sick and bringing hope to the poor. You did none of that on this test. You did, however, do well. You got 87 percent." This little experience has prompted me to wonder how I, as a mathematics teacher, can do some good for my students and for my world—how I can be an agent for peace. It is my responsibility as a teacher to set tasks for my students in such a way that, if they follow my instructions well, they will be doing something good.

REFERENCES

Borba, M., & Skovsmose, O. (1997). The ideology of certainty in mathematics education. *For the Learning of Mathematics, 17*(3), 17–23.

D'Ambrosio, U. (1994). Cultural framing of mathematics teaching and learning. In R. Biehler et al. (Eds.), *Didactics of Mathematics as a Scientific Discipline* (pp. 443–455). Dordrecht, NL: Kluwer.

D'Ambrosio, U. (1998). Mathematics and peace: Our responsibilities. *Zentralblatt für Didaktik der Mathematik, 30*(3), 67–73.

Gerofsky, S. (1996). A linguistic and narrative view of word problems in mathematics education. *For the Learning of Mathematics, 16*(2), 36–45.

Skovsmose, O. (2000). Aporism and critical mathematics education. *For the Learning of Mathematics, 20*(1), 2–8.

This chapter was originally published as:

Davis, B. (2005). Emergent insights into mathematical intelligence from
cognitive science. *delta-K: Journal of the Mathematics Council of the
Alberta Teachers' Association, 42*(2), 10–19.

CHAPTER 43

EMERGENT INSIGHTS INTO MATHEMATICAL INTELLIGENCE FROM COGNITIVE SCIENCE[1]

Brent Davis

Editor's Note: Brent Davis is professor and Canada research chair in mathematics education and the ecology of learning with the Department of Secondary Education at the University of Alberta. He taught junior high mathematics and science through the 1980s after completing his undergraduate work and before beginning his graduate studies (all at the University of Alberta). He currently researches and teaches courses in mathematics education, cognition and curriculum.

In this article, I point to a handful of recent developments in cognitive science in an attempt to highlight how they might contribute to a rethinking of the nature of mathematical intelligence. In the process, I also offer some preliminary speculations on what these developments might mean for the teaching of mathematics.

I must begin with a disclaimer: Cognitive science is a burgeoning field. It is really only a half-century old, and it has just taken off in the last decade,

Selected Writings from the Journal of the Mathematics Council of the Alberta Teachers' Association, pages 357–375
Copyright © 2014 by Information Age Publishing
All rights of reproduction in any form reserved.

spurred along by the invention of technologies that enable researchers to peer into brains in real time. Some surprising observations have been made—ones that have compelled researchers to question and reject an array of deeply entrenched assumptions about how people learn, how brains work, what thinking is, and what intelligence is all about.

Cognitive science isn't actually a field. The phrase is an umbrella term that stretches across certain research in artificial intelligence, linguistics, cultural studies, philosophy, experimental psychology, neurology, neurophysiology, ecology, cybernetics, and complexity science—to mention a handful of the more prominent areas. In brief, the emergence of cognitive science as a domain of research might be taken as recognition that investigations into such phenomena as learning and intelligence require a transdisciplinary approach. None of the above-mentioned fields on its own has the capacity to answer the big questions about human cognition.

With regard to education, this move toward transdisciplinarity is a significant development. For most of the past century, educators relied almost exclusively on psychology for their formal definitions of intelligence, the tools to measure it and advice on how to nurture it. As it turns out, much of that advice was good, despite some troublesome assumptions. But much of it was also a bit problematic. In particular, the reliance on psychology has contributed to some deeply ingrained and unfixable dichotomies—between, for example, skills-based and understanding-oriented instruction, or between teacher-centred and learner-centred instruction. Most of what we've borrowed from psychology compels us to take one side or the other, or to live with some uncomfortable compromise.

But, as John Dewey (1910) noted a century ago, we never solve such radical splits. We simply get over them. So none of what I present here should be taken as an argument for or against, for example, skills-based or student-centred instruction. Rather, I'm actually arguing that recent cognitive science provides us with a way of sidestepping these sorts of quagmires and opening spaces for more interesting and productive discussions.

Before going too much further, it's important to be clear about how cognitive science defines intelligence—and let me emphasize that this definition represents a break with popular and psychology-based orthodoxies. For instance, for the cognitive scientist, intelligence is not what IQ tests measure, as might be inferred from the fact that some patently unintelligent machines are able to perform at the genius-level on most IQ tests. As well, an individual's IQ score can vary by as much as 50 points, depending on the time of day, warm-up activities, hunger, thirst, and so on.

Cognitive science uses a much broader definition: Intelligence is the capacity to respond to new situations in ways that are not only appropriate, but that open up new spaces of possibility. Intelligence, then, is not merely about getting the right answer to a trick question. It is about coming up with solutions to real problems, with answers that go beyond routine responses and that enable the person to go further than he or she could before taking on the problem. Intelligence, in these terms, is about breeding new possibilities, opening up new vistas, not about responding to mind-twisters devised by others.

POINT #1: CONSCIOUSNESS IS SMALL.

One solid, rigorously demonstrated conclusion of the research out of 19th- and 20th-century psychology was that human intelligence is greatly constrained by some rather severe biological limitations on consciousness. In particular, a frequently cited factoid is that humans are capable of juggling a maximum of 6 or 7 details in their heads at a time, but can only do that for about 15 seconds before some or all fall away. This 6–7 limitation is especially interesting when considered against the total number of sensory receptors in an average human body, which is estimated to be somewhere in the 10 to 20 million range. (Some researchers contest that the total is in the order of 10^{10}; see Norretranders, 1998.) To drive that point home, fewer than one in every million sensory events (and the number may be closer to one in a billion events) ever rises to consciousness.

This insight is actually an old one, thoroughly demonstrated in the 1800s. It was a key tenet in the emergence of discourses as diverse and incompatible as B. Skinner's behaviourist psychology and Sigmund Freud's psychoanalysis, both of which were under development about a century ago.

A brief demonstration might be useful here. First read the following instruction, then follow it. Close your eyes and imagine two dots, then three, then four, then five, then 20, then 100.

Chances are that your image of three was arranged in a triangle, that your four was a square, your five was either a pentagon or a square with a dot in the middle. You shouldn't have been able to imagine 20 or 100, but you might have invoked a strategy like a grid to think of these quantities in terms of smaller, more readily imagined amounts.

Now repeat the tasks, this time with all of the imagined objects in a single row—no grids, polygons or subgroupings allowed. You will likely max out at five. I know of no one who can imagine six side-by-side, ungrouped objects.

There is some compelling evidence that the capacity to imagine small quantities might actually be built in. It's been established that very young babies can discern between one object and two objects, likely between two and three, and perhaps between higher quantities (see Gopnik, Meltzoff, & Kuhl, 1999). It also seems that we share that ability with lots of mammals, some birds, and a few other species.

The realization that consciousness is so tremendously limited is one of the principles that undergirds the highly parsed structure of modern curricula, especially mathematics curricula, which have been the subject of more psychologically based research than any other topic area. (In fact, math curricula have been the focus of more research than all the other areas combined.) The practice of structuring a lesson around one small topic, such as adding integers, long division, or factoring a trinomial, originated in part from the embrace of the factory model of schooling, but the bolt that holds it in place is research into the limitations of consciousness.

In fact, that research is so compelling that I have structured this article around it. My psychologist colleagues tell me that the best I can hope for is that you'll retain at most six or seven bits of information. So I've limited my foci to seven points.

Before moving on to the second of those seven, I want to nod to a few implications of this first point for our efforts to nurture mathematical intelligence. Two implications:

- We have to limit the amount of new information in any given learning event.
- We have to use design learning in ways that help learners focus their attention on what really matters.

We've already mastered the first point. The second one is a little more complex than it might appear.

There is a connection between intelligence and discernment. In fact, intelligence was originally conceived as the capacity to discern what is really important in a situation. As it turns out, there are teaching strategies that can support people's discernment-making abilities—that is, that help them be intelligent.

Anne Watson of Oxford University and her husband John Mason of the Open University in the United Kingdom have done considerable work on this issue. An example based on their work is the following:

Compare the two lists here:

$$3 : 3$$
$$1.7 : 1.7$$
$$x : x$$
$$e^{\pi i} : e^{\pi i}$$

and

$$3 : 3$$
$$6 / 4$$
$$2 \text{ to } 9$$
$$\underline{1.2n}$$
$$0.36n^2$$

The point Anne intends through this sort of comparison may seem counterintuitive. She argues that the first list might be a better pedagogical tool because it is designed to assist the learner to make a key mathematical discernment. In contrast, the second list obscures the discernment. Too much is going on. Her argument is that if there is not much variety, we generalize. If there is too much variety, we categorize. And for the most part, the intelligent mathematical action is about making the sorts of discernments that enable generalization, not categorization.

The first list sets itself up for questions like, What's the same? What's different? Is it always, sometimes, or never true? Are there examples that don't fit the pattern? In other words, even though it might look like there is less there, it's much easier to strike up a conversation about what is presented— that is, to pinpoint and emphasize what really matters.

The fact that consciousness is limited also points to the need for repetition and practice, which is something that traditional mathematics teaching has done well and that reform teaching has often done less well. Let me underscore this point.

POINT #2: INTELLIGENCE RELIES ON THE CAPACITY TO ROUTINIZE KNOWLEDGE AND PROCEDURES SO THAT CONSCIOUSNESS IS FREED UP TO WORK ON OTHER TASKS

Consider this sequence of numbers:

$$1, 11, 21, 1211, 111\,221, 312\,211$$

What comes next?

The following discussion will be more meaningful if you actually try to respond to the question.

When you first take on this sort of problem, your brain activity spikes and continues to do so until you either find a solution or give up on it. If you do in fact come up to a solution, your brain very quickly works to routinize things by delegating the task to a sub-regions or clusters of sub-regions while the rest of the brain returns to its usual near-resting state.

The realization of the importance of routinization for intelligence is quite a recent development. Or, at least, the proof for it is recent. Now that we can watch the brain in action, we can see that brains respond in different ways to novel situations. When presented with an unfamiliar problem or context, all brains begin to fire rapidly. And the whole brain fires when it meets a novel problem, not just parts of it (see Calvin, 1996). I'll return to this point later.

The quality that most distinguishes the intelligent brain from the unintelligent brain is that it quickly settles on what's important, routinizes it, and assigns it to subconscious processes. So, in terms of the profile, there's an initial spike of whole-brain activity that settles very quickly into lower-level, region-specific activity. By contrast, the unintelligent brain continues at a high level of whole-brain activation, apparently groping for what's important.

The happy thing is that the brain can improve its abilities to make vital discernments. One key is practice. Let me tell you a quick story.

Each week for the past three years, I've been meeting with Krista, an adolescent, about her mathematics. When I first met her, she was in grade 9 and was unable to see patterns in lists of numbers like

$$1 \quad 4 \quad 9 \quad 16 \quad 25 \quad 36 \quad 49 \quad 64\ldots$$
$$1 \quad 1 \quad 2 \quad 3 \quad 5 \quad 8 \quad 13 \quad 21\ldots$$

It didn't take much probing to discover that a large part of the problem lay in the fact that she couldn't work with even single-digit numbers reliably. Calculations like $6 + 7$ and $5 \cdot 3$ were problems for her.

This meant that she was failing mathematics badly, and had been doing so since grade 1. The school board had been testing her annually and she had had at least eight years of focused help with special needs teachers. Yet in grade 9 she couldn't do things that are routinely expected of children in grades 2 and 3. I decided to work with her because I thought she might be one of those interesting cases of people with location-specific brain injuries, which I imagined could be a fascinating thing to study from the point of view of an educational researcher.

It turned out that I was quite mistaken in this suspicion.

The first year of our association was spent on what I thought of as educating her intuition—a phrase that refers to engagement with processes and situations intended to help one develop a feeling for quantities and manipulations of quantities. For instance, we spent a total of about six hours (a month's work together) figuring out different ways to estimate the number of grains of rice in a bowl. We spent considerably more time on paper-based activities, such as folding, cutting, assembling, and dismantling. We did anything I could think of that might be interpreted in terms of basic operations on whole numbers, integers, and rationals.

Significantly, I insisted on practice. Krista had daily homework exercises, which included flashcard drill on multiplication facts, writing out explanations of why things seem to work how they work, spending time on non-routine problems and so on. Six months into our work together, the psychometrician who had worked with her for three years was surprised to note that her score on the mathematics portion of the test he used had soared from grade 2.3 (at the end of grade 8) to grade 1 0.8 (in the last half of grade 9).

I cite those statistics cautiously. Krista really was not working at a grade 10.8 level. (I had no access to the test, so I cannot comment on what was really being assessed.) But the numbers do suggest that something important had happened. At the time of this writing, she is enrolled in the grade 12 applied stream mathematics course. Her average in mathematics is consistently in the 80 percent range. That's gratifying, but what is really exciting is that while she's writing an exam she can now tell whether or not she's doing well. Two years ago, she couldn't tell you what sort of grade she might get on a test. If she passed (which was not often), she attributed it to luck. Now she can predict her score with a high degree of accuracy.

I recently asked her about her new capacity to predict her exam results and how she could feel so sure of her predictions. She responded that a few years ago, her brain would "just go crazy in math exams." She couldn't focus, she couldn't remember. Now, in her exact words, her "brain just goes calm" when she realizes she can respond to the questions.

I haven't had a chance to monitor her brain activity, but I'm fairly confident in the assertion that two years ago, in a test situation, her brain was spiking throughout the test, to no avail. Now, it's spiking and settling in— just as an intelligent brain is supposed to do.

As for teaching implications, a central point is one that we all know deeply—if we want to be proficient in an activity, much of it has to be routinized. Be it playing hockey, playing the piano or adding fractions, certain levels of practice are needed not only to develop the basic mechanical competencies but to get a feel for what one is doing.

There is one caveat here. Practice must be contextualized. The brain resists learnings that lack context or that are not anchored in purposeful activity for reasons that I will develop later. But first, I want to make one more point on the role of practice.

POINT #3: MATHEMATICAL GENIUS (IN FACT, ANY CATEGORY OF GENIUS) IS, IN GENERAL, MUCH MORE ABOUT FOCUS AND PRACTICE THAN IT IS ABOUT INNATE, BIOLOGICALLY ROOTED TALENTS OR GIFTS

Rena Upitis of Queen's University often asks audiences to do the following: Think about something you're really, really good at. Now answer two questions: Do you practise it? And did you learn it at school?

You probably said yes to the first and no to the second.

The fact of the matter is that talent and genius are dependent on practice. So long as the basic biology of the brain isn't compromised, an otherwise typical person can obsess his or her way toward genius in some domain of activity because the brain is what neurologist and psychologist Merlin

Donald (2001) describes as a "superplastic structure" whose resources can be co-opted and reassigned through dedicated practice. If those resources are focused on mathematics—or golf, or the cello, or plumbing—otherwise ordinary individuals can achieve quite extraordinary feats after years of focused effort.

One interesting statistic in this regard is that the rates of mental illness, particularly obsessive compulsive disorders, are several times higher among elite mathematicians, musicians, athletes, and other high performers (Richardson, 1999). This is not to say that obsession is a good thing; it is merely to underscore that, biologically speaking, most of the super geniuses of the world began life with capacities that were very similar to the ones the rest of us were born with.

I'm not suggesting that people are all born with the same cognitive architectures or that there's no such thing as natural mathematical talent. Clearly, such notions are misguided. The point is that most of the differences that we observe among adults have more to do with habits of mind than with raw horsepower. A person who begins with typical ability but who is obsessive about mathematical concepts can be a much better mathematician than a person with considerable natural ability but no inclination to develop his or her own capacities.

I return to Krista here. Two years ago, she was mathematically inept. She is far from a mathematics genius, but she is now mathematically capable. And just being capable means that her mathematical intelligence has skyrocketed.

The claim here is that one can become more intelligent, and it is an assertion that flies in the face of some deeply engrained beliefs and practices. IQ tests, for instance, are developed around the assumption that something innate is being measured, not something that can be honed through practice. Howard Gardner's theory of multiple intelligences is anchored in the assumption that differences in human capacities for mathematics, interpersonal relations, music, and so on are all rooted in variations among inborn brain structures. And we are confronted with tale after tale based on the assumption—and that advances the belief—that mathematical talent is innate. Consider the popular Hollywood films *Good Will Hunting*, *Little Man Tate*, *A Beautiful Mind*. The implication in these stories often seems to be that education is supposed to stay out of the way of a genius.

But the fact of the matter is that there are no documented cases, anytime or anywhere, of a full-blown mathematical genius who became that way without extensive practice and some formal education. It simply doesn't happen. By contrast, there is no shortage of evidence to support that assertion that mathematical intelligence is not fixed. We can make ourselves smarter.

Teachers can play an important role here. Emotions like curiosity and pleasure can be infectious. In fact, all emotions are. We humans are prone to being caught up in others' emotional expressions. So it's worthwhile asking yourself what emotions are you expressing in your classroom toward the mathematics? Enthusiasm? Indifference? Amusement? Obsession? (See Damasio, 1994 for a discussion of the relationship between emotion and logical competence.)

On the issue of making ourselves and our students smarter, it turns out that there are critical moments in life for nurturing one's intelligence.

POINT #4: BRAINS ARE CONSTANTLY CHANGING—
AND THEY CHANGE MOST RAPIDLY IN THE FIRST FEW
YEARS OF LIFE AND DURING EARLY ADOLESCENCE

What do you see in the inkblot below?

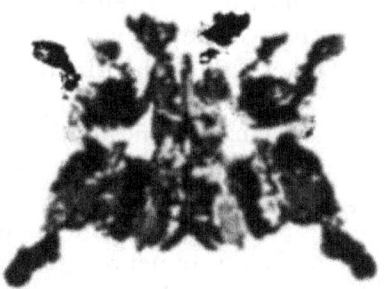

Now, what if I tell you that this is actually a picture of two people squatting back-to-back holding ducks in their laps?

Once you apply this interpretation into the image, you can't help but see what you were told to see.

In other words, I have affected your brain structures by imposing a specific interpretation. That interpretation is compelling because your brain immediately went to work to activate the associations necessary for you to perceive the image as described. That is, your brain is physically different because of my intervention. Every lived experience entails a physical transformation of your brain.

Now consider such common turns-of-phrase as "taking things in," "attaining one's personal potential," and "brain as computer." We have dozens of such expressions, all of which assume and assert a fixed brain architecture—as though the brain were some kind of preset and unchanging receptacle. Nothing could be further from the truth. Some details:

1. Brains account for about 5 percent of the body's weight, but consume about 20 percent of the body's energy. In other words, they're incredibly physically active, and when I say physical, I mean physical.

Things are actually moving about up there. On an MRI the brain looks vastly more like an anthill than it does a computer.

2. Infant and adolescent brains operate in overdrive, consuming two to three times as much energy as a typical adult brain. The claim has been made that if only a three-year-old could have an adult's knowledge and experience, all of the great problems of science would be solved in short order (Gopnik, Meltzoff, & Kuhl, 1999). They're geniuses. Super geniuses. And we might expect as much. They have to develop language, put together a theory of how the world works, and master the complexities of interpersonal relationships in just a few years. None of us adults can do that.

   Until about five years ago, it was believed that brain activity undergoes a gradual and steady decline from toddlerhood to adulthood. But some recent research has demonstrated that there's a renewed surge of brain activity in early adolescence, especially around junior high age. They're geniuses again.

   Some, including Pinker (1997), theorize that this second surge in brain growth and activity is an evolutionary response to the need to cope with some new and fairly significant distractions. Whether or not that's the case, it would seem to make sense to take advantage of their amplified cognitive powers.

3. One of the differences between intelligent brains and not-so-intelligent brains is the density of neurons. Einstein's brain is pretty normal in size. There are no odd bulgy areas. However, Einstein's neurons were more tightly packed and more intricately interconnected than typical brains.

   It turns out that neuronal interconnections can be grown. In fact, whole new neurons can be grown. These things happen in response to experience and need. As Canadian neurologist Donald Hebb (1949) wrote 50 years ago, "Neurons that fire together, wire together." A key here is, once again, contextualized and rich practice.

Considered together, the above points underscore an important conclusion: Your brain, at this moment, is different from the brain that you had when you started reading this article. Every experience you have contributes to the ongoing restructuring of the brain. Put in somewhat different terms, the brain isn't hardware and knowledge isn't data or information. These popular and pervasive ways of talking about learning and knowledge are way, way off.

In terms of implications for teaching, the sorts of things that contribute to increased neural density and interconnectivity are the sorts of things that force learners to think outside the box. Such activities include sustained engagements with mathematical puzzles, attending to the different ways that concepts can be interpreted and doing things that are unfamiliar and non-routine. In particular, for a learner to develop mathematical intelligence and robust mathematical understandings, she or he has to be aware of how mathematical concepts can be interpreted in different ways. I turn to an example of this presently.

POINT #5: HUMAN THOUGHT AND LEARNING ARE MAINLY ASSOCIATIVE, NOT RATIONAL—THAT IS, ANALOGICAL, NOT LOGICAL. MATHEMATICAL INTELLIGENCE AND CREATIVITY ARE ROOTED IN THE CAPACITY TO SELECT AND BLEND APPROPRIATE ASSOCIATIONS

What is multiplication?

It turns out that this question has at least a dozen distinct responses, all of which are correct. In a recent workshop with a group of K–12 teachers, the following list was generated:

- Repeated addition: $2 \times 3 = 3 + 3$ or $2 + 2 + 2$
- Grouping process: 2×3 means "2 groups of 3"
- Sequential folds: 2×3 can refer to the action of folding a page in two and then folding the result in 3
- Many-layered (the literal meaning of multiply): 2×3 means "2 layers, each of which contains 3 layers"
- Grid-generating: 2×3 gives you 2 rows of 3 or 2 columns of 3
- Dimension-changing: a two-dimensional rectangle of area 6 units—2 can be formed when one-dimensional segments of lengths 2 units and 3 units are placed at right angles to one another
- Number-line-stretching or -compressing; $2 \times 3 = 6$ means that "2 corresponds to 6 if a number-line is compressed by a factor of 3"
- Rotating: for example, multiplication by –1 means rotate the number line by 180°—which reverses its direction

This list is far from exhaustive. It could easily be extended to include interpretations that are needed to make sense of the multiplication of vectors, matrices, and other familiar mathematical objects.

It's important to emphasize that all of these interpretations point to distinct actions. They can be mapped onto one another, but they cannot be reduced to one another. And it's important that they're distinct. The power of mathematical processes like multiplication is not that they can be reduced to a single definition or process, but that they actually consist of clusters of interpretations.

There are some major teaching implications here. For most of the past four centuries, school mathematics has been organized around the assumption that mathematical learning proceeds logically and sequentially, like the construction of a building. Think of some of the metaphors that tend to be used: solid foundations, the basics, a cornerstone of logic, the structure of knowledge, and building and constructing ideas.

There is a popular assumption that the history of mathematics unfolded logically and sequentially as well. Nothing could be further from the truth. The more recent histories of mathematics underscore this point (for example, Mlodinow, 2001; Seife, 2000). The great leaps in the emergence of mathematical knowledge didn't occur through moments of logical insight, but through the development of new analogies. The concept of multiplication, for instance, has grown over the centuries as new interpretations have been proposed and blended into the existing definition (see Lakoff & Nunez, 2000; Mazur, 2003).

What does this mean for mathematical intelligence? Let me preface my answer to that question with a quick visit to the field of artificial intelligence (AI) research. AI started in the 1950s when computers were beginning to outperform their programmers on some difficult mathematical tasks. Based on this early success, computer scientists and science fiction writers began to make confident predictions about the future of machine intelligence, forecasting that electronic intellects would soon dwarf flesh-based intellects.

Fifty years later, we see that they were spectacularly wrong. The reason for the collective error is instructive: They assumed—as did the original IQ-test inventors, many curriculum designers, and writers of *Star Trek*—that logic is the root of intelligence. The belief was supported by their own experiences. Like most people, they found logical tasks very difficult.

And there is a reason why they're difficult—it's because our brains are analogical. That is, the root of intelligence is not logic, but the capacity to make new associations among experiences—through storying, analogy, metaphor, and other figurative devices. Ours is an intelligence that is capable of logic, but that capacity rides on top of very different sorts of competencies.

There's a rather shocking implication here—our current mathematics curriculum might be stifling mathematical intelligence, not supporting it, an assertion that might be linked to Point #2. Brains resist decontextualized, overly abstract constructs. When the brain meets something new, it works very hard to weave the experience into the web of existing associations. But if the new topic comes without obvious associations, then it can't be learned on any level other than the mechanical. But human brains are notoriously unreliable when it comes to rigidly procedural knowledge.

One of the major implications for teaching is something that we can't do much about at the moment. Mathematics curricula are structured after the model of the logical proof. You begin by developing the premises or basics and proceed by assembling those premises into more sophisticated truths. In terms of the analogical nature of human cognition—and, in fact, in terms of the emergence of mathematical knowledge—this instructional sequence amounts to putting the cart before the horse. Logical justification has always come after the development of a new way of interpreting things.

Speaking of the model of formal logic, did you know that Euclid's five axioms aren't sufficient for his geometry? He missed some necessary axioms because he was thinking analogically, not logically. About a century ago, David Hilbert (1988/1899) identified several others that are needed for Euclid to be logically complete. It took more than two millennia for mathematicians to notice the gap. Why? Because humans are much more analogical than logical.

But, of course, we can't wait for full-scale curriculum restructuring. In the meantime, to nurture your students' mathematical intelligence, I recommend that you work with them to try to uncover the associations that have been built into mathematical concepts. Start with addition. What are some of the ways we interpret adding? (If you want one answer to that question, you might check Lakoff & Nunez, 2000.)

Let me re-emphasize that robust understandings and flexible applications of mathematical ideas—that is, the underpinnings of mathematical intelligence—are completely dependent on access to the range of meanings that are knitted together in a concept.

POINT #6: THE REAL POWER OF MATHEMATICS ARISES IN CLEVERLY STRUCTURED SYMBOLIC TOOLS, WHICH COLLECT TOGETHER BUT CONCEAL THE ARRAYS OF INTERPRETATIONS AND EXPERIENCES THAT UNDERLIE CONCEPTS

Close your eyes and imagine $\sqrt{15}$.

It's not so easy. And yet, as it turns out, $\sqrt{15}$ is utterly imaginable. Barry Mazur, a Harvard University mathematician, explains how in his 2003 book *Imagining Numbers*. Space prohibits an adequate summary of his discussion, but I can mention that to imagine $\sqrt{15}$, you have to know that the concept relies on the notion of multiplication-as-rotation. That is, multiplication by a negative means a 180°-rotation, and multiplication by two negatives means a 360°-rotation (which takes you back to the starting orientation). One more detail is needed: one might think of a square root as half of a multiplication, as indicated by the exponent of 1/2. If you blend these ideas—as mathematicians did a few centuries ago—you get the root of a negative is a half of a 180°-rotation, which is a 90°-rotation, which generates the complex plane. The roots of −15, then, are the points that are just beneath the +4 and just above the −4 on the i axis of the complex plane.

Lakoff and Nunez (2000) take this sort of thinking even further and demonstrate how it's possible to imagine Euler's formula: $e^{\pi i} + 1 = 0$. Even more significantly, they attempt to impress that this very complex notion is rooted in bodily action, like moving forward, spinning and so on.

My point here is not really that such imaginings are doable nor that we should be doing them in our math classes—although I do believe that they are doable and that we should be doing them in our math classes. It is, rather, that knitted into these symbols are an incredible array of experiences and possibilities. They are intelligently designed tools that greatly expand what we are able to do.

To put a finer point on it, tools such as language, mathematical symbols, and calculators aren't just the product of human intelligence—they are bestowers of intelligence. Humans with language are much more intelligent than humans without language. And, although I don't have nearly the raw intelligence of Archimedes or Newton or other mathematical giants of history, I can do things that they didn't even imagine doing because of the tools they helped to build.

Now, by psychologistic definitions of intelligence, you might argue, the fact that I can solve an unsolved differential equation by typing it into Maple does not make me a mathematical genius. And according to measures of IQ, that's true. But going by the cognitive science definition of intelligence (that is, intelligence is the capacity to respond to new situations in ways that are not only appropriate, but that open up new spaces of possibility), intelligence is about an ever-growing horizon of possibility, not the capacity to master what's already been established. What's more, intelligence is obviously not an individual phenomenon. Not only can we make ourselves smarter, we can contribute to the intelligence of others by giving them access to the tools of our intelligence. On this point, it's important to emphasize that we're routinely asking high school students to perform mathematical operations that were accessible only to the geniuses of a few centuries ago.

Now, to be clear, I'm not suggesting that technology on its own makes us smarter. Giving an iMac to a caveman would be a bit of a waste. And we have probably all seen people grab a calculator in order to add 0 or to multiply by 1. Those are decidedly unintelligent acts.

The point is, rather, that intelligence is not some mysterious quantity that's locked in our heads. Intelligence is about appropriate and innovative action, and to be intelligent in mathematics in this day and age requires more than a mastery of the conceptual tools that have been developed by our forebears. Intelligence is greatly enabled by a facility with contemporary tools. That's certainly true among research mathematicians. Our mathematics pedagogy hasn't adapted to take that into account, even though electronic technologies have contributed to dramatic reshaping of the landscape of mathematics research. We have to think about ways of incorporating these technologies to amplify possibilities, not just to brush aside tedious calculations as we cling to a curriculum that hasn't much changed in 400 years.

POINT #7: THE CLINICALLY BASED RESEARCH THAT SUPPORTS POINT #1 IS FLAWED, AND THE FLAWS ARE INSTRUCTIVE

Most of the consciousness research that was conducted through the 20th century was undertaken in laboratories. And it turns out that if we isolate people in a room without any of the tools we use to extend intelligence, their conscious capacities will turn out to be not just amazingly limited, but amazingly equal, whether they are Nobel Prize laureates with something to prove or six-year-old brats who couldn't care less.

Now, it seems to me that this fact should have prompted curriculum developers to hesitate a little before structuring programs of study around the limitations of consciousness by parsing up concepts into small, 45-minute-lesson-sized concepts. But it didn't. It seems that no one thought to ask what it might be that enables some people, with essentially the same conscious capacities, to achieve such remarkable feats. Inborn ability is certainly part of a factor, but the range of inborn abilities is simply too limited to explain the variations in achievement that we see. Obsession is a huge factor, too, but we all know that obsessing about something doesn't necessarily lead to great insight.

A major clue into the difference between ordinary and extraordinary performances has emerged over the past few decades, as we've developed the technical abilities to study humans in contexts that are a bit more natural than the laboratory setting. Some surprising things have been shown. One of them is that humans have the capacity to "couple their consciousnesses" (Donald, 2001); that is, to link their minds, to coordinate the rhythms and cycles of their brains' activities. In the process, they can form grander cognitive unities. One common sort of coupled consciousnesses is a "conversation."

It turns out that, in the context of a conversation, humans are able to collectively juggle not 7 ideas, nor 7 + 7 ideas, but more in the order of 7×7 ideas. And some of those ideas can endure not for 10 or 15 seconds, but for minutes and hours.

This point is critical to the production of mathematical knowledge. The image of the focused and still mathematician labouring alone in a locked chamber is not at all representative of how research mathematicians work.

There may be moments when they're on their own, but like anyone in any domain who is concerned with the development of new insights, they surround themselves with others and others' ideas. No mathematician is an island.

Elaine Simmt, also of the University of Alberta, and I have been trying to understand the sorts of collective structures that support the work of mathematicians. Drawing from complexity science (see, for example, Kelly, 1994), we have identified a handful of conditions that are common to such intelligent collectives (see Davis & Simmt, 2003). This thinking is still in its infancy, but I can report briefly on what is involved in prompting the emergence of an intelligent collective in the classroom—a collective that, in turn, supports the development of each individual's mathematical intelligence.

Over the past 20 years, complexity scientists have been labouring to identify the sorts of conditions that enable the emergence of complex systems—how, for example, ants interact to form anthills, species couple together within ecosystems, cells knit themselves into organs, and organs into individuals, and individuals into societies, and so on. Among the necessary conditions for these happenings, the following six seem to have a particular relevance to the work of the mathematics teacher:

- *Internal Diversity*—Internal diversity refers to the pool of possibilities that a system has to choose from when it's faced with a novel circumstance. It is the basis of the collective's intelligence. A system in which all of the components are expected to do the same thing at the same time will not be an intelligent one.
- *Internal Redundancy*—That being said, it's important that the agents in a system have enough in common to be able to work together, whether talking cells, birds, people, or social systems. Redundancy is also necessary for a robust system. If one agent fails, another can step in.

  Some redundancies among participants in a collective have a lot to do with actions and competencies that are automatized. This is where traditional mathematics teaching has focused. The only way that a system's diversity can be a source of intelligence is if its agents are sufficiently alike for the bit of diversity to be embraced and elaborated.
- *Neighbour Interactions*—This condition might seem ridiculously obvious. Of course the agents in a system need to interact if that system is to become a system.

  But in the context of the classroom, the agents that need to interact aren't necessarily people. They can also be ideas or interpretations. As already mentioned, mathematics knowledge emerges as new ideas are blended with old ones. These blendings open up spaces for more powerful notions. So the phrase *neighbour interactions* doesn't refer to pod seating or group work, but to ensuring there is a sufficient density of diverse thought represented for the possibility of new ideas, as in the example of the varied interpretations of multiplication.

- *Liberating Constraints*—Consider these three tasks:
  - Write down all that you know about three fourths.
  - Write down two fractions equal to three fourths.
  - Write down three things that you know about three fourths.

  In most classroom contexts, the first of these is much too broad to generate much that is interesting. The second suffers from being much too narrow, but has the same result—it likely won't generate much that's interesting either. But the third, like Baby Bear's porridge and bed, might be just right. It's open enough to allow for diverse possibilities, but sufficiently constrained to ensure that ideas won't be too diverse to prevent them from working together. (Of course, whether it is suitable depends on the collective.)

  Complex systems have to maintain this delicate balance between so much structure that they lock into place and so little structure that they decay into chaos. And the tasks that you set will determine whether or not intelligent—once again, appropriate and innovative—action can emerge.

- *Organized Randomness*—With a complex system, there's always a bit of randomness. Some of that randomness is ignored by the system—which is to say, it doesn't really affect what the system does. Other bits of randomness come to be really important—the unexpected observation, the sudden insight, the fact that this student's father is a painter and he knows the world doesn't work the way the question about ratios says it should work. Really intelligent systems, it seems, take advantage of more of these random events, and they're able to do so because they have strategies to organize such events.

- *Decentralized Control*—One of the big changes at Microsoft, Apple, IBM, Hewlett Packard, and other locations of cutting-edge knowledge production has been an abandonment of the top-down model of centralized management in favour of a more distributed sort of control. Intelligent collective action can't be orchestrated into existence, at either the individual or the collective level. Space to negotiate the parameters and possibilities is needed.

All this being said, we have a long way to go before we'll be able to give much more direct advice on how to nurture mathematical intelligence. However, we can be quite specific about the opposite—on how to militate against the emergence of intelligent action. For instance, if diversity (among interpretations and among people) is suppressed, if ideas aren't plentiful and not permitted to bump against one another, if tasks are too open or too narrow, if control of the outcomes is strictly in the hands of the teacher, then chances are that intelligence will be stifled—intelligence of not just the collective, but of the individuals in the collective.

NOTES

1. Some of the research data reported in this article are drawn from studies supported by the Social Sciences and Humanities Research Council of Canada. The article itself is a modest revision of a presentation made at the NCTM Regional Conference in Edmonton on November 22, 2003.
2. The reviews of *Where Mathematics Comes From* have been varied, especially with regard to the issue of whether Lakoff and Nunez actually succeed in explaining the bodily basis of Euler's formula. Nevertheless, most reviewers have acknowledged that their discussion of the analogical substrate of our logical abilities is compelling and has significant implications for the teaching of mathematics.

REFERENCES

Calvin, W. H. (1996). *How brains think: Evolving intelligence, then and now*. New York, NY: Basic Books.

Dewey, J. (1910). The influence of Darwin on philosophy. In J. Dewey, *The influence of Darwin on philosophy and other essays* (pp. 1–19). New York, NY: Henry Holt.

Damasio, A. R. (1994). *Descartes' error: Emotion, reason, and the human brain*. New York, NY: G. P. Putnam's Sons.

Davis, B., & Simmt, E. (2003). Understanding learning systems: Mathematics education and complexity science. *Journal for Research in Mathematics Education, 34*(2), 137–167.

Donald, M. (2001). *A mind so rare: The evolution of human consciousness*. New York, NY: W. W. Norton.

Gopnik, A., Meltzoff, A. N., & Kuhl, P. K. (1999). *The scientist in the crib: What early learning tells us about the mind*. New York, NY: Perennial.

Hebb, D. (1949). *The organization of behavior: A neuropsychological theory*. New York, NY: Wiley.

Hilbert, D. (1988). *Foundations of geometry* (Leo Unger, Trans.). Chicago, IL: Open Court. (Original work published in 1899)

Kelly, K. (1994). *Out of control: The new biology of machines, social systems, and the economic world*. Cambridge, MA: Perseus.

Lakoff, G., & Nunez, R. (2000). *Where mathematics comes from: How the embodied mind brings mathematics into being*. New York, NY: Basic Books.

Mazur, B. (2003). *Imagining numbers (particularly the square root of minus fifteen)*. New York, NY: Farrar Straus Giroux.

Mlodinow, L. (2001). *Euclid's window: The story of geometry from parallel lines to hyperspace.* New York, NY: The Free Press.

Norretranders, T. (1998). *The user illusion: Cutting consciousness down to size* (J. Sydenham, Trans.). New York, NY: Viking.

Pinker, S. (1997). *How the mind works.* New York, NY: W. W. Norton.

Richardson, K. (1999). *The making of intelligence.* London, UK: Weidenfeld & Nicolson.

Seife, C. (2000). *Zero: The biography of a dangerous idea.* New York, NY: Penguin Books.

BIBLIOGRAPHY

Armstrong, S. (executive producer), & van Sant, G. (director). (1997). *Good Will Hunting* [Motion picture]. U.S.A.: Miramax.

Hallowell, T. (executive producer), & Howard, R. (director). (2001). *A beautiful mind* [Motion picture]. U.S.A.: Universal.

Rajski, P., Rudin, S., & Stone, R. (producers), &Foster, J. (director). (1991). *Little man Tate* [Motion picture]. U.S.A.: Orion.

This chapter was originally published as:

Glanfield, F. (2006). Secondary mathematics education curriculum developments: Reflecting on Canadian trends. *delta-K: Journal of the Mathematics Council of the Alberta Teachers' Association, 43*(2), 17–19.

CHAPTER 44

SECONDARY MATHEMATICS EDUCATION CURRICULUM DEVELOPMENTS

Reflecting on Canadian Trends

Florence Glanfield

Editor's Note: A former Alberta mathematics educator, past president of MCA-TA and winner of MCAT's award for outstanding contribution to mathematics education in 1995, Florence Glanfield is currently a teacher educator and researcher in the College of Education, University of Saskatchewan. Her current research interests are teachers' mathematical understanding, mathematics education in the Aboriginal community, and Aboriginal students' perspectives of mathematics.

I recently attended a professional meeting where a conversation focused on the potential changes in the Western and Northern Canadian Protocol (WNCP) secondary mathematics curriculum. The suggestion was made that Alberta drop its applied mathematics courses in favour of returning to the Mathematics 10-20-30, Mathematics 13-23-33 and Mathematics 14-24 programs of the early 1990s. I was saddened to hear this conversation.

I am a mathematics teacher educator in Saskatchewan, a province that does not have an applied or consumer mathematics program. There is one mathematics program for all secondary students. Most of the teacher candidates in the teacher-education program at the University of Saskatchewan are graduates of the provincial system. I am saddened when I hear about their experiences in the secondary mathematics program—experiences that have led many of them to believe that they cannot "do" mathematics. Yet, the teacher candidates who tell these stories are the elementary mathematics teachers of the future, the teachers who will be expected to excite children and help them develop a passion for mathematics.

A 1981 report by the Council of Ministers of Education, Canada (CMEC) provided the impetus for provinces and territories to consider how they might humanize mathematics (Wheeler, 1982) through their curricula. O'Shea (2003) describes the recommendations of the CMEC report as follows:

> The use of applications as a motivation for learning mathematics and as an experience in using mathematical models received increased emphasis. Respondents asserted that this reflected an effort to link the mathematics of the classroom to the real world in the interest of helping students see the practical uses of mathematics, understand the relationship of mathematics to other disciplines, and understand more clearly the mathematical ideas themselves.

O'Shea continues,

> In general, the trends at this level reflected a reduction of emphasis on rigor and formal structure. The reason usually given for this was that students might not be ready for formalization until they had developed concepts through informal approaches, including finding patterns, working with physical models of mathematical concepts, and using other manipulable aids. The result was an evolving mathematics curriculum that included emphasis on topics such as numerical skills, applications, and problem solving, and an accompanying deemphasis on abstract topics such as set theory and algebraic manipulations.

Across Canada we see mathematics curricula that include a focus on problem solving, applications, and the development of mathematical concepts from a concrete approach; the design of programs intended for students who are postsecondary-bound but are not going into science and mathematics; and the inclusion of computer technology (O'Shea, 2003). Below are examples of how the provinces and territories have responded to the CMEC report and taken a more humanistic approach to mathematics.

APPLIED MATHEMATICS

To address the needs of postsecondary-bound (but not necessarily science- or mathematics-bound) students, the Western Canadian Protocol (WCP) (now the WNCP) developed a series of three courses called applied mathematics. The curriculum document states that the "applied clusters ... emphasize applications of mathematics rather than precise mathematical theory. The approaches used are primarily numerical and geometrical" (WCP, 1996, p. 19). Throughout the applied mathematics courses, students are engaged in projects and activities that explore the mathematics in a given context. The textual resources that support the courses are filled with projects and laboratory activities. Students are encouraged to work collaboratively throughout the courses, and the mathematical ideas are expected to be developed from the numerical and geometrical approaches.

An example of an applied mathematics expectation is "Use properties of circles and polygons to solve design and layout problems" (WCP, 1996, p. 107). Students might work on a problem such as the following:

> The pattern on a piece of vinyl flooring consists of a square and four equilateral triangles. Each equilateral triangle has as its base one side of the square. Circles are inscribed in each triangle and in the square.
>
> a. Start with a square of side length 6 cm. Draw the design, full size.
>
> b. Determine the ratio of the area of the small circle
>
> c. to the area of the large circle. (p. 107)

In the Atlantic Canada collaboration, the applied mathematics courses start in grade 10, similar to the WCP collaboration. In Ontario, an applied mathematics course has been developed for grade 9.

CONSUMER MATHEMATICS

A second area in curriculum development for students who may or may not be university- or collegebound is the development of consumer or basic mathematics courses for grades 10–12. Currently, three provinces (British Columbia, Manitoba, and Ontario) and the three territories have adopted such courses. An example of a grade 12 consumer mathematics curriculum outcome is "Develop, use, and justify mathematical strategies by analyzing puzzles and games" (Manitoba Education, 2004). (As a side note, Manitoba Education has negotiated with its postsecondary institutions to accept consumer mathematics as a general admission requirement into postsecondary programs that are not mathematics intensive.) In Ontario, the courses are called *Mathematics for Everyday Life* and are referred to as workplace-preparation courses (Craven, 2003).

ROLE OF COMPUTER TECHNOLOGY

The use of graphing calculators has been evident in the curricula of most provinces and territories since the early 1990s. More recently, however, computer technology has played a more prominent role in secondary mathematics. Across Canada, secondary mathematics students are expected to learn to use spreadsheet programs and geometry software. You will see curriculum expectations such as "Students will solve problems using spreadsheet functions and templates" and "Students will use geometry software to develop the geometric properties of circles."

REFLECTION

The mathematics education community and curriculum developers must not give up on the idea of investigating mathematics from a variety of perspectives. I believe that when we offer secondary students a variety of ways to see mathematics, they will begin to see themselves as mathematical beings and recognize that they, too, are part of the mathematical community. The development of programs such as applied mathematics and consumer mathematics and the use of computer technology in mathematics classes invite students to be engaged in mathematics from multiple perspectives.

Kissane (2002, p. 191) notes Wheeler's (1982) prophetic suggestion of more than 20 years ago that "mathematics teachers were in the midst of three major educational upheavals: mass secondary education, the rise of new and available technologies, and the revolution of humanizing mathematics." In Canada, we could say that in the past 20 years, secondary mathematics curriculum development has also been in the midst of these three upheavals. I believe that secondary mathematics curriculum development has been addressing "the rise of new and available technologies" and "humanizing mathematics" in order to address "mass secondary education." Generally, we now see secondary mathematics curricula that include the use of technology (not just computer and calculator technology) to study mathematics and the introduction of courses, such as applied and consumer mathematics that humanize mathematics.

We must continue to look for ways, through secondary mathematics curriculum development, to humanize mathematics. We must work hard as a profession and engage in problem solving to overcome issues around implementation and postsecondary acceptance of mathematics courses other than those labeled as pure mathematics in order to begin to make a difference in how our students view themselves within the mathematics community.

REFERENCES

Council of Ministers of Education, Canada (CMEC). (1981). *Mathematics: A survey of the provincial curricula at the elementary and secondary levels.* Ottawa, ON: Author.

Craven, S. (2003, May). *The state of mathematics education in Ontario: Where we came from and where we are.* Paper presented in the Presentation of Regional Situations panel, Canadian School Mathematics Forum, Montreal. Retrieved from www.math.ca/Events/CSMF2003/panel/OntarioReport.pdf

Glanfield, F. (2003, May). Western Canada Regional Perspective on Mathematics Education. Paper presented in the Presentation of Regional Situations panel, Canadian School Mathematics Forum, Montreal. Retrieved from www.math.ca!Events/CSM F2003/panei/WestReport.pdf

Kissane, B. (2002). Three roles for technology: Towards a humanistic renaissance in mathematics education. In A. Rogerson (Ed.), *Humanistic Renaissance in Mathematics Education: Proceedings of the International Conference* (pp. 191–196). Palermo, Sicily: Mathematics Education into the 21st Century Project. Retrieved from http://math . unipa.it/-grim/SiKissane.PDF

Manitoba Education. (2004). Senior 4 Consumer Mathematics (40S) Outcomes by Unit. Retrieved from www.edu.gov.mb.ca/ks4/cur/math/s4consumer_outcomes.html

O'Shea, T. 2003. The Canadian mathematics curriculum from the new math to the NCTM standards. In G. M. A. Stanic & J. Kilpatrick (Eds.), *A history of school mathematics* (Vol. 1, pp. 843–894). Reston, VA: National Council of Teachers of Mathematics.

The Western Canadian Protocol (WCP) for Collaboration in Basic Education. (1996). *The common curriculum framework for K–12 mathematics: Grade 10 to Grade 12.* Np: WCP.

Wheeler, D. (1982). Mathematisation: The universal capability. *Australian Mathematics Teacher, 38*(4), 23–25. Cited in Kissane, 2002.

This chapter was originally published as:

Long, J. S. (2006). Noticing as a form of professional development in teaching mathematics. *delta-K: Journal of the Mathematics Council of the Alberta Teachers' Association, 44*(1), 5–7.

CHAPTER 45

NOTICING AS A FORM OF PROFESSIONAL DEVELOPMENT IN TEACHING MATHEMATICS

Julie S. Long

Editor's Note: Julie S. Long is a doctoral student in the Faculty of Education at the University of Alberta. Her area of specialization is elementary mathematics education. Julie has taught at various levels from kindergarten to postsecondary. Her research interests centre on care and mathematics in the elementary classroom, but also include preservice teachers, metaphors of mathematics, mathematical reasoning in young children, discourse analysis of school textbooks, and ways of knowing in academia.

As a teacher, I constantly change my practice. I make changes not to correct something but to respond to students and to answer my own questions about teaching mathematics. These changes are a form of professional development because they are often derived from readings or working with other teachers. This article focuses on how John Mason's book *Researching Your Own Practice: The Discipline of Noticing* (2002) can enrich professional development. In particular, I will look at *accounts-of* and *accounts-for* experience, professional development, and connections to mathematics.

ACCOUNTS-OF AND ACCOUNTS-FOR

Mason's (2002) work centers on developing sensitivities for attending to, or noticing, aspects of unexamined and habitual practice, so that choices in moments of teaching practice might be better informed. Mason's research has shown the importance of reflection in developing professional practice by offering a "detailed, structured, systematic" (p. 25) way to record and act on reflections.

Mason differentiated between an account-of and an account-for an experience. An account-for an experience includes explanations, judgments, and evaluations surrounding an event; an account-of an experience minimizes these aspects. The idea is to write up the account so that others recognize the experience. Mason (2002, p. 41) wrote that collecting these accounts-of "is one step towards...identifying a type of situation, tension, issue or interaction which is exemplified in several different incidents or experiences." I decided to try it by writing an account-of a teaching moment.

ACCOUNT-OF FRACTIONS

While James presented his ideas about dividing fractions, he drew circles on the whiteboard. He explained his method for dividing one-half by one-quarter, and his classmates asked him questions. James re-explained his method using different words and the same drawings. When he stopped talking, students turned their heads from James and looked at me. James sat down and I elaborated on his strategy. Students then discussed fractions in groups and made drawings of the fractions.

In this account-of a teaching moment, I distilled the experience into a short paragraph. I avoided emotional words and explanations of my thinking, which was difficult because emotions were important in my decisions that day. I also had difficulty identifying the essence of the experience. I had to work at stressing certain aspects, such as what I could remember about the physical situation, and ignoring others, such as my emotions. At first I thought that I was writing about listening to students, but I realized that the essence in his account-of was a moment of taking authority in the classroom. This is different from my original account-for.

ACCOUNT-FOR FRACTIONS

One day I talked about dividing fractions and how simply knowing the procedure is not helpful. It's easy to forget what to do to which fraction. I explained that if you understand it, the procedure is meaningful. I drew an example on the board.

While students were working on a problem, I circulated and chatted. James explained his thinking about the division of fractions to me a couple of times, and I had difficulty understanding him. The students at his table were also confused. We all asked many questions until some students began to lose interest. We were off task, but I thought that the exploration was important.

I asked James to record his thinking for me so that I could consider it some more. I puzzled over his ideas and fraction circles until next class and then asked him for more clarification. When James presented his ideas to the class, his classmates had lots of questions. Students were getting annoyed because the method made no sense to them. Although James was good at explanations and answering questions, I needed to intervene. I told him that I didn't understand his thinking. James sat down and I presented a similar strategy. The class was focused and silent as I spoke and wrote. I was nervous about adding to the confusion, but I could almost hear that audible *aha* from students. They began to excitedly talk in groups. I hoped that meant that they were sharing their understanding and not getting bored or confused.

A few days later a student mentioned that she had never understood the division of fractions until that class. I wondered how those teaching moments came about. A lot had led up to that moment, including positive and negative feelings. I'm not sure that the moment would have been as meaningful if the students hadn't struggled to understand a classmate's unfamiliar idea, if there hadn't been time to think and discuss, if they hadn't been emotionally involved or if they hadn't already spent time listening to each other's ideas.

In this account-for, I skipped ahead of simply describing the incident to explaining my actions and trying to draw a lesson out of the experience. If I shared this account-for with others, it might be difficult for them to support me in my re-examination of the experience because I have already explained it and there are no alternatives to explore. Sharing accounts-of (not accounts-for, though, the line between the two is unclear) experiences might be "used explicitly to foster and sustain professional development in others" (Mason, 2002, p. 139).

PROFESSIONAL DEVELOPMENT

Although Mason (2002) mainly focused on how to use noticing for one's own practice, he also wrote about how to use accounts-of to support professional development in others.

> A good way to expose people to alternative practices without pressuring them to suddenly adopt one and to act differently is to arrange that one person gives a brief-but-vivid account of some problematic situation, and then others recount situations which they think have some similarities. In the process, different practices will be revealed, but in a non-threatening manner. ... It is a matter of offering a brief-but-vivid account without the intention of "offering a solution." (p 146)

Following this suggestion, preservice and inservice teachers could write accounts-of their teaching and share them with others, but they must be open to this sort of inquiry and look to change their own practice. Developing trusting and collegial relationships would also be important, though Mason did not write about this explicitly. Assuming that these conditions are met—which is no small feat—this could be a fruitful way of interacting with teachers. Blending teachers' practical concerns with professional development is possible. These concerns might also be used as a basis for research, whereby theory and practice overlaps. This research might be done by teachers from the inside of practice or by researchers in conjunction with teachers from the outside of practice.

CONNECTIONS TO MATHEMATICS

The discipline of noticing and Mason's previous work *Thinking Mathematically* (Mason, Burton, & Stacey, 1985) are parallel. This helps me to better understand how professional development of teaching in general is connected to teaching and learning mathematics.

The acts of stressing and ignoring are part of both mathematical thinking and professional development through the discipline of noticing. When thinking mathematically, I often stress one part of the question while ignoring other parts; for example, looking at a geometric shape and stressing the global characteristics (it looks like a diamond), while ignoring the specific characteristics, such as the angle measures. What I stress and what I ignore can be described as habitual and depends on the situation's context. By stressing and ignoring, I can first specialize and then generalize; both are essential features of mathematical thinking (Mason, Burton, & Stacey, 1985). The discipline of noticing calls me to attend to what is stressed and ignored in my own mathematical work as well as in my teaching practice. In addition, I am invited to stress the essence and ignore the emotions in writing accounts-of experience. Stressing and ignoring are mathematical ways of examining teaching practice.

Accounts-of experiences are described as "brief but-vivid" (Mason, 2002, p. 47). A problem of teaching practice is distilled into a few sentences. In this discipline of noticing, a number of these accounts are examined for relationships and inconsistencies. This is similar to the work of mathematicians, which might be characterized as compressing information, and of teachers, which can be thought of as unpacking this condensed knowledge (Ball & Bass, 2003). Through noticing as a form of professional development, teachers condense experiences and then reconceptualize their accounts to transform practice.

QUESTIONS AND ISSUES

This article explains how to consciously notice. Practices can be transformed by attending to experiences, recording them systematically, questioning the accounts, and then acting deliberately. Though Mason's description of the discipline of noticing resonates with my own process of reflecting, I also wonder about the power of accounts-for in this discipline. I have used accounts-for in my writing, talking, and thinking with meaningful results. Though I resist the focus on the accounts-of (as opposed to the accounts-for), part of this resistance comes from the difficulty and work involved in writing a compelling account-of an experience. Writing and using both accounts-for and accounts-of have been fruitful ways to engage in professional development in my teaching of mathematics.

REFERENCES

Ball, D. L., & Bass, H. (2003). Toward a practice-based theory of mathematical knowledge for teaching. In E. Simmt & B. Davis (Eds.), *Proceedings of the 2002 Annual Meeting of the Canadian Mathematics Education Study Group* (pp. 3–14. Edmonton, AB: CMESG/GCEDM.

Mason, J. (2002). *Researching your own practice: The discipline of noticing.* New York, NY: Routledge.

Mason, J., Burton, L., & Stacey, K. (1985). *Thinking mathematically* (Rev. ed.). Harlow, UK: Addison-Wesley.

CHAPTER 46

TEXTBOOKS IN MATHEMATICS LEARNING

The Potential for Misconceptions

Ann Kajander and Miroslav Lovric

Editor's Note: Ann Kajander is an associate professor in the Faculty of Education at Lakehead University, in Thunder Bay, Ontario. She is interested in facilitating creative exploration in mathematics at the elementary level. She is particularly interested in giftedness and creativity in mathematics. She has designed a program called kindermath and runs this as an afterschool mathematical program for elementary children. Miroslav Lovric is an associate professor in the Department of Mathematics and Statistics at McMaster University, in Hamilton, Ontario, and has a long-time interest in mathematics education. He is the recipient of a number of teaching awards, in particular for his work supporting graduate tutors.

INTRODUCTION

As textbook authors ourselves (for example, Kajander, 2007; Lovric, 2007a) we have found ourselves face to face with the tension between the historical use of the text as the mathematical authority and the notions of mathematics reform as described in the Principles and Standards for School Mathematics (National Council of Teachers of Mathematics [NCTM] 2000), which position the learners as prominent participants in their own learning. "Rather than the textbook and the teacher acting as major sources of authority, this intended curriculum encourages students to rely on their own mathematical reasoning and evidence when discussing mathematical solutions" (Herbel-Eisenmann, 2007, p. 345).

As we began to critically examine sample secondary and postsecondary mathematics texts, we saw evidence of textbook formats that might have been intended to simplify the learning process for students. In our examination we found many examples of such attempts that, in fact, potentially introduced mathematical misconceptions. This article summarizes some of the types of issues we have found related to mathematical exposition and suggests areas requiring attention in textbook writing, design, and use.

BACKGROUND

Both older and more recent research seems to highly privilege the role of the textbook in classroom mathematics learning. McKnight, Crosswhite, and Dossey (1987) reported that more than 95 percent of grade 12 teachers indicated that the textbook was their most commonly used resource, and recent research in Ontario indicates that this situation has changed little (Kajander & Mason, forthcoming). More generally, "Commercially published, traditional textbooks dominate mathematics curriculum materials in US classrooms and to a great extent determine teaching practices" (Clements, 2007, p. 55). Yet even with all this emphasis on textbooks in learning, we are becoming convinced that research about the textbooks themselves, and particularly the implications for mathematical content interpretation they impart, is modest at best.

Some effort has been put into content analysis and exploring the ways in which textbooks are used in classrooms and beyond (for example, Love & Pimm, 1996; McCrory, 2006). Recent work (Herbel-Eisenmann & Wagner, 2007; Herbel-Eisenmann, 2007) has examined how the textbook positions the learner, particularly with respect to discourse. However, market research, rather than research based on mathematics education sources, is usually used to determine textbook content and approach (Clements, 2007).

Project 2061 is a long-term project, aimed at evaluating resources in science and mathematics, supported by the American Association for the Advancement of Science. According to one of their studies, "the majority of textbooks used for algebra . . . have some potential to help students learn, but they also have serious weaknesses" (Project 2061, 2000). More than half of the 12 textbooks evaluated by the project team were considered adequate, but none were rated highly; three were rated as highly inadequate for student learning. According to the Project 2061 findings, authors of textbooks generally ignore the research on how students acquire new mathematical ideas and concepts as described in the Standards (NCTM, 2000).

Our work is grounded in the theory of conceptual change (Davis, 2001; Biza, Souyoul, & Zachariades, 2005), which provides a framework for the study of potential student misconceptions related to learning from textbooks. The theory describes learning processes of adults as well as children and hence is appropriate in addressing high school and university students.

THE STUDY

We have begun to take a closer look at textbooks commonly used at the secondary and first-year university levels, particularly in Ontario, to determine how mathematics textbooks might contribute to the development of students' conceptions and misconceptions about mathematics, particularly with respect to the development of conceptual understanding. We begin by providing our current working model, which may be helpful to others wishing to undertake critical mathematical examinations of texts, and follow this with an example.

OUR WORKING MODEL

As we worked through a number of textbook examples, we found issues that appeared to fall into a number of somewhat overlapping groupings. We have attempted to make sense of these by offering the following framework for such analyses, which has emerged from our work thus far. Roughly speaking, these groupings are

1. the use of colloquial, reader-friendly language beyond an intuitive introduction, and in places where mathematical precision was warranted
2. incorrect generalizations, often taken out of context
3. diagrams as sources of misconceptions (sometimes supporting oversimplified definitions)
4. oversimplification, often leading to inaccuracy
5. discussion of concepts not yet properly defined
6. design issues such as summary boxes

We are continuing to use and refine this rough framework as we examine more sources.

AN EXAMPLE

As one of many examples we have encountered, consider the concept of the tangent line to the graph of the function $y = f(x)$. From our own experience, common ideas held by many students entering first year university include incomplete or inaccurate conceptions, such as "the tangent is the line that touches the graph of $y = f(x)$ at only one point," or "a tangent line cannot cross the graph of $y = f(x)$." Accordingly, we began exploring how such conceptions might arise from previously studied written sources. When we examined a grade 12 calculus textbook currently in use in some Ontario schools, we easily found evidence that could support (or at best fail to correct and clarify) such misconceptions. For example, in one source, an initial explanation near the beginning of the chapter on tangents states clearly (with accompanying diagrams), "In the graphs of the circle and the parabola, a tangent line touches exactly one point of the graph, P. For other curves, such as the one in the third diagram [an example of a tangent that also crosses the curve at two other points] a tangent line touches the graph at the point of tangency, P, but may pass through other points on the graph as well" (Kirkpatrick et al., 2002, p. 183).

While this explanation is thorough, a subsequent coloured summary box contains the incomplete definition of a tangent as follows: "A tangent is a line that touches exactly one point on the graph of a relation" (Kirkpatrick et al 2002, p. 190).

In fact, this oversimplified statement is not true in all cases, and emphasizing it in a summary box may increase the likelihood of its being the definition remembered by the student, rather than the concept being deeply understood. This oversimplification is an example of Issue 4 in our working model, and is further emphasized by the design issue (Issue 6 in the model) of placing the text in a coloured summary box on the page. While the summary is meant to make things easier for students, highlighting such oversimplifications, which are not always true, introduces the potential for incomplete understanding and misconceptions. In a traditional classroom environment, definitions are often presented by the teacher rather than developed by the students themselves as meaning is made. If the definition used by a teacher is the page 190 summary version on its own, and students are not encouraged to investigate the concept more fully with suitable prompts and questions, students will likely be left with an incomplete understanding. These and other examples made it painfully obvious to us that students' common misconceptions in firstyear calculus may often have their roots in previously studied materials.

Similarly, standard university calculus texts contain numerous illustrations of tangent lines. However, in a majority of cases, the tangent is shown in the generic position where it touches the curve at one point and does not cross it. The concept of touching might be further suggested in examples where students are given the graph of a function $f(x)$ and are asked to sketch the graph of its derivative $f'(x)$. Although the tangent is defined as the limit of secant lines, these examples do not attempt to encourage drawing the tangent or thinking about it (either in illustration or in accompanying text) as limiting the position of secant lines.

Continuing with this example, to address this potential misconception, a textbook (or an instructor) could ask students to create illustrations that show relationships between curves and lines and identify which are (or are not) tangents (Lovric, 2007b). These illustrations should include cases such as a tangent that crosses the graph at the point of tangency; a tangent line that touches the graph at more than one point; a line that touches the graph, does not cross it, but is not a tangent (cusp); and so on. By analyzing such situations, students could gain a more accurate understanding.

DISCUSSION

Given that textbooks appear to remain the resource of choice for many teachers, we believe that more attention needs to be focused on the quality of their content. To date, "we know little about the impact of curriculum materials on the . . . relationships developed between these readers and the curriculum materials themselves . . . [thus] we must first examine the relationships that the textbook materials themselves encourage" (Herbel-Eisenmann, 2007, pp. 345–346).

In our work to date, we have been struck by how the issues we identified as problematic for learning from textbooks—such as generalizations that were incomplete and that lacked connections to intuitive understanding, ideas that were separated from connections to context, and incomplete summaries placed in specially designed coloured boxes implying that they were to be remembered—echoed a traditional pedagogy. On the other hand, the treatments that we found more effective—based on a gradual shift from informal language, good visuals and identification of misconceptions to more formal, rigorous, and precise generalizations—echoed a more reform-based philosophy as defined in the Standards (NCTM, 2000).

We believe that as students move through the secondary curriculum and into the postsecondary realm, they might obtain deeper and more accurate conceptual understandings of fundamental concepts if more attention were paid to textbooks that support rather than undermine such growth. Rather than criticizing the predominant use of textbooks by teachers, better success might be achieved by actively ensuring the quality of these texts in promoting deep and accurate student understanding.

NOTE

Ann Kajander has written books of activities for elementary and early secondary students, while Miroslav Lovric is the author of undergraduate textbooks.

REFERENCES

Biza, I., Souyoul, A., & Zachariades, T. (2005, February). *Conceptual change in advanced mathematical thinking.* Discussion paper presented at the Fourth Congress of the European Society for Research in Mathematics Education, Sant Feliu de Guíxols, Spain.

Clements, D. H. (2007). Curriculum research: Toward a framework for "research-based curricula." *Journal for Research in Mathematics Education, 38*(1), 35–70.

Davis, J. (2001). Conceptual change. In M. Orey (Ed.), *Emerging perspectives on learning, teaching, and technology.* Retrieved from www.coe.uga.edu/epltt/ConceptuaiChange.htm

Herbel-Eisenmann, B. (2007). From intended curriculum to written curriculum: Examining the "voice" of a textbook. *Journal for Research in Mathematics Education 38*(4), 344–369.

Herbel-Eisenmann, B., & Wagner, D. (2007). A framework for uncovering the way a textbook might position the mathematics learner. *For the Learning of Mathematics, 27*(2), 8–14.

Kajander, A. (2007). *Big ideas for growing mathematicians.* Chicago, IL: Zephyr.

Kajander, A., & Mason, R. (forthcoming). Examining teacher growth in professional learning groups for in-service teachers of mathematics. *Canadian Journal of Mathematics, Science and Technology Education.*

Kirkpatrick, C., McLeish, R., Montesanto, R., Suurtamm, C., Trew, S., & Zimmer, D. (2002). *Advanced functions and introductory calculus.* Scarborough, ON: Nelson.

Love, E., & Pimm, D. (1996). This is so: A text on texts. In A. I. Bishop, M. A. Clements, C. Keitel, J. Kilpatrick, & C. Laborde (Eds.), *International handbook of mathematics education: Part 1* (pp. 371–409). Dordrecht, NL: Kluwer Academic.

Vector calculus. Hoboken, NJ: Wiley and Sons.

Lovric, M. (2007b, January). *Mathematics textbooks and promotion of conceptual understanding.* Presentation to the Education Forum of the Fields Institute for Mathematical Sciences, Toronto, Ontario.

McCrory, R. (2006). Mathematicians and mathematics textbooks for prospective elementary teachers. *Notices of the American Mathematical Society, 53*(1), 20–29.

McKnight, C., Crosswhite, F. J., & Dossey, J. A. (1987). *the underachieving curriculum.* Champaign, IL: Stipes.

National Council of Teachers of Mathematics (NCTM). (2000). *Principles and standards for school mathematics.* Reston, VA: Author.

Project 2061. (2000, April 26). Algebra for all—Not with today's textbooks, says AAAS [Press release]. Retrieved from www.project2061.org/about/press/pr000426.htm

CHAPTER 47

RATIONALE GONE MISSING

A Comparative and Historical Curriculum Search

Lynn McGarvey

Editor's Note: Lynn McGarvey is an associate professor at the University of Alberta. She teaches courses in mathematics education for preservice and inservice teachers. Her research is focused on how young children (particularly ages 3 to 5) understand mathematics and how parents and educators support mathematical thinking in the home and in early childhood settings. Related interests are international public policy for preschool curriculum development and the rationale for mathematics learning with young children, pedagogical expectations, and the range of definitions, descriptions, and outcomes related to mathematical concepts and processes.

Whether it is appropriate or not, there is an implicit hierarchy in the subject areas taught in public schools. Educational institutions are notoriously underfunded and have limited resources. Educators, particularly those in nonacademic subjects such as fine arts or physical education, need to simultaneously defend the importance of their subject areas and actively lobby for their fair share of time, money, equipment, and personnel. As a mathematics educator, I often feel I am letting my subject area down. I usually leave my soapbox tucked neatly under my desk and keep quiet. Yet, even in my silence there is an assumed value and importance given to school mathematics by educators, parents, policymakers, and most citizens. I have listened patiently to people explain how horrible they are in math but in the next breath say how important mathematics is for such things as everyday calculations, postsecondary institution admission, future job prospects, and future income possibilities. Not satisfied with any of these reasons, I examined the front matter of the *Alberta K–9 Mathematics Program of Studies* (Alberta Education, 2007)[1] in search of a rationale for teaching mathematics. Perhaps it would provide me with a thoughtful reason to dust off my soapbox.

As I read, I was bemused by the fact that the introduction included the heading, "Purpose of the Document," but the purpose of the content or of the entire subject was not clear. I read through the program of studies, asking the question, "Why is mathematics important for children?" Nothing in the early pages answered the question. Instead, the document provides goals for students such as, "to prepare students to use mathematics confidently to solve problems" (p. 4). But what are the problems to be solved? If the main goal is to solve the problems that mathematics educators give students in the course of learning mathematics, then this goal is a self-referencing and self-serving. The goals chosen must be based on a rationale of some sort, but I couldn't find it.

I continued to sift through the document, asking myself, "Why is mathematics important?" I found one sentence under "Nature of Mathematics" that was at least a partial response: "Mathematics is one way of trying to understand, interpret and describe our world" (p. 10). This statement could form the basis for a rationale. It is, in fact, similar to the rationale provided in the science program of studies (1996), which appears under the explicit heading, "Rationale." It states: "Learning about science provides a framework for students to understand and interpret the world around them" (p. 1). The statement in the mathematics program of studies is very similar, but it was buried on page 10 and no further explanation is given. Nowhere in the document is an explicit rationale provided.

On one hand, I was relieved that no attempt was made to pinpoint a rationale. Mathematics educators certainly do not agree on a singular, appropriate reason to teach mathematics—especially to all children (e.g., Davis, 2001; Huckstep, 2000; Noddings, 1994). If a rationale had been written, it surely would have been criticized by several stakeholders. On the other hand, I was dismayed. With no clearly defined vision for why mathematics is important, what assumptions are hidden beneath the choice of goals for students, content inclusion and exclusion, authorized resources, and so on?

Perhaps an explicit rationale in a program of studies is passé. I continued to think about my colleagues in other subject areas and their ability to put forth a united front regarding the purpose of their subject areas and their insistence that their subject is a vital component of schooling—at least in elementary schools. Out of curiosity I opened the Programs of Study (Core Curriculum) site from Alberta Education.

Because I am a former physical education teacher, I opened the K–12 Physical Education Program of Studies (2000b) to see if it still included a rationale. It does. In fact, in contrast to the mathematics curriculum, the physical education document starts with "Program Rationale and Philosophy" in big bold letters. I asked the question, "Why is physical education important for children?" Part way through the first page I read:

> Physical activity is vital to all aspects of normal growth and development, and the benefits are widely recognized. Students do not develop automatically the requisite knowledge, skills and attitudes that lead to active, healthy lifestyles. Such learning should begin in childhood. Schools and teachers can be prime facilitators in providing opportunities for the development of the desire for lifelong participation in physical activity. (p. 1)

Although we can critically dissect this statement and find faults, unlike the mathematics curriculum, the physical education document places a strong rationale for the subject up front. Not only does the document provide a justification for physical education, it also addresses the vital role that schools and teachers can (or should) play.

I also opened the programs of study for other marginalized subjects, including drama (1985b), art (1985a), music (1989) and health (2002). Each one started with "Program Rationale and Philosophy." Other academic subject areas begin the same way. Science, as I mentioned, starts with an explicit rationale. Social studies (2005), the program of studies that has been most recently revised, also provides extensive information on rationale, a program vision, a definition, and the role of social studies in schools.

The remaining core subject, English language arts, arguably tops the hierarchy of subjects, particularly in elementary schools. As I had done with other subjects, I opened the document and started searching with the question, "Why is learning language arts important?" The word rationale is not used in the *K–9 English Language Arts Program of Studies* (2000a). Instead, the document begins with "The Importance of Language" followed by the subheading, "The nature of language."

> Language is the basis of all communication and the primary instrument of thought. Composed of interrelated and rule-governed symbol systems, language is a social and uniquely human means of exploring and communicating meaning. As well as being a defining feature of culture, language is an unmistakable mark of personal identity and is essential for forming interpersonal relationships, extending experiences, reflecting on thought and action, and contributing to society. (p. 1)

Although the subheadings "the nature of language" and "the nature of mathematics" are parallel, their treatment and placement in the two documents are not at all equal.

Regardless of where the subject area was on the perceived hierarchy, all current curriculum documents, except mathematics, provide a (more or less) persuasive response to the question of why that subject matter is important.

Why, then, is no rationale for mathematics education provided? Is the justification so obvious that inclusion is not needed? Is it simply an oversight? Or were the writers not able to reach an agreement? Has an explicit rationale for teaching mathematics to children ever existed? This last question prompted an examination of the 1996 program of studies, then the 1982 curriculum, and compelled me to look through the dusty historical collection of almost a century of curriculum guides for elementary mathematics.

A CENTURY OF RATIONALES FOR MATHEMATICS IN SCHOOLS

Beginning in 1914, most Alberta programs of study for elementary schools begin with a general introduction and aims or guiding principles for education in schools. The introduction is then followed by pages devoted to individual subject areas, usually with their own set of aims that are usually aligned with and reference the general goals of education previously stated. In arithmetic/mathematics, several common aims seemed to resurface throughout the nearly 100-year history that I reviewed. The following discussion highlights a few recurrent aims and assumptions pertaining to the importance of mathematics.

Perhaps a place to start is with the assumption that mathematics sits at or near the top of the hierarchy of school subjects. In 1918, the *Course of Studies for the Public Schools Grades I–VIII* includes "academic subjects" along with nature study, art, manual and household arts, physical training, hygiene, and music. The teacher is advised to follow the outline in each subject, "but should give each pupil a thorough training in Reading, Writing, Spelling, Oral and Written Composition, and Arithmetic, as these subjects form the basis for future progress of education" (J. T. Ross, introductory note).

The supposed supremacy of literacy and numeracy has existed without change since the inception of public schooling. This viewpoint is expressed clearly in 1918, but similar views are expressed several times in the historical documents: "From the earliest times mathematics has occupied an honored place in the courses of study that have been pursued" (Department of Education, 1924, p. 144). Also, "[f]rom time immemorial mathematics has played, and will continue to play, an important role in the history of man's existence and for this" and other reasons provided, "mathematics is considered to be one of the 'basics' of education" (1982, p. 2). In many respects these statements claim that mathematics is and has always been important in education, but without explicitly stating why. Why does mathematics occupy an "honored place in the courses of study"? The reason that mathematics is more important than hygiene or physical training, for example, is never explained or questioned.

Most Alberta curricula through the years do provide some form of rationale for why learning mathematics is important. Perhaps the most common reason stated is the utility of mathematics to solve everyday problems. However, there appeared to be a shift with respect to what and whose problems were important. The early years of public schooling emphasized "training" in and "mastery" of the four operations to solve "concrete problems" (1918, p. 7) and "problems ordinarily met within the activities of life" (1924, p. 144). Also, in grade 8 (1924), the pinnacle of education for many students, the aim was training in "the genuine problems of life met within ordinary occupations of the community" (p. 158). Throughout the years, mathematical skills are valued as a means to solve concrete problems, everyday problems and problems related to community occupations. However, I wonder why the problems related to mathematics are more important than the problems of, say, manual and household arts or hygiene?

Occasionally, the perspective on problem solving in mathematics shifted to a focus on problems arising from the children's present needs and interests. This trend was particularly evident in the "progressive" era of the 1930s, which emphasized a Deweyan perspective. That is, learning should be interesting and relevant for children "in the life they are living as boys and girls" (1936, p. 4). A similar emphasis also appears in 1982: "The [mathematics] program should be focused on the child's world. . . . An awareness of some real-world applications of mathematics and some of the technological advances *which will directly affect the child's life*, should be imparted to the student" [italics added] (p. 5).

The 2007 mathematics curriculum is sufficiently vague in its position on the utility of mathematics to solve problems. While students are expected to be confident problem solvers, the type of problems and their relevance are never specified.

Another common assumption is that learning mathematics is justified because it is required for future learning of mathematics. For example, "[o]f equal importance [to the growth and development of the child] is the aim of providing pupils with the background they will require for the study of mathematics in the later years of their school life" (1962, p. 31; 1963, p. 24; 1968, p. 19). Even in grade 1, "[t]he major objective of a primary number curriculum is to provide through enriching experiences a background of attitudes, appreciations, facts, and skills that will aid in the understanding of the formal Arithmetic of later grades" (1936, p. 89; 1940,

p. 255). Rationalizing the learning of mathematics in the present for the learning of mathematics in the future is self-serving and rarely convincing.

Mathematics is also said to be important for future careers. In 1997, mathematics is justified because "a greater proficiency in using mathematics increases the opportunities available to individuals" (p. 2). Although this rationale appears only once in the documents, it is one of the most common notions I hear from others who explain why they think mathematics is important. However, if "[w]e are teaching for a changing world" (1936, p. 89), how can we know what the opportunities will be available in a "rapidly advancing, technological society" (1997, p. 1)?

A final assumption appearing throughout the decades is the focus on the individual learner. In the first half of the last century, educational goals focused exclusively on individuals. Consider these statements from the *Programme of Studies for the Elementary School*, written in 1947:

> The ultimate goal of education is the happiness of the individual. Accordingly, the teacher's purpose is to assist each child in the class to unfold as fully as possible his unique potentialities. (1947, p. 10)

The arithmetic "point of view," in 1947, also focuses on individual learners, but I didn't see anything suggesting that the mathematics program was purposefully contributing to the children's happiness. The view that education should foster "the fullest development of each child's potentialities" occurred repeatedly (1963, p. 7; 1968, p. 4), particularly in mathematics.

In 1975, the educational goals shift radically to suggest that "education must provide opportunities to meet individual and societal needs" (p. 1). However, the mathematics general objectives make little mention of society and continued to read "the aim of Elementary School Mathematics is to foster continuous and maximum development of each child's potentialities" (p. 21). In fact, it emphasizes, "the growth of a mature individual who thinks and acts effectively, and who, as a result *may contribute to society*" [italics added] (p. 21). The hedge on making a contribution to society seems particularly odd. In 1978, the educational goals also emphasize "fulfilling personal aspirations while making a positive contribution to society" (p. 3), but the mathematics goals in that year focus exclusively on individuals and make no mention of society or social relationships.

FINAL THOUGHTS

My quest to locate a convincing rationale for teaching mathematics through almost a century of curriculum documents has been wholly unsuccessful. To my mind, learning mathematics for the purposes of future progress of education, for its utility, for later mathematics learning, for future careers, and for individual growth are, at best, inadequate and, at worst, inappropriate. Why do we teach mathematics to children?

The current program of studies says that "Mathematics is one way of trying to understand, interpret and describe our world" (2007, p. 10). I mentioned previously that this is perhaps one part of a rationale, but it, too, is inadequate. The statement neglects the fact that humans who use mathematics play an essential role in creating our world—a mathematical world. Much like the "nature of language" in the English language arts curriculum (2000a), mathematics is also a "primary instrument of thought" and a "defining feature of culture" (p. 1). How does mathematics shape our thinking? How has it defined our culture? What has mathematics allowed us to create? How has it constrained or perhaps disallowed possibilities for knowing, seeing, doing, and being? Why is mathematics the only "academic" school subject that does not include a critical thinking component or address issues within the discipline? How does our uncritical stance on the value of mathematics prevent us from moving beyond individual development towards cultural implications of a mathematical world? Mathematics is important for children, but there are many more questions that need to be raised and discussed before I'll be comfortable dusting off and stepping on my soapbox.

NOTE

1. All references to programs of studies are to programs produced by the Department of Education (1914–1968), Alberta Education (1975–1985; 2000–2008) and Alberta Learning (1997).

REFERENCES

Alberta Education. (1975). *Program of studies for elementary schools.* Edmonton, AB: Author.

Alberta Education. (1978). *Program of studies for elementary schools.* Edmonton, AB: Author.

Alberta Education. (1982). *Elementary mathematics curriculum guide.* Edmonton, AB: Author.

Alberta Education. (1985a). *Art (elementary).* Edmonton, AB: Author. Also available at www.education.alberta.ca/media/3 12998/elemart.pdf

Alberta Education. (1985b). *Drama (elementary).* Edmonton, AB: Author. Also available at www.education.alberta.ca/media/313001/elemdram.pdf

Alberta Education. (1989). *Music (elementary).* Edmonton, AB: Author. Also available at www.education.alberta.ca/media/313004/elemusic.pdf

Alberta Education. (2000a). *English language arts*. Edmonton, AB: Author. Also available at www.education.alberta.ca/media/450519/elak-9.pdf

Alberta Education. (2000b). *Physical education kindergarten to grade 12*. Edmonton, AB: Author. Also available at www.education.alberta.ca/media/450871 / phys2000.pdf

Alberta Education. (2002). *Health and life skills kindergarten to grade 9*. Edmonton, AB: Author. Also available at www.education.alberta.ca/media/313382/ health.pdf

Alberta Education. (2005). *Social studies kindergarten to grade 12*. Edmonton, AB: Author. Also available at www.education.gov.ab.ca/k_12/curriculum/bySubjectlsocial/ sockto3.pdf

Alberta Education. (2007). *Mathematics kindergarten to grade 9*. Edmonton, AB: Author. Also available at www.education.alberta.ca/ media/645594/kto9math.pdf

Alberta Education. (2008). *Programs of study (Core Curriculum)*. Edmonton, AB: Author. Also available at www.education.alberta.ca/ teachers!core.aspx

Alberta Learning. (1997a). *Mathematics kindergarten to grade 6*. Edmonton, AB: Author. Also available at www.education.albertaca/media/450754/elemmath.pdf

Alberta Learning. (1997b). *Science*. Edmonton, AB: Author. Also available at www. education.alberta.ca/media/654825/elemsci.pdf

Davis, B. (2001). Why teach mathematics to all students? *For the Learning of Mathematics, 21*(1), 17–24.

Department of Education. (1918). *Course of studies for the public schools: Grades I–VIII*. Edmonton, AB: J W Jeffery, King's Printer.

Department of Education. (1924). *English, citizenship and arithmetic. Part I of the programme of studies for the elementary schools of Alberta. Grades I to VIII inclusive*. Edmonton, AB: J W Jeffery, King's Printer.

Department of Education. (1936). *Programme of studies for the elementary school. Grades I to VI*. Edmonton, AB: A Shnitka, King's Printer.

Department of Education. (1940). *Programme of studies for the elementary school. grades I to VI*. Edmonton, AB: A Shnitka, King's Printer.

Department of Education. (1947). *Programme of studies for the elementary school. Bulletin II. The Enterprise, Language, Reading, Arithmetic, Music*. Edmonton, AB: A Shnitka, King's Printer.

Department of Education. (1962). *Program of studies for elementary schools of Alberta*. Edmonton, AB: Author.

Department of Education. (1963). *Program of studies for elementary schools of Alberta*. Edmonton, AB: Author.

Department of Education. (1968). *Program of studies for elementary schools of Alberta*. Edmonton, AB: Author.

Huckstep, P. (2000). The utility of mathematics education: Some responses to scepticism. *For the Learning of Mathematics, 20*(20), 88–103.

Noddings, N. (1994). Does everybody count? Reflections on reform in school mathematics. *The Journal of Mathematical Behavior, 13*(1), 89–104.

CHAPTER 48

PERCEPTIONS OF PROBLEM SOLVING IN ELEMENTARY CURRICULUM

Jennifer Holm

Editor's Note: Jennifer Holm is currently a doctoral student at Lakehead University, where she teaches a Math for Teaching course for junior/intermediate preservice teachers. She has a master of education from Lakehead University, as well as a bachelor of arts in elementary education from the University of West Florida. Jennifer has previously worked for elementary school boards in both Tennessee and Florida in the United States. Her research is focused on furthering mathematics education while supporting educators in utilizing reform-based methodologies for teaching mathematics.

INTRODUCTION: A LOOK AT THE MEANINGS OF PROBLEM SOLVING

Recent research with intermediate teachers[1] indicates that the phrase *problem solving* often evokes multiple meanings in mathematics teaching and learning (Kajander & Mason, 2007). While some teachers support the vision of problem solving espoused by the National Council of Teachers of

Mathematics (NCTM) (2000) as "engaging in a task for which the solution method is not known in advance" (p. 51), others view it as something to be done after students are taught and only if there is time (Kajander & Mason, 2007). The goal of this article is to examine the usefulness and intent of the various meanings of this "problematic" phrase, while shedding light on the best way to engage in effective problem solving with students.

The inclusion of problem solving in mathematical learning is not new. Polya's (1957) famous model is even included in some provincial curricula (for example, Ontario's Ministry of Education, 2005), as shown in Figure 48.1, and is one of the best known outlines of the possible processes involved in problem solving. What is perhaps new in many classrooms is that effective problem solving should be more than having students solve a problem using formulas or methods that the teacher has previously shown.

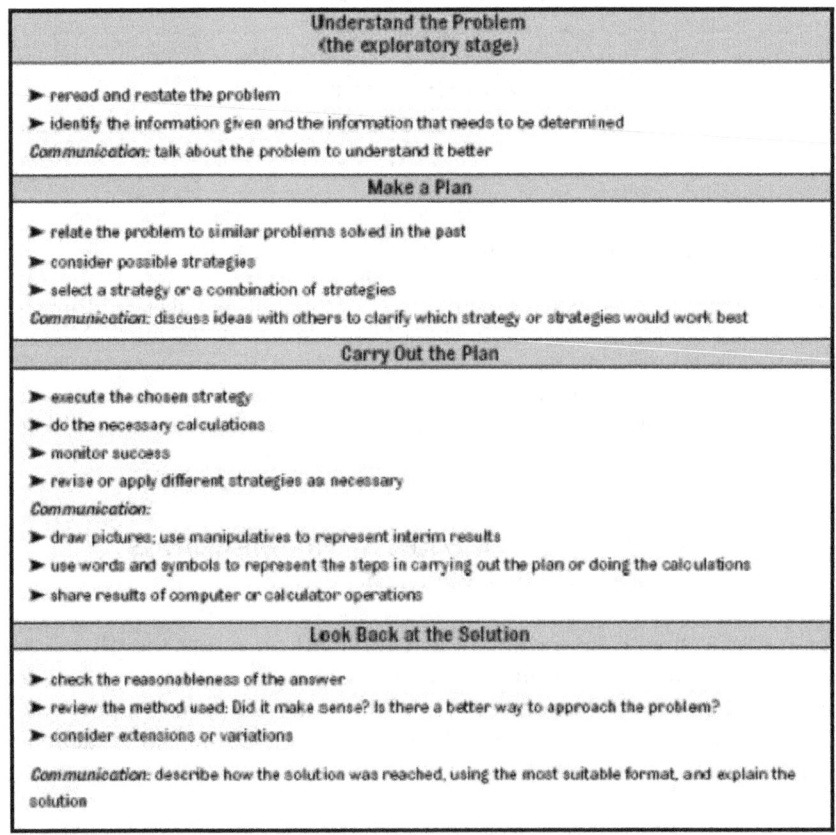

Figure 48.1 Polya's problem-solving model. (Ontario Ministry of Education, 2005, 13).

As a grade 2 teacher in the United States, I was able to experience first-hand the effects of new legislation that required us to drill mathematics facts and algorithms into the minds of our students so that they could survive testing. Each year the same concepts had to be reviewed, because retention was minimal if at all. We met each year as a school staff to discuss ways to better meet the needs of the students with information handed to us from our school board. We discussed at length using problem solving to improve our mathematics test scores. Yet what this actually entailed, according to what we were told, was handing students a sheet of word problems to solve using the algorithm the teachers had already given them to use. It was suggested that we do a similar word problem with the students so that they would follow a similar process. Similar understandings of the implementation of problem solving have also been found in some Canadian classrooms (e.g., Kajander & Mason, 2007; Kajander & Zuke, 2008).

Elementary mathematics curricular goals may refer to problem solving without exploring what the phrase really means. In Alberta, for example, the curriculum endorses the importance of using a problem-solving approach noting that "students need to explore problem-solving strategies in order to develop personal strategies and become mathematically literate" (Alberta Education, 2007, p. 1). Ontario, as well, mentions that "problem solving forms the basis of effective mathematics programs and should be the mainstay of mathematical instruction" (Ontario Ministry of Education, 2005, p. 11). These statements could be interpreted in multiple ways. From these statements in the mathematics curriculum guides, teachers could assume that problem solving means having students read a problem and find a correct solution using a given, previously taught method. Research has shown that this is not the most effective way to use problem solving in mathematics classrooms (Bay-Williams & Meyer, 2005; Boaler & Humphreys, 2005; Buschman, 2004), nor is it the most effective method for teaching mathematics (Askey, 1999; NCTM 2000; Van de Walle & Lovin, 2006). This article will further examine some alternatives to this common interpretation.

PROBLEM SOLVING AS LEARNING

I take the stance that true problem solving involves students really learning something new and not just applying a previously taught strategy to a new example or task. This position underscores the importance of problem solving as learning. As Bay-Williams and Meyer (2005) note, "teacher-directed instruction may help a teacher feel that more topics have been covered, but it reduces the chances that students are (1) making connections with other mathematical ideas and (2) understanding the concepts related to the skill" (p. 340). In fact, students should engage in rich problem-solving tasks *in their daily mathematics classroom experience* in order to construct new knowledge and understanding by connecting it to their previous knowledge.

This interpretation of effective problem solving differs from the belief that students must be taught the concepts before they can engage in problem solving (Kajander & Mason, 2007; Kajander & Zuke, 2007, 2008). One view might be that by assigning the problem-solving questions in the textbook for homework, exercises, and on tests, students have the opportunity to problem solve. Typically, such problems are really applications of known formulas or methods to new examples. However, the goal with problem-solving tasks should be to allow students to figure out *how* they will solve the problem. The importance is placed on the *method for determining the solution,* as opposed to the solution itself. As McGatha and Sheffield (2006) point out, in problem-solving classrooms "students are pushed beyond simply *finding a right answer to questioning the answer*" (p. 79; emphasis in original). The one single right answer is no longer the singular goal of the mathematics classroom; rather, the process taken to find an answer is where the real learning lies. Students should subsequently be given opportunities to discuss how they solved the problem so that they can learn from each other and see different ways of arriving at a possible solution. This is very different from classrooms in which the teacher tells the students how to go about solving problems so that they can arrive at the single right answer in this so-called correct way. True problem-based learning involves students constructing new ideas based on their experiences with appropriate problems, *not* applying known methods to new contexts.

HOW EFFECTIVE PROBLEM SOLVING IS ACCOMPLISHED

Effective problem-solving tasks can be implemented as part of a three-part lesson plan (Van de Walle & Lovin, 2006). It is important to consider that the actual lesson may take more than a single mathematics class period to finish, depending on the students. In the first part of the lesson, the teacher sets up the current problem to be worked on. The teacher acquaints the students with any previously unknown vocabulary at this time. This portion of the lesson does not include the teacher showing the students a similar problem and how to solve it. After the teacher sets the stage for learning, the students begin to explore the given problem.

The second stage of a problem-solving lesson requires teachers to set up an environment and procedures that are conducive to exploratory learning. While exploring the problem, students may work individually, in pairs, or in groups. Students need to be arranged in a way that allows them to share their ideas with each other. During this phase of the lesson, the students work with the problem to figure out a solution method that makes sense to them. As students work with the mathematics concepts embedded in the problem, they should record their thoughts to share during the final portion of an effective problem-solving lesson—the discussion.

Discussion is an absolutely essential phase of the problem-solving method because it allows students to come together and share while explaining their thinking. As Boaler and Humphreys (2005) note,

> students are not asked to present their answers; they are asked to show representations of their ideas and to justify why they make sense. None of the audience members will have the exact same answer, and all the students have a role. (p. 50)

Not only are students more engaged while discussing ideas with their peers, they also learn more from each other and discover new ways of thinking about a problem. Students need to be able to put their solutions into words and discuss how they solved the problem so that they can explain their methods to others. This forces students to get at *how* their solution was found, not just what they decided was the correct answer. It is important that students learn "to *question the answers* by posing additional questions when solving the original problem [because this] is one way that teachers and students can develop mathematical power" (McGatha & Sheffield, 2006, p. 79; emphasis in original). It is this power that helps further students' understanding of and learning in mathematics. Boaler and Humphreys (2005) suggest using the method of "convincing a skeptic" when trying to explain the solution the students came up with (originally from Mason, Burton, & Stacey, 1982). Their belief is that "this strategy... helps place responsibility on the person who is explaining to make his [sic] explanations understandable and gives permission for anyone who doesn't understand yet to play the role of being unconvinced rather than being just slow to catch on" (p. 67; emphasis in original). Students are given the opportunity to question each other and refine their thought processes until everyone sees why the solution method works. Seeing alternative solutions is important because "if their knowledge is limited to the computational procedure without any idea why the procedure works, this is also not enough to build on. Students need both" (Askey, 1999, p. 3). Through exploring a problem and discussing the solution, students learn how and why their method and procedures work and gain deeper mathematical understanding. At this point, teachers can help students see the generalizations or the procedures that are being developed through examining the students' solutions. Teachers play an important role in fostering this development of ideas. Since students are sharing their knowledge and understandings, or even misunderstandings, during this portion of the lesson, the teacher must create an environment where all contributions are valued and allowed to be expressed.

In order to use the problem-solving method effectively, students must be given opportunities to share their solution methods so that the teacher can see where any misunderstandings or confusions lie. These essential discussions also allow students to learn from each other. This very important aspect can be the deal breaker for the success of problem-based lessons if the teachers do not allow time for sufficient sharing of ideas. Sometimes issues or difficulties that arise during the discussion can prompt the teacher to suggest a new problem for the next class.

In a problem-solving lesson as just described, problem solving is the vehicle for knowledge and learning instead of simply the way that students showcase what they have learned. One issue with doing problem solving after the teacher has taught a concept is that students have trouble switching from a teacher-directed lesson one day to a lesson in which they control the learning path the next (Van de Walle & Folk, 2007). Also, if students are to truly engage in problem solving, they need to know that the teacher is not about to step in and tell them the strategy or the answer eventually. If they feel that this will happen, my experience is that some students will simply wait for the instruction or answer to come from the teacher and therefore will not deeply engage in the task. In other words, they feel that their work will be devalued eventually when the teacher provides the right answer or method, and they have simply learned to wait for this to happen. By engaging in problem-solving lessons as the main curricular vehicle, students learn that their thoughts and ideas are important and are correct ways to solve a problem.

Teachers should choose problems that allow students to explore and construct knowledge for the big ideas or the overall expectations of the grade level. This allows the teacher to address multiple curricular expectations in one problem while, at the same time, addressing the needs of different learners. The benefit of well-chosen problems is that they "can be solved at different levels of sophistication, enabling all children to access the powerful mathematical ideas embedded in the problem" (English, Fox, & Watters, 2005, p. 156). For example, a problem like the handshake problem could be used with a class: "There are 20 students in a class. On the first day, the teacher asks each student to shake hands with each other student. How many handshakes were there?" (Small, 2008, p. 567; similar problem in Kajander, 2007). In order to make the problem accessible to all students, the teacher could make several different size classes, starting with 5 students and ranging up to 20. Students who are more advanced could begin looking for patterns in the different numbers in order to make the problem more challenging, while students who are struggling can simply tackle what would happen if five students shook hands. By having students solve the problems from their ability level, all students are engaged and learning from each mathematics lesson.

The problem-solving approach also allows all students to be included in the discussions. Students choose methods to solve the problem that make sense to them, which is more meaningful than just repeating what the teacher has said. Using problem solving in the classroom allows all students to reach mathematical understanding at a level that they are comfortable with. Since the goal is to have students use their prior knowledge, all students will be able to work with the problems using what they already know to build their own new ideas. As the Alberta curriculum asserts, "students learn by attaching meaning to what they do, and they need to construct their own meaning of mathematics" (Alberta Education, 2007, p. 1). Teachers can also use this baseline knowledge to help students to come up with new ideas and more effective solution methods instead of teaching formulas that students apply without really understanding. For example, the solution to the handshake problem could be arrived at in many different ways, including drawing a picture, acting out the problem with children, looking for patterns, or even solving algebraically. Students would be able to solve the problem with their own solution methods, but during the discussion would be exposed to all the different methods and thereby learn from the other students. The problem itself can be used to teach or review addition, multiplication, division, geometric patterning, numeric patterning, pattern rules, and iterative patterns (Kajander, 2007), depending on how the teacher guides the students through the discussions and what areas are highlighted as students present their solution methods.

One caution does need to be made in choosing effective problems to solve in order for students to gain the most benefits. Teachers should avoid forcing too much content into a single lesson; therefore, each lesson "focuses on investigating one rich problem, probing deeply into a different mathematical content strand each day" (McGatha & Sheffield, 2006, p. 79). By narrowing the focus to one main concept each day, teachers can allow students to look further into the problem in order to reach a deeper understanding. For example, simply introducing a simple problem like the handshake problem with different-sized classes would allow the students to explore the solution methods; the teacher could then guide discussions to accomplish the necessary curriculum goals. By focusing on one problem, students are not overwhelmed by a worksheet full of problems and could be challenged to come up with multiple solution methods. Another important consideration is that, as one teacher said, "too much choice could be overwhelming for the children and difficult for me to manage" (Whitin, 2004, p. 181). Putting too much into one lesson is not only hard for a teacher to organize and observe, but it can confuse students and prevent them from delving deeply into the topic being explored.

Another benefit of using problem solving extends beyond the mathematics classroom. Since the goal is not for teachers to show students a formula and the exact method to solve the problem, students use their own problem-solving skills to solve the problem. This can affect students' lives—not only do students learn mathematical concepts with deep understanding, they also gain skills that enable them to solve problems in their daily lives. The benefits of using problem solving and allowing students to learn how to solve a problem in their own daily lives are great. I turn now to showing how this can fit within the mathematics curriculum.

EXAMPLES OF USING PROBLEM SOLVING WITH ONTARIO AND ALBERTA CURRICULUM

It is my experience that curriculum guides mention problem solving while not explicitly laying out how to use problem solving in a classroom. For example, in the curriculum I am most familiar with, the Ontario Ministry of Education sets out several mathematical processes that should be included in the elementary curriculum: "problem solving; reasoning and proving; reflecting; selecting tools and computational strategies; connecting; representing; communicating" (Ontario Ministry of Education, 2005, p. 11). These processes are listed as separate entities; yet an effective problem-solving approach to teaching would encompass all of these processes and would, therefore, be the only method necessary to accomplish these curricular goals. After being given a problem, students would have to *select tools* and the *computational strategies* needed to solve the problem. Since the students would be using prior knowledge to pursue a solution, they are *connecting* the new concept to previous knowledge and skills. Students would then be required to *reason* through their solution and *prove* that it works to the class and teacher. Through the discussion of the solution, students would have to *reflect* on whether or not their method makes sense in order to solve the problem. By sharing their solution with others, students would be required to *communicate* their thought processes and *represent* the solution so that others can see how they solved the problem. Using problems with a similar focus on different days would allow students to practise their skills and create more in-depth conceptual knowledge. By using a problem-solving approach to teaching, teachers are able to simplify their planning while meeting all the goals of the curriculum.

In Alberta, the curricular goals identified are that students will "use mathematics confidently to solve problems; communicate and reason mathematically; appreciate and value mathematics; make connections between mathematics; make connections between mathematics and its applications; commit themselves to lifelong learning; [and] become mathematically literate adults, using mathematics to contribute to society" (Alberta Education, 2007, pp. 2–3). As with the Ontario curriculum, the curriculum guide provides the goal of using problem solving in the classroom as part of the routine. Effective problem solving would also help to accomplish the other goals by giving students ample opportunities to use mathematics in meaningful ways that will benefit students throughout their lives. Where Alberta's curriculum differs from Ontario is that it explicitly states that students "must realize that it is acceptable to solve problems in a variety of ways and that a variety of solutions may be acceptable" (p. 1). This statement lends itself more to the problem-solving methods described within this article, but teachers should keep in mind that this should not mean giving students different solution methods but allowing them to discover multiple solution methods. The two curriculums mention the importance of problem solving, but it is my experience that teachers are often left to their own devices to locate the problems that would address these goals. Some resources that I have used and found successful with students include *50 Problem-Solving Lessons: Grades 1–6,* by Marilyn Burns, and *Big Ideas for Small Mathematicians,* by Ann Kajander. The problems in these books mention different curriculum strands addressed by the problems, and a single problem could be used in a lesson to give students a chance to delve deeply into the topics. In order to encourage more problem solving, the Ontario Ministry of Education has also created its own lessons that relate to the curriculum for grades 7 to 10, called *Targeted Implementation and Planning Supports* (TIPS). Textbooks might also be a helpful tool—teachers could choose or create a single problem from the lesson that students could explore on their own in order to determine their own solution methods.

In Ontario, the Ministry of Education (2005) does provide a valuable framework that can be used with students while exploring a problem. During the exploratory phase of problem solving, the Ministry of Education suggests using Polya's problem-solving model (see Figure 48.1) to guide the students in thinking about how to solve a problem. The belief is that teachers should guide students in grades 1 and 2 through the model without directly teaching the steps, whereas students in grade 3 and above should be taught the terminology of each step of the model directly. For grades 1 and 2, a simpler way of remembering the steps can be beneficial.

Thomas (2006) has suggested the use of the THINK strategy to get students organized in their thinking, which could be a useful mnemonic for grades 1 and 2. First, students talk about the problem. Second, students look at how the problem could be solved. Third, students identify a strategy for solving the problem. Fourth, students notice how the strategy helped solve the problem. Finally, they keep thinking about the problem. As students continue working on the problem, they may need to cycle through this framework several times until they arrive at a solution that makes sense to them. According to the research study, Thomas notes that "students who used THINK demonstrated greater growth in problem solving than students who did not use the frame- work" (p. 86). The use of a model is beneficial because a teacher who is aware of the model and who uses it to guide his or her questioning and prompting during the problem-solving process will help students internalize a valuable approach that can be generalized to other problem solving situations, not only in mathematics but in other subjects as well. (Ontario Ministry of Education, 2005, p. 13)

Students could be coached using this model while they are exploring the problem. The Polya model should be directly taught to higher grade levels (Ontario Ministry of Education, 2005). While older students are working with the problem, the first step is for them to understand the problem. According to Outhred and Sardelich (2005), "understanding the problem requires children: to be able to read the problem; to comprehend the quantities and relationships in the problem; to translate this information into mathematical form; and to check whether their answer is reasonable" (p. 146). Students begin by rereading the problem and deciding what the problem is asking them to figure out. Next, students make a plan for deciding how to solve the problem through examining different strategies to solve it. As Askey (1999) discovered when working with teachers, "the teachers argued that not only should students know various ways of calculating a problem [solution] but they should also be able to evaluate these ways to determine which would be the most reasonable to use" (p. 6). Third, students enact the plan that they decide to use to solve the problem. Finally, students assess whether or not the solution is reasonable through reexamining the problem. If the solution is determined to be unreasonable, students would then go back through the model. This model is important because when students are used to traditional instruction they typically do not have the skills and strategies developed to effectively problem solve (Van de Walle & Folk, 2007; Kajander & Zuke, 2007). Once students are given a problem, they need to be given a way of organizing their thinking in order to solve the problem (Figure 48.1).

SUMMARY

Problem solving may have varying definitions for different teachers, but effective problem solving should allow students to explore a problem for themselves to find a solution. I have argued that problem solving does not involve giving students a method or formula for how to get the answer; rather, it involves giving them a framework to think through the problem and work to develop their own method. Students need a structure to develop problem-solving skills, and this must be supported by peer and teacher-facilitated discussion at certain points in the learning. Neither of these can take place when problem solving is attempted in isolation as homework or on tests. True problem solving cannot happen for most students (or even most mathematicians!) in a time-limited situation such as a test. Students need time to reflect, discuss, and try possibilities. Tests are simply not good places to attempt problem solving. While tests might play a role in efficiently assessing procedural skills, learning and assessment tasks are much better vehicles for learning through problem solving. Teachers in an effective problem-solving environment are no longer disseminators of knowledge but facilitators and coaches who help students create their own knowledge. In my experience, one of the most rewarding experiences can be watching students grapple with a problem and come to a solution after they have worked on the concepts within the problem. The excitement and feelings of accomplishment that accompany the final product can be empowering to their mathematical abilities, as well as foster the idea that they, too, can do mathematics! While using problem solving and discussion may be uncomfortable at first, the long-term benefits for both student learning and engagement are phenomenal (NCTM, 2000). The goal of mathematics classrooms is to have students learn and understand mathematics, and engaging in effective problem-solving tasks is the best way to accomplish these goals.

ACKNOWLEDGEMENTS

I would like to acknowledge the contributions of Dr Ann Kajander at Lakehead University for her invaluable input and guidance in writing this article. This work was funded by the NSERC University of Manitoba CRYSTAL grant for the project "Under- standing the Dynamics of Risk and Protective Factors in Promoting Success in Science and Mathematics Education."

NOTE

1. This research was conducted by the University of Manitoba CRYSTAL subproject located at Lakehead University, funded by the National Science and Engineering Research Council of Canada.

REFERENCES

Alberta Education. (2007). *Mathematics Kindergarten to Grade 9. Program of studies.* Edmonton, AB: Alberta Education.

Askey, R. (1999). Knowing and teaching elementary mathematics. *American Educator, 23*(3). Retrieved from www.aft.org/pubs-reports/american_educator/fall99/amed1.pdf

Bay-Williams, J. M., & Meyer, M. R. (2005). Why not just tell students how to solve the problem? *Mathematics Teaching in the Middle School, 10*(7), 340–341.

Boaler, J., & Humphreys, C. (2005). *Connecting mathematical ideas: Middle school video cases to support teaching and learning.* Portsmouth, NH: Heinemann.

Burns, M. (1996). *50 problem-solving lessons: Grades 1–6.* Sausalito, CA: Math Solutions Publications.

Buschman, L. (2004). Teaching problem solving in mathematics. *Teaching Children Mathematics, 10*(6), 302–309.

English, L. D., Fox, J. L., & Watters, J. J. (2005). Problem posing and solving with mathematical modeling." *Teaching Children Mathematics 12*(3), 156–163.

Kajander, A. (2007). *Big ideas for small mathematicians.* Chicago, IL: Zephyr.

Kajander, A., & Mason, R. (2007). Examining teacher growth in professional learning groups for in-service teachers of mathematics. *Canadian Journal of Mathematics, Science, and Technology Education, 7*(4), 417–438.

Kajander, A., & Zuke, C. (2007). *Factors contributing to success for intermediate students of mathematics: the needs of the teacher and the characteristics and needs of students at risk.* Final Report. Thunder Bay, Ont: Northern Ontario Educational Leaders (NOEL). Retrieved from http://noelonline.ca/depo/ fd/les/Kajander-NOEL%20FINAL%20report%20APRIL%2030%202007.pdf

Kajander, A., & Zuke, C. (2008). Listening for their voices: A study of intermediate students at risk in mathematics. *delta-K, 45*(2), 21–24.

Mason, J., Burton, L., & Stacey, K. (1982). *Thinking mathematically.* London, UK: Addison-Wesley.

McGatha, M. B., & Sheffield, L. J. (2006). Mighty mathematicians using problem posing and problem solving to develop mathematical power. *Teaching Children Mathematics, 13*(2), 79–85.

National Council of Teachers of Mathematics (NCTM). (2000). *Principles and standards for school mathematics.* Reston, VA: Author.

Ontario Ministry of Education. (2005). *The Ontario curriculum grades 1–8: Mathematics* (Rev. ed.). Ontario: Queen's Printer.

Ontario Ministry of Education. (n.d.). *Targeted implementation and planning supports for revised mathematics.* Retrieved from www.edu.gov.on.ca/eng/studentsuccess/lms/tips4rm.html

Outhred, L., & Sardelich, S. (2005). "A problem is something you don't want to have": Problem solving by kindergartners. *Teaching Children Mathematics, 12*(3), 146–154.

Polya, G. (1957). *How to solve it* (2nd ed.). Garden City, NY: Doubleday.

Small, M. (2008). *Making math meaningful to Canadian students, K–8.* Toronto, ON: Nelson Education.

Springer, G. T., & Dick, T. (2006). Connecting research to teaching: Making the right (discourse) moves: Facilitating discussions in the mathematics classroom. *Mathematics Teacher, 100*(2), 105–109.

Thomas, K. R. (2006). Students THINK: A framework for improving solving. *Teaching Children Mathematics, 13*(2), 86–95.

Van de Walle, J., & Folk, S. (2007). *Elementary and middle school mathematics: Teaching developmentally* (2nd Canadian ed.). Toronto, ON: Pearson.

Van de Walle, J. A., & Lovin, L. H. (2006). *Teaching student-centered mathematics: Grades 3–5*. Boston, MA: Allyn & Bacon.

Whitin, P. (2004). Promoting problem-posing explorations. *Teaching Children Mathematics, 11*(4), 180–186.

This chapter was originally published as:

Sterenberg, G., et al. (2010). To become wise to the world around us:
   Multiple perspectives on relating indigenous knowledges and math-
   ematics education. *delta-K: Journal of the Mathematics Council of the
   Alberta Teachers' Association, 47*(2), 21–29.

CHAPTER 49

TO BECOME WISE TO THE WORLD AROUND US

Multiple Perspectives of Relating Indigenous Knowledges and Mathematics Education

**Gladys Sterenberg, Liz Barrett, Narcisse Blood,
Florence Glanfield, Lisa Lunney Borden, Theresa McDonnell,
Cynthia Nicol, and Harley Weston**

Editor's Note: Liz Barrett is a teacher and a First Nations Outreach and Training Consultant for JUMP Math. Narcisse Blood, Ki'naksaapo'p, Iitsitssko'pa, was a previous Coordinator of the Kainai Studies program at Red Crow College, and has been recognized as an Eminent Scholar (K.Ph.D.) by Red Crow College. He currently teaches for the Kainai Studies program, the Department of Education at Lethbridge University, and the International Indigenous Studies Department at the University of Calgary. Narcisse has served as Chair for the Mookaakin Cultural and Heritage Foundation of the Blood Tribe and served on the Blood Tribe Chief and Council. He has principal transfer rights in the Niitsitapi knowledge disciplines of Iitskinaiksi and Ninnaimsskaiksi. Florence Glanfield is of Métis ancestry and an Associate Professor in mathematics education in the Department of Secondary Education at the University of Alberta. Her research interests include mathematics teacher development, community-based research, narrative inquiry, and Aboriginal curriculum perspectives. Lisa Lunney Borden is an Assistant Professor in the Faulty of Education at St. Francis Xavier University. Her research interests include culturally responsive mathematics education and Aboriginal education. Theresa McDonnell is a teacher at Siksika Nation High School in Alberta. She is interested in curriculum development and learning from place. Cynthia Nicol is an Associate Professor in the Department of Curriculum and Pedagogy at the University of British Columbia. Her research activities include mathematics teacher education, mathematics and culturally responsive pedagogy, Aboriginal education, case-based and problem based learning, and issues and practices related to conducting participatory action research, community-based research and self-study research. Gladys Sterenberg is an Assistant Professor in the Department of Elementary Education at the University of Alberta. Her research interests include mathematics and story, teacher professional development, learning from place, and curriculum studies. Harley Weston is a Professor Emeritus in the Department of Mathematics and Statistics at the University of Regina. He is the director of Math Central (MathCentral.uregina.ca).

INTRODUCTION

Significant curricular initiatives in mathematics have been undertaken across Canada to appropriately and respectfully consider Indigenous knowledges and perspectives. For example, the Common Curriculum Framework for K–9 Mathematics: Western and Northern Canadian Protocol (Alberta Education, 2006) now describes Aboriginal learners in its Program of Studies:

> Aboriginal students in northern and western Canada come from diverse geographic areas with varied cultural and linguistic backgrounds. Students attend schools in a variety of settings including urban, rural and isolated communities. Teachers need to understand the diversity of cultures and experiences of students.
>
> Aboriginal students often have a whole-world view of the environment in which they live and learn best in a wholistic way. This means that students look for connections in learning and learn best when mathematics is contextualized and not taught as discrete components.
>
> Aboriginal students come from cultures where learning takes place through active participation. Traditionally, little emphasis was placed upon the written word. Oral communication along with practical applications and experiences are important to student learning and understanding. It is also vital that teachers understand and respond to non-verbal cues so that student learning and mathematical understanding are optimized.
>
> A variety of teaching and assessment strategies is required to build upon the diverse knowledge, cultures, communication styles, skills, attitudes, experiences and learning styles of students. (p. 3)

These initiatives, which call on teacher educators, teachers, administrators, and students to reconsider their received understandings of Indigenousness, challenge inherited conceptual frames derived from the lengthy colonial process of looking for mathematics in cultural activities and by doing so validating the activities as mathematical from a Euro-Western perspective. We wonder about possibilities of enacting an Aboriginal perspective within our educational contexts.

In May 2009, a group of mathematicians and mathematics educators of non-Aboriginal and Aboriginal descent participated in a study group on Indigenous knowledges at the Canadian Mathematics Education Forum. For three days, we engaged in authentic conversations (Clark, 2001) as we investigated the following questions: What do Indigenous knowledges offer for teaching and learning mathematics by both Aboriginal and non-Aboriginal students and teachers? What role does language play? What role does place, community, and culture play? Narcisse Blood, recognized as an Iitsitsskopa (emplaced-for-a-reason, or elder), led our conversations and offered guidance and wisdom in our endeavour to better understand the relationship between Indigenous knowledges and teaching mathematics.

This paper continues these conversations as we draw on our educational experiences to consider the following question: What is the relationship between Indigenous Knowledges and mathematics education in our current research/teaching projects? We have organized stories of our multiple perspectives into four sections: Culturally responsive education, Language and culture, Learning mathematics from place, and Relationships. Our hope is that you may be inspired to appropriately and respectfully consider Indigenous Knowledges and perspectives in your teaching and learning.

CULTURALLY RESPONSIVE EDUCATION

Narcisse Blood

What has culture come to mean today? Culture came from anthropologists and the old country. The view of people back then was that any society other than ours is inferior. And anthropologists started saying, "No, that's not the case. As a matter of fact, some societies are every bit as complex if not more complex than us." That's where the term 'culture' started from. But today, it's changed. It's not that. It has a different connotation, especially where I work and in the community. Culture has evolved into a definition of right. We're going to work with the [Universities] and they say, "Oh yeah, let's not leave out the cultural component." You know, as an afterthought. "Let's get Ryan and Narcisse in there to appease." It sounds good. Because when I was in high school, Indian Affairs came up with that same kind of thing: "We'll throw in culture. We're going to teach them how to bead." And they brought Mrs.

Rosie Day Rider and the late Louise Cropped Ear. They knew what they were up to. In the classes, you'd be doing beadwork, and they'd be telling stories. It was not what Indian affairs wanted, you know, "We'll throw a bead and feather in there to shut them up."

Cynthia Nicol

For the past few years I have had the honour of working with researchers at the University of British Columbia along with teachers, students and community members in two rural communities and one urban community. In each community we are exploring what it means to live culturally responsive mathematics education. Our work is guided and informed by Indigenous knowledges in multiple ways. First our project brings together both Aboriginal and non-Aboriginal scholars, school-based teachers and researchers, students and community members to re-imagine mathematics teaching and learning. Second our work is informed by and guided by the framework of Indigenous storywork developed by co-reseacher Jo-ann Archibald. Working with teachers we are exploring the seven principles of reverence, respect, reciprocity, responsibility, synergy, interrelatedness, and holism in terms of our research and mathematics pedagogy.

We are exploring our awareness of these principles, how they can be used to connect mathematics, community, culture, and Indigenous Knowledges and what this could mean for improving students' mathematical experiences and emotions. For example, Indigenous storywork as a methodology focuses our attention on the importance of building relationships that are respectful and mindful of community protocols, of deeply appreciating the stories and experiences shared, of giving back and paying forward our work so that future teachers and students may learn from our experiences, and considering our multiple responsibilities in the research process. Experiences with the land and nature emphasize the value of and connection of the inner passion, feel, and heart with the environment. For us, Indigenous epistemologies can be characterized as being experiential, storied, relational, contextual, and wholistic.

Our goal is transforming mathematics education so that it serves the diverse interests and aspirations of all Aboriginal and non-Aboriginal students within our project communities and beyond. Our goal is actualized by the following processes: (1) embracing a wholistic and interconnected view of mathematical knowledge; (2) exploring practical approaches for using local and traditional knowledges as resources for mathematics pedagogy; and (3) using local cultural values to extend possibilities for pedagogical practices, connecting with students, and learning from students.

Project teachers are exploring culturally responsive mathematics education in multiple ways. One is by developing mathematics problems inspired by traditional legends. For example, teachers designed a lesson from an audio recording of Raven Brings the Light as told by community elders and youth. In this story the Raven brings light to the world after taking it from a series of nested bentwood boxes. The lesson teachers designed and are currently piloting in their classrooms invites students to listen to the story and then explore the mathematics of building bentwood boxes out of paper. A second way of exploring culturally responsive education is through developing digital storybooks. Teachers and students have collected a series of images of the land and activities connected to the land. Together they have designed three mathematics photobooks the involved developing, adapting, and writing mathematics problems inspired by the images. A third context for exploring culturally responsive education has involved collaboration with the local museum. In this context teachers are exploring the mathematics of the process of canoe building from sapling to sea. Teachers are researching their own experiences of culturally responsive education within their classrooms and are meeting once every couple months to share their ideas and stories.

Harley Weston

During the past three years I have had the opportunity to develop a relationship with a number of Aboriginal Elders and other leaders in the Aboriginal community in Western Canada. I feel extremely privileged to have benefited from these leaders sharing their knowledge. Their knowledge has guided Judi MacDonald and me in the development of resources for use by teachers, students, and parents in the Aboriginal and non-Aboriginal communities. These resources, which are being written by education students at the University of Regina and the Gabriel Dumont Institute, grow out of the audio and video capture of activities and conversations with people in the Aboriginal community. The use of audio and video material in this initiative facilitates the inclusion of a cultural context as an integral part of the resources as well as making the content accessible to students in younger grades.

The following are three examples of the activities/conversations for which teaching resources are being prepared.

- Elder Glen Anaquod from the Piapot First Nation led the students from Kitchener School through a tipi raising. This is an inner school in Regina where many of the students are Aboriginal and through this activity Elder Anaquod shared traditional Saulteaux teachings with the students.

- Birch bark biting is an art form produced by folding a paper-thin sheet of birch bark and placing a design on it through perforating the birch bark by biting. The artist, Rosella Carney is a Cree woman form La Ronge Saskatchewan. Rosella is also a Cree language instructor, and her interview includes a discussion of the number words in Cree.
- Cassandra Opikokew was a student in the Indian Communications Arts program at the First Nations University of Canada and the School of Journalism at University of Regina School from which she graduated in 2009. In her interview, Cassandra talks about how she uses the skills she gained in both programs and the importance of Aboriginal issues and voices in the media.

LANGUAGE AND MATHEMATICS

Narcisse Blood

What we're talking about is a language that was built over thousands of years. So one of the barriers to learning is also time, concepts of time, constructs of time. We work with the museums or some of the parks like Writing on Stone Provincial Park. We have something similar about preserving sacred sites, but their concept of preserving is it's in the past. By virtue of us being there and speaking the language and looking at the petroglyphs, takes it out of that model and brings it to the present. And to get that across is quite difficult but there's movement along that line. And therefore, when we open a bundle, it's, if you want to put it this way, it's history happening right there. So the language is that, it's evolved over thousands of years.

Lisa Lunney Borden

I began my teaching career in We'koqma'q First Nation on Cape Breton Island in Nova Scotia, which is a Mi'kmaw community. Mi'kmaw communities in Nova Scotia have a unique jurisdictional agreement with the Government of Canada that gives them control over their education system and collective bargaining power. These schools live within the tension of wanting to provide culturally responsive and language-rich programs for their students while at the same time being required by law to offer provincially approved curricula.

Disengagement from mathematics and science is a concern for many teachers in these schools as they grapple with the tensions between school-based mathematics and Mi'kmaw ways of reasoning about things seen as mathematical. Having taught secondary mathematics for ten years, I had experienced these tensions myself. It was this experience, and the related learning from community members, that brought me to my doctoral work. My goal was to work towards the development of culturally responsive mathematics curricula with participant schools, yet it soon became evident that a necessary first step was to investigate the tensions between Mi'kmaw cultural ways of knowing and school-based mathematics.

My research took place in two schools and involved after-school discussion with teachers in the form of *mawikinutimatimk* (coming together to learn together). This traditional community practice values the contributions of all participants and acknowledges that we each have something to teach and something to learn. Through our conversations, four key areas of concern emerged as themes: (1) the need to learn from Mi'kmaw language, (2) the importance of attending to value differences between Mi'kmaw concepts of mathematics and school-based mathematics, (3) the importance of attending to ways of learning and knowing, and (4) the significance of making ethnomathematical connections for students. Although interconnected and interdependent, each of these themes can be tied to the need to learn from Mi'kmaw language.

The important role of indigenous language in understanding mathematics was demonstrated by Denny (1981) who used a "learning from language" approach while working with a group of Inuit elders in Northern Canada to explore mathematical words in the Inuktitut language. Rather than developing curriculum and translating it into Inuktitut, they used the mathematical words to develop the curriculum and associated mathematics activities. More recently Bill Barton (2008) has shared the stories of his similar struggles in translating mathematics concepts into the Maori language. He has argued that mathematics evolves with language and as such claims that:

> A proper understanding of the link between language and mathematics may be the key to finally throwing off the shadow of imperialism and colonialisation that continues to haunt education for indigenous groups in a modern world of international languages and global curricula. (p. 9)

This connection between language and decolonization of education was certainly evident in the conversations with participants during my doctoral work.

The principal of one of the participant schools arrived at our session one day and told me that I should call my dissertation "lost in translation." She proceeded to talk about the difference between *lnuitasi* (Our people's ways of thinking) and *aklasiweitasi* (Anglophone ways of thinking). She argued that many conflicts arise for children when their ways of thinking (*lnuitasi*) come into conflict with teachers who have different ways of thinking (*aklasiweitasi*). There is something being lost in the translation of worldviews, of ways of thinking and styles of communication. She claimed that she sees the conflicts arising on a daily basis as the students in her school struggle to find their way through a colonizing curriculum. Something is being lost in translation, and she is searching for ways to resolve this challenge. The Mi'kmaw language holds the key to *lnuitasi*.

Three key pieces emerged in our discussion about the role that language plays in understanding the link between Indigenous Knowledge and mathematics. First, there was an expressed need to include more Mi'kmaw language in the mathematics classroom. This group in particular stressed the importance of reclaiming mathematical words and supporting Mi'kmaw speaking teachers to develop a lexicon of words that could be used in their classes. Many participants shared their belief that using the Mi'kmaw language as much as possible can only benefit students. Several participants expressed how much more responsive students were when they were asked to complete a task or do more work in Mi'kmaq. In particular, one teacher participant commented on the way in which her students often don't understand what she means when she says "how many" but noted, "say '*Tasikl* (how many—inanimate)?' and they get it."

Second, there is value in asking questions such as "What's the word for...?" or "Is there a word for...?" through raising such questions, our groups began to gain new insight into the ways of thinking (*lnuitasi*) embedded in the language. It was often most interesting to discover that a certain word that was taken for granted to be understood by children in school-based mathematics perhaps did not even exist in the Mi'kmaw language. "Flat" is one such word. There is no word for *flat* in Mi'kmaw, yet it is commonly used in mathematics. When I asked elders in the community how they would describe the bottom of the basket which is flat (based on an Anglophone worldview), I was told it was "just the bottom of the basket. It's what lets it sit still." And when I asked about a flat tire I was told that it was "out of air." If *flat* is not a word that is used in Mi'kmaq, what happens when this word is used in mathematics classrooms and it is often taken for granted that there is a shared understanding of its meaning?

Third, there is a sense of motion embedded in the Mi'kmaw language that is not apparent in school-based mathematics. Shape and space words act as verbs in Mi'kmaw and are dynamic. For example, to describe something that it straight, Mi'kmaw speakers might say "*pektaqtek*," which means "from here to there it pretty much goes straight." This dynamic nature of mathematical ideas in Mi'kmaq comes into conflict with the tendency in school mathematics towards nominalization—turning actions and processes into nouns. Our group wondered what might happen if we drew on the verb-based discourse of Mi'kmaq rather than the noun-based discourse of school mathematics to engage our Mi'kmaw students in the learning of mathematics. Such "verbification"—utilizing the verb-based discourse—may provide Mi'kmaw students with increased understanding of and connection to mathematical concepts (see Lunney Borden, 2009, for a more detailed discussion).

While there is much more work to be done to explore the connections between Aboriginal languages and Indigenous Knowledge for mathematics education, this certainly seems to be an important area of investigation to support the development of culturally responsive mathematics education.

Liz Barrett

Something truly unique has happened in Mount Currie, British Columbia. Xit'olacw Community school is proud to be one of the oldest Band-operated schools in Canada, and a group of dedicated elders and parents have worked tirelessly to ensure that it is one of the leading schools in First Nation Language Education. Reclaiming and protecting the language of the community is a high priority of the school. They have one of the few immersion programs here, so students can learn and understand Lilwat, St'at'imcets, which is the Lilloet language.

The teacher at the Clao'alcw (Raven) School, Terri Williams invited me into her class so that I could model some math strategies that are very effective with teaching young children. The idea struck me that we should teach the same math program being used by other classes at the school, but we should translate this into the language. Everyone was excited by the idea and so we contacted the author of the math program, who was very supportive of the idea and gave us permission to start the translation of the grade 1 books.

This was the start of a language and math adventure. Funding was received from First Voices and so Lois Joseph, team leader of the Lil'wat'ul Culture Centre used the funding to get Elders together to start the intense discussions as to what language would be best used to describe the mathematical concepts that were in the books. New language had to be created to describe things like squares and rectangles, triangles and core patterns, and so the process was a collaborative living process of embracing and agreeing on new words that could and should be used.

What was unique for me was watching the process unfold and observing the respect that the Elders showed to each other throughout the discussions. This was and is still a mammoth team effort and I wish to pay tribute to the team made up of Mary James, Laverne James, Priscilla Ritchie, Georgina Nelson, Veronica Bikadi, Jean Andrew, Vera Edmunds, Theresa Jones, and Dixie Joe. Tanis Grandbois helped to capture the data electronically and Burt Williams is helping to edit the text. We thank Lois Joseph for all her support of the project and we look forward to other communities following the lead of Mount Currie.

NITAWAHSIN-NANNI: LEARNING MATHEMATICS FROM PLACE

Narcisse Blood

> [My colleague and I] were asked to come and share [at the University]. But what stayed with me is, here are the buildings and there's a parking lot. And here are trees that go down to that environment. What struck me is the dissonance in terms of learning. What transformation will take place for the University to acknowledge the place and how tragic that can be. Because we're sitting up there overlooking the harbour. And you have all these ships and all these raw materials that are going to be shipped away somewhere. And how unsustainable that is.

Theresa McDonnell

Infusing the curriculum with Aboriginal perspectives is what Alberta teachers are now required to do. Over the last two years, I have taken on this challenging but important endeavour in my middle school mathematics classes. I teach in Siksika, Alberta but myself, I am of Cree and Irish-Canadian decent. Siksika is one of the member nations of the Blackfoot Confederacy.

In my classroom, I have observed silent struggles with identity creating the greatest barriers to success among my students. They long deeply to learn more of their proud history and accumulated knowledge while being profoundly involved and invested in a global world.

In this setting, my experience has taught me that teaching from an Aboriginal perspective is best accomplished by focusing on the perspectives of the Siksika people. As their knowledge has accumulated over the millennia in one place, it is fair to suggest that Blackfoot perspectives arose from their interaction with the land now labelled Southern and Central Alberta, Eastern Saskatchewan, and Northern Montana. This knowledge is rich in mathematics. I came to see my job as a teacher as empowering Siksika students in recognizing their intuitive mathematical abilities. This was partly done by making time in the schedule for learning from place.

Learning from place continues to be a valid and meaningful method of interpreting and understanding the world; including mathematics. As most mathematicians know, mathematics can be found everywhere, and nature is no exception. Therefore, I decided not to rely solely on mathematic textbooks as reference and guide. Instead we explored mathematics in the world as well as the textbook. They complimented each other well. For example, I taught part of both the Shape and Space units using the spokes, angles, and circular geometry of the Majorville medicine wheel site. Students learned concepts with help from the textbook, but moreover, concepts were explored in activities at the site and in the creation of models afterward in the classroom.

By taking students to sacred sites within the Blackfoot Confederacy and bringing some lessons outside, we soon realized that the land became a very present third party. Lessons don't follow the same schedule that a teacher carefully plans when taught outside. Suddenly other factors arise such as weather, animals, and other unexpected guests, as well as an abundance of student curiosity. This alternative method of teaching mathematics is challenging because it requires a plethora of planning, a field trip budget, awareness of protocols, support from culturally knowledgeable staff or community members, flexible and supportive administration, and a willingness to relinquish some power and structure in teaching.

How each teacher accomplishes meeting the new goals set out by the province of Alberta will be varied and unique. For me, "infusion" is the wrong term. Teaching from an Aboriginal perspective is simply finding what is meaningful and relevant to the students, that honours the ancestors of the host territory in which teachers live and teach. It means teaching the curriculum and addressing silent identity issues simultaneously by revering the land and people from which the students came. This can be accomplished by continuing to find meaning in places and inviting students to see the world mathematically and intuitively.

Gladys Sterenberg

Working alongside my Aboriginal colleagues and friends, I am coming to understand my unique role as a person who can provide bridging experiences among teachers of federally and provincially funded schools. Perhaps this is best illustrated by a story of one such experience.

In my work as a mathematics teacher educator, I have been invited to work in partnership with teachers in several First Nation communities. At one Nation High School, we are investigating how learning from place might be enacted. To date, we have taken the students on two field trips to sacred Blackfoot sites, one of which was to the Big Rock. Located west of Okotoks, Alberta, the Big Rock is the largest glacial erratic in the world and is part of a series of boulders stretching from Utah to Jasper. This site is significant to the Blackfoot community and the splitting of the Big Rock by Napi's actions is a familiar story. On our field trip to the Big Rock, the elder who was with us began our visit with an offering and stories of Napi, a Blackfoot trickster. While we were there, elementary students from the Big Rock School in Okotoks arrived to investigate the meaning of their school name. I knew one of the teachers of this group and once I found out why they were there, I suggested that perhaps her students might want to hear the Napi stories told by the elder. Following protocol, I asked the elder if he could share the stories with them and he was very excited about the opportunity to do this. The elementary students were highly respectful when listening to him. The principal of Big Rock School wrote an affirming editorial response about this experience, and this was published in a local newspaper. In this particular situation, my comfort with federally and provincially funded school communities facilitated a bridging experience. For me, this was a profound experience as I better understood the importance of creating opportunities for sharing knowledge between Aboriginal and non-Aboriginal communities and the possible role I might have in this process.

For me, relating Indigenous Knowledges and mathematics education is focused on the act of listening. As a non-Aboriginal visitor, I am respectful of protocols, always mindful of my place within this place. Learning from place recognizes the intimate relationship that Indigenous people have with the land. In our visits to sacred places, the land has spoken to me as I have listened. Stories of place have grounded our visits and our response to knowing the land has informed our acts. Carbaugh (1999) describes listening as dwelling-in-place. He suggests that Blackfeet listening is a "highly reflective and revelatory mode of communication that can open one to the mysteries of unity between the physical and spiritual, to the relationships between natural and human forms, and to the intimate links between places and persons" (p. 250). The sacred places have spoken to me as I have listened. Through the stories told to us by cultural Elders, I am learning to dwell-in-place.

RELATIONSHIP—RESPONSIBILITY—RECURSION

Narcisse Blood

Really what it is for my people, and I think all of us and sometimes we forget, really what it is all about is this relationship. Ultimately, that's who we are. It's relationship. If you want to learn about the Blackfoot, it's about relationship and relationship and relationship. About everything. Those that we see, and those that we don't see. So with that, I'm starting to get to know a lot of you in here, it's good to see you. And that's important. In our world, we don't take that for granted.

Florence Glanfield

"Florence, ohmigosh this is FOIL! Why didn't anyone ever tell me that there was a relationship? I knew how to multiply... but could never figure out FOIL!" These were the words of Cindy, a student in a mathematics education class for elementary preservice teachers, approximately seven years ago. Cindy and her peers had been exploring the underlying concepts behind the multiplication of two two-digit numbers. In Cindy's work with using base-10 blocks and the subsequent numerical representations, she came to realize that the multiplication of two two-digit numbers was the distributive property of multiplication, or in Cindy's words, 'FOIL.' I now 'hold' Cindy's words with me in my teaching and learning, and have 'held' Cindy's words as I've come to know, and acknowledge, myself as an Aboriginal person in teaching mathematics.

When I consider Aboriginal perspectives in a mathematics classroom I think about the importance of Relationship. In my experiences as an Aboriginal person I've come to understand the importance of acknowledging relationships that we have in all aspects of my life. For example, people have relationships with the land (some call it Mother Earth) as human beings draw sustenance from the land, and once we acknowledge the relationship with the land we then have a responsibility to sustaining the relationship and sustaining the land. A second example is that we have relationships with our family—we draw sustenance from our family and when we acknowledge that we draw sustenance from our relationship with family then we also have a responsibility for sustaining the relationships and sustaining the family. Once we acknowledge the relationships that we have with the land and the family and we act on our responsibility, as being a part of that relationship then our initial relationship is now different than it was. Hence we are in a continuous cycle of Acknowledging Relationship; Acting Responsibility; and Re-living or Re-telling the Relationship (Recursion).

What does this mean for a mathematics education class? Most of my teaching is now teaching preservice teachers about what it means to teach mathematics. The focus of the planning work that I do in my classes is around this notion of Relationship—Responsibility—Recursion. In a classroom there are multiple perspectives on relationship that I believe a mathematics teacher 'holds' in their practices. To name a few—a teacher has a relationship with mathematics; with the experiences they had in learning mathematics; with colleagues in the school; with the community in which the school resides; with their worldview about what it means to learn, teach, and specifically teach and learn mathematics; with their family; with their students; and with themselves. I also believe that a teacher must acknowledge that within each of these relationships there are other relationships that can be named—for example within the relationship one has with mathematics one might be aware that there are relationships within the content of mathematics itself. My development as a mathematics teacher and mathematics teacher educator then becomes a life-long journey as I will continue to come to know about the multiple relationships that exist within my classroom and that I continue to have multiple responsibilities. In other words as I attend to or notice (Mason, 2002) the complexity, and the responsibility, of the relationships that exist in my classroom then my teaching continues to focus on sustaining relationships; my life as a teacher is a series of recursive acts in sustaining relationships.

What might be an example? For example, suppose I am teaching about the multiplication of two binomials in a grade 9 or a grade 10 mathematics class. To focus on relationship, I might ask myself questions such as

- How is it that I understand the multiplication of two binomials?
- In what way(s) is the multiplication of two binomials related to multiplication of two two-digit numbers?
- How is it that I can invite my students to see the relationship between the multiplication of two binomials and the multiplication of two two-digit numbers?
- What were my experiences in learning about the multiplication of two binomials? How is it that those experiences have informed me in how I understand the multiplication of two binomials?
- How is it that learning the multiplication of two binomials contributes to my students' deeper understanding of mathematical ideas?
- In what way will be my actions as a teacher in teaching about the multiplication of two binomials sustain my relationship with my students?
- In what way will my actions as a teacher in teaching about the multiplication of two binomials sustain the relationship my students are developing with mathematics?
- In what way does how I see myself as a learner of mathematics impact the decisions I make as a teacher of mathematics and the tasks that I ask my students to engage in which learning about the multiplication of two binomials?
- How is it that I share my understanding of the relationships within mathematics with my students in a way that sustains the relationship that I have with my students?
- How is it that I share my understanding of the relationships between the multiplication of two binomials and other areas of mathematics with my colleagues?
- How is it that I participate in the development of a shared understanding of these mathematical ideas within the community in which I teach?

The ideas that I write about here are a way that mathematics teachers might begin to focus on the relationships or the interconnectedness of our content, our lives as teachers, the lives of our students, and the context in which we teach. Bopp et al. (1988) write that "the great lesson of the sacred circle is always that separate entities, when seen in light of the universe, are equal and necessary parts of the larger whole. It brings out the ancient teaching of the interconnectedness of all things" (p. 62).

CONCLUSION

Colonization was an intentional act. If we try to convince our students that the only mathematics that exists and is worth studying has its roots in Euro-Western traditions, then we are engaged in such an intentional act. By honouring multiple perspectives of mathematical thinking and knowing, we can come to know mathematics and ourselves in a different way.

The *Common Curriculum Framework for K–9 Mathematics: Western and Northern Canadian Protocol* (2006) states, "The strategies used must go beyond the incidental inclusion of topics and objects unique to a culture or region, and strive to achieve higher levels of multicultural education." While we agree that token gestures of inclusion are inappropriate, we believe that relating Indigenous Knowledges and mathematics education is context specific. Multiple perspectives are necessary and embraced. Moreover, any project must be community initiated.

At the end of our time together, we were left with many questions: How do we teach in ways that are responsive to students and challenging of the systemic inequities? How do we invite others to think with us in similar kinds of critical reflexive research? How do we engage in our own decolonizing? In this report, we offer our stories and challenge others and ourselves with our questions. As we reflect on our experiences within this working group, we contemplate the next steps to be taken and invite others to walk alongside us as we attempt to live our lives differently:

> Take our stories. They're yours. Do with them what you will. Use them in your planning. Tell them to other educators. Forget them. But don't say in the years to come that you would have lived your life differently if only you had heard our stories. You've heard them now. (adapted from Thomas King, 2003)

ACKNOWLEDGMENTS

Liz Barrett and Narcisse Blood were supported by the Pacific Institute for the Mathematical Sciences to attend CMEF.

Lisa Lunney Borden was supported in part by the Pacific Institute for the Mathematical Sciences and the University of New Brunswick to attend CMEF. Funding for her doctoral work came from the Social Sciences and Humanities Research Council.

Theresa McDonnell and Gladys Sterenberg's participation in CMEF was supported by the Alberta Advisory Committee for Educational Studies. Their work on learning from place was supported by the Canadian Council on Learning.

Harley Weston's participation in CMEF was supported by the Imperial Oil Foundation through the foundation's support of Math Central.

REFERENCES

Alberta Education. (2006). *Common curriculum framework for K–9 mathematics: Western and Northern Canadian protocol.* Edmonton, AB: Alberta Education.

Barton, B. (2009). *The language of mathematics: Telling mathematical tales.* New York, NY: Springer.

Bopp, J., Bopp, M., Brown, L., & Lane, P. (1988). *The sacred tree: Curriculum guide.* Lethbridge, AB: University of Lethbridge Four Worlds Development Press.

Carbaugh, D. (1999). "Just listen": "Listening" and language among the Blackfeet. *Western Journal of Communication, 63*(3), 250–270.

Clark, C. M. (2001). Good conversation. In C. M. Clark (Ed.), *Talking shop: Authentic conversation and teacher learning* (pp. 172–182). New York, NY: Teachers College Press.

Denny, J. P. (1981). Curriculum development for teaching mathematics in Inuktitut: The "Learning-from-Language" approach. *Canadian journal of anthropology, 1*(2),199–204.

Lunney Borden, L. (2009). *Transforming mathematics education for Mi'kmaw students using mawikinutimatimk.* Unpublished dissertation, University of New Brunswick, Fredericton, NB.

Mason, J. M. (2002). *Researching your own practice: The discipline of noticing.* London, UK: Routledge Farmer.

King, T. (2003). *The truth about stories: A Native narrative.* Toronto, ON: Anansi Press.

This chapter was originally published as:

Chapman, O., Letts, K., MacLellan, L. (2009). Developing inquiry-based teaching through lesson study. *delta-K: Journal of the Mathematics Council of the Alberta Teachers' Association, 46*(2), 27–33.

CHAPTER 50

DEVELOPING INQUIRY-BASED TEACHING THROUGH LESSON STUDY

Olive Chapman, Krista Letts, and Lynda MacLellan

Editor's Note: Olive Chapman is a professor of mathematics education and an assistant dean in the Faculty of Education at the University of Calgary. She is associate editor of the *Journal of Mathematics Teacher Education.* Her research interests include self-study approaches to inservice teachers' professional development and preservice teacher education, mathematics teacher thinking, mathematical problem solving, inquiry-based teaching and classroom discourse. She has published in all of these areas. Krista Letts is an experienced elementary school teacher. She has taught grade 2 and currently teachers grade 1 at Edgemont School in Calgary, Alberta. She is interested in inquiry-based teaching and teaching through problem solving. Her love for literature has led her to explore ways to incorporate it into her teaching to support mathematics literacy. She has co-led professional development sessions in learner-centered mathematics pedagogy for colleagues and made presentations at mathematics teachers' conferences. Lynda MacLellan is an experienced elementary school teacher. She currently teaches grade 6 at Arbour Lake Middle School in Calgary, Alberta. She served as a mathematics curriculum leader at Edgemont School and an AISI mathematics learning leader for the Calgary School Board. She is interested in inquiry-based teaching,

Selected Writings from the Journal of the Mathematics Council of the Alberta Teachers' Association, pages 433–445
Copyright © 2014 by Information Age Publishing

teaching through problem solving, and mathematical connections through literature, other school subjects and real-world situations. She has co-led professional development sessions in learner-centered mathematics pedagogy for colleagues and made presentations at teachers' conferences.

Inquiry-based teaching is widely promoted in mathematics education to help students develop a deep understanding of mathematics and mathematical thinking. In inquiry classrooms, students construct mathematical meaning through reasoning, communicating, exploring, and collaborating with peers and teachers while working on inquiry-oriented tasks, including non-algorithmic problems and investigations (NCTM, 1991). However, adopting an inquiry perspective is challenging for teachers because they were taught mathematics that way. This can create a difficult situation when preparing today's students and tomorrow's workers for the demands of the 21st century unless teachers are provided with meaningful learning opportunities to deepen their understanding of inquiry-oriented pedagogy. This article describes such a learning opportunity based on the experience of a group of elementary teachers. Specifically, it discusses a lesson-study approach and the type of knowledge of inquiry teaching and actual teaching resulting from the experience.

LESSON STUDY

Japanese teachers use lesson study as professional development to systematically examine their teaching to become more effective in teaching mathematics (Fernandez & Chokshi, 2002; Stigler & Hiebert, 1999). It offers teachers a structure to transmit, reformulate and share craft knowledge through practice and collaboration with peers (Shimahara, 2002). It has recently been introduced to North America and is becoming popular for promoting teacher-driven instructional change. Teachers use it to enhance their mathematical knowledge and pedagogical skills, which leads to student-centred teaching in the classroom (Stiggler & Hiebert, 1999). It fosters improvement in increased development of instructional knowledge and understanding of subject matter (Lewis, Perry, & Hurd, 2004).

Lesson study involves recursive cycles of planning, teaching, analyzing, and revising of lessons that contribute to a continuous improvement model of professional development for teachers. A small group of teachers work collaboratively to plan, teach, observe, and analyze the lessons. They start by identifying a goal or problem to explore. This is followed by a recursive cycle composed of four phases: collaboratively developing a lesson plan, implementing the lesson while colleagues and other experts observe, analytically reflecting on the teaching and learning that occurred, and revising the lesson for reimplementation (Curcio, 2002; Stigler & Hiebert, 1999). During each cycle of implementation, a different teacher teaches the lesson to students in a normal classroom setting while the other group members observe and take notes. Finally, the teachers produce a report on what they have learned from the study lessons, particularly with respect to their goal.

Finding the time to engage in lesson study is a challenge. Unlike North American teachers, Japanese teachers build time into their weekly teaching schedule for professional development. However, the experience of the group of teachers described below shows that lesson study can be adopted effectively.

THE STUDY GROUP

In spring 2002, 14 Calgary elementary teachers formed a mathematics study group as part of their school's requirement to develop and implement a professional growth plan. Their broad goal was to develop and foster a commitment by all staff to participate in professional growth opportunities to improve the quality of teaching and learning in classrooms. To achieve this, the initial intent was to reflect on their teaching during group meetings. Divisions I and II teachers were represented in the group. Three teachers assumed the role of group leaders to organize meetings and activities. A mathematics education professor joined the group as a mentor in fall 2002. The group met once every three weeks for about one and a half hours at the end of the day.

To highlight the role of the group leaders, the professor will be referred to as the *mentor,* the group leaders as the *team* and the participants as the *group.* The mentor's role was to provide support, theoretical validation, and help in abstracting general ideas from the teachers' thinking and experiences. In general, the mentor acted as a colleague and group participant in sharing ideas and learning from the experience as opposed to being in authority. The mentor had no previous experience with lesson study or teaching at the elementary school level but had taught mathematics methods courses for prospective elementary teachers.

The mentor introduced the idea of lesson study to the group as an approach that can be used to design collaborative learning experiences and encourage growth and continuous improvement. After reading the description of lesson study in Stigler and Hiebert (1999) and with the support of the principal, the group decided to try it. Because this was a new idea for all involved and time was an issue, they decided that the team would participate in the first round of the lesson study and provide feedback to the group. If the approach proved effective, the others would participate in future rounds. The team was granted two days of release time from teaching during the 2003 winter term to implement the first round. This time was allocated as follows:

1. Half day for initial planning to identify a focus, process, and schedule
2. Full day for researching, developing relevant ideas, and planning
3. Half day for classroom implementation, observation, and briefing

These sessions occurred at different times during the term to accommodate the mentor's schedule and arrangements for substitute teachers for the team.

The positive result of this first round of lesson study on the team's learning led to planning the other rounds to include the whole group. The after-school meetings were used to take the group through the process. In addition, all the teachers received a half-day release time from teaching, during which they planned the mathematics lessons to be studied. Each team member worked with a subgroup of teachers to plan the lessons. The teachers covered each other's classes to obtain the release time to observe the lessons. This way, over three years, 22 teachers had the experience and learned from it. This number was a result of teachers leaving because of other commitments and others entering.

THE LESSON-STUDY APPROACH

This section describes the different stages of the lesson-study approach in the initial round and summarizes the follow-up rounds.

Identify a Topic of Study

The team members were interested in inquiry-based teaching but wanted to focus on a specific feature for the lesson study. They identified this feature by examining the philosophy section of the mathematics curriculum document *Alberta Program of Studies for K–9. Western Canadian Protocol for Collaboration in Basic Education* (Alberta Education, 1996), which incorporates a perspective of mathematics learning that supports inquiry-based teaching. This document states: "Students learn by attaching meaning to what they do; and they must be able to construct their own meaning of mathematics. The meaning is best developed when learners encounter mathematical experiences that proceed from the . . . concrete to the abstract" (p. 2). The goals for students include using "mathematics confidently to solve problems and communicate and reason mathematically" (p. 2). The curriculum also emphasizes seven mathematical processes (p. 4): communication, connections, estimation and mental mathematics, problem solving, reasoning, technology, and visualization, all of which can play an important role in inquiry-based teaching if they are interpreted as intended. Thus, the curriculum provided validation and a basis for the starting point to identify a topic of study.

The team members focused on the seven mathematical processes in the curriculum and, after a lengthy discussion and reflection on their teaching, concluded that communication in an inquiry-teaching context was the key process that they would like to study in the first round. As stated in the curriculum, "Students need to communicate mathematical ideas clearly and effectively, orally and in writing" (Alberta Education, 1996, p. 4). But this communication is different from the traditional mathematics classroom where the focus is on transmitting information, copying notes, and recording solutions to exercises or routine problems. The team was interested in focusing on other features of communication that allowed students to think and actively engage in their learning. Once a topic for the lesson study was established, the next stage was determining specific features of communication and inquiry-based teaching to use in designing the lesson to be studied.

Video Case Study

The team and mentor discussed possible ways to obtain relevant information about specific features of communication and inquiry-based teaching. Instead of beginning with reading theory on these topics, the team members preferred to study a video as the basis of their learning. The mentor suggested the DVD series, *Mathematics: With Manipulatives* (Burns, 1988). The team chose *Pattern Blocks* and *Cuisenaire Rods,* two of the six DVDs in the series. To have focus in studying the videos, the team and mentor attended to the following: lesson goal, students' role, teacher's role, specific questions posed by the teacher to stimulate and extend students' thinking, classroom environment, nature of inquiry tasks, and key features of the inquiry lesson.

After studying the videos, the following key components for inquiry-based lessons were identified:

1. Three modes of communication: oral, written, and nonverbal (for example, observing, listening, and acting/gesturing)
2. Tasks that allow for open and guided exploration, predication, discussion, and evaluation
3. Student roles, teacher roles, and environment (specific features identified are provided in Figure 50.1)
4. Questions and prompts (examples are provided in Figure 50.2)

Student Role	Teacher Role	Classroom Environment
Communicate in all three modes	Pose questions and prompts	Manipulatives
Engage in inquiry	Allow time for exploration	Small groups
Collaborate	Select appropriate tasks	Whole-class sharing
Take risks	Allow time for communication	Supportive/risk free
Show curiosity	Observe and listen to	Lots of mathematical talk
Reflect	Support/encourage	

Figure 50.1 Roles and environment.

1. What do you notice?
2. What else do you notice that is different?
3. Who can explain how (or why) this makes sense?
4. What do you think the answer (or pattern or outcome) could be? How do you know?
5. How do you know it will (will not) work?
6. Where (or when) would you use this _____?
7. Suppose I want to_____, how can I start?
8. Who can describe it so that I can do it?
9. Present your idea.
10. Explain the problem to your partner (the class).
11. What do you know about _____ (for example, this topic)?
12. Can you make a general statement about _____?

Figure 50.1 Questions and prompts.

Following the video study, the team later read and discussed the National Council of Teachers of Mathematics (2001) standards on discourse. However, the outcome of the video study formed a key basis for planning the experimental inquiry-based lesson.

Planning the Experimental Lesson

The grade 1 teacher on the team volunteered her classroom for the first round of lesson testing; thus, the team chose a topic from the grade 1 curriculum to develop an inquiry-based lesson. The topic "explore and classify 3-D objects according to their properties" was selected to correspond with the teacher's class schedule. The team brainstormed different approaches to teach the topic. Each teacher on the team described what she might do in her class.

Teacher One
- Observe objects in classroom
- Discuss why these objects have certain shapes
- Post pictures of objects in the real world around the classroom and time to identify shapes

- Name geometric objects
- Link to objects in class
- Refer to chart with formal names
- Investigate attributes
- Relate to real world—why things have certain shapes

Teacher Two

- Describe geometric objects in groups or pairs
- List names of objects and descriptive words on a chart
- Build a model of one object (a skeleton representation)
- Discuss, comparing skeleton and actual object
- Introduce formal names

Teacher Three

- Pose a problem—for example, build a house with this object
- Discuss attributes
- Explore attributes
- Classify attributes
- Describe common features

Reflecting on these approaches and the outcome of the video study, the team and mentor identified the following set of key components for inquiry-based teaching to use to structure the experimental lesson:

Goal

- Prerequisite
- Free exploration and discussion
- Prediction of properties of concept
- Application of concept
- Testing predictions
- Evaluation of knowledge of concept
- Extension of concept to new situations

Using this structure and key questions to promote inquiry-based communication, the team designed a plan for the experimental grade 1 lesson on introduction to geometric solids. Figure 50.3 provides an abbreviated outline of the plan.

Conducting and Observing the Experimental Lesson

The grade 1 teacher of the team taught the lesson to her students in her classroom. The other team members and the mentor observed and made

- Free exploration objects (10 minutes) (Talk, experiment and observe in small groups.)
- Discussion (5 minutes). (What did you notice? Record answers.)
- Prediction (5 minutes). (Will objects roll or slide? Record individually.)
- Pose problem/real world application (5 minutes). (Suppose I want to build a house on a mountain, what would I need to know about objects?)
- Test prediction, record results (5 minutes).
- Comparison (5 minutes). (Discuss solutions with partner and support answers.)
- Evaluation (10 minutes). (Venn diagram, sort shapes and make general statements about "What I know about 3D objects.")
- Extension (homework). (Look for things at home and around school that roll or slide.)

Figure 50.3 Outline of experimental lesson plan.

notes on an observation sheet (Figure 50.4). The intent was to focus on how the lesson was conducted and how effective the questions and prompts were for communicating the components of the lesson to facilitate meaningful and worthwhile inquiry and learning of the mathematics concepts.

Analyzing the Experimental Lesson

The lesson was analyzed immediately after it concluded. The observers and the teacher of the lesson shared notes, focusing on communication and the components of the lesson. For the most part, the lesson went as planned.

Notice?	
Make sense?	
Predict?	
How know?	
Make connections?	
Describe/explain	
Generalize/summarize	
Other	

Figure 50.4 Observation sheet.

It consisted of the following sequence of activities: brief introduction to set the tone; free exploration (in small groups) of 11 3-D geometric objects; whole-class discussion; individual prediction using worksheet with pictures of the 11 3-D objects and columns for rolls only, slides only, and rolls and slides: comparison with a partner; whole-class discussion of an application (think of self as builder); predication if all will agree; focused exploration to test predictions; comparison and discussion of findings with others; whole-class discussion of findings; 3-D vocabulary of objects and building of Venn diagram on whiteboard with pictures; evaluation/reflection/generalization in relation to goal of lesson; and an application task for homework. These components and activities were effective in creating a learner-centred classroom and promoting inquiry and rich mathematical communication. The children were actively involved throughout the lesson doing mathematics, and thinking and communicating mathematically. The teacher of the lesson kept posing questions that stimulated or revealed their thinking as she circulated during the lesson and during the whole-class discussion. This allowed her and observers to learn from and about the children based on their ways of thinking. The team was amazed and impressed with what the children were able to do and the richness of their thinking.

The only change that was identified was to clarify two more components to an inquiry lesson: comparison and reflection. The team members concluded that their inquiry-teaching model was workable and they felt comfortable and confident implementing it in their teaching. They emphasized that the structure of the lesson is flexible as to which components are used and how they are sequenced. Thus they decided to represent the approach in the form of a jigsaw puzzle, which they called the Jigsaw Teaching Model. The experimental lesson was one way of applying this model.

The Jigsaw Teaching Model

Figure 50.5 is the final version of the Jigsaw Teaching Model developed by the team to represent the key components of an inquiry-based classroom and key questions and prompts. The model is in the form of a jigsaw because the pieces do not have to follow a linear pattern but can be organized differently, combined or omitted depending on the topic being taught. The pieces represent different components that are important in inquiry teaching.

The model requires the teacher to (1) identify learning goals that include conceptual understanding, (2) expose students' prerequisite knowledge/conceptions for the concept being taught in an inquiry way, (3) have student make predictions about possible outcomes related to the concept, (4) allow students to engage in free exploration of the concept (through discourse and/or using manipulatives), (5) engage students in focused

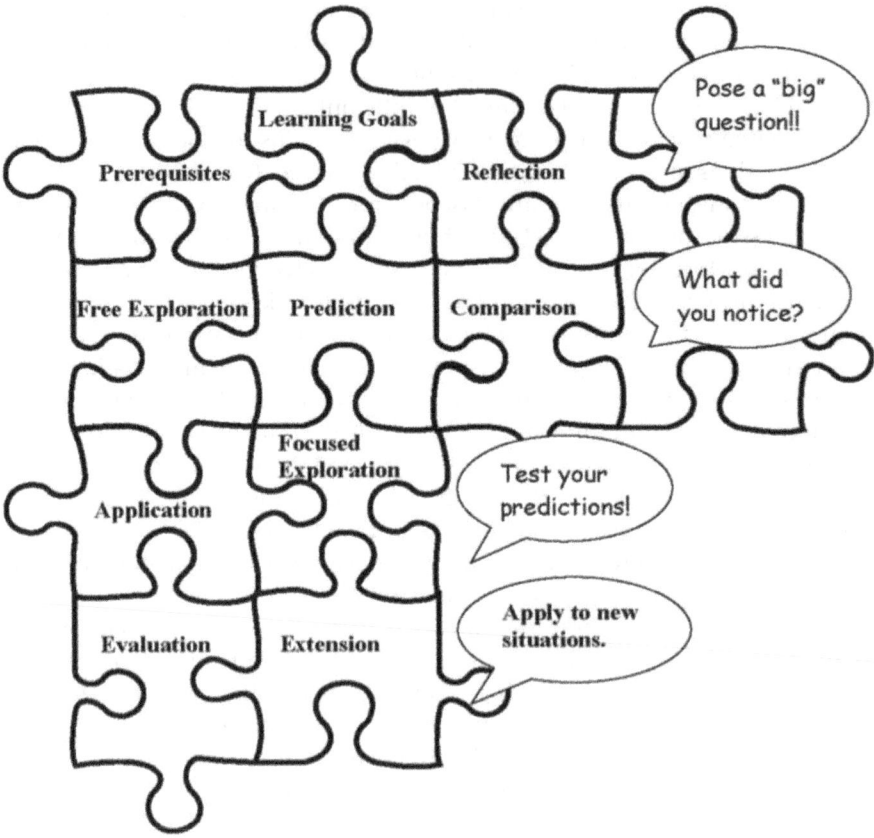

Figure 50.5 The jigsaw teaching model.

exploration, (6) have students work on application of concept, (7) engage students in comparison, evaluation and reflection of their learning, and (8) suggest extension of concept to other situations or related concepts.

Follow-up Rounds of Lesson Study

The team did not follow the recursive cycle of the lesson study in terms of revising and teaching the same lesson. Instead, they made the Jigsaw Teaching Model the basis for subsequent rounds of the lesson study with the group. These rounds of experimental lessons included (1) the meaning of five (kindergarten); (2) estimation with mass (grade 3), representing a multidigit number in different ways (grade 3) with goals that students will understand why the value of a digit changes depending on its position in a number and the meaning of regrouping among hundreds, tens and ones; (3) mental arithmetic (grade 5); and (4) measurement (grade 6)—for example, perimeter and area.

Later, the teachers also worked on exploring lessons on problem solving in their classrooms Figure 50.6 describes a lesson taught four years later by the grade 1 teacher who taught the first experimental lesson, which shows that the approach was being sustained. The teacher explained:

> The students' thinking started us off on a new game of trying to stump the rest of the class with our ordering rules. Students began to use mass, capacity, width, height, cost, temperature, time and so on to place the objects in order. The discussions that followed were lively and informed. We all learned more than the original activity, and it was a fabulous culminating activity. The following day students were presented with a puddle question about how they could measure a puddle after a rainy day. They recorded their answers in their math journals in the form of a mind map or brainstorming diagram. They used words, pictures diagrams and so on to show their thinking. Finally they shared their ideas with several partners and discussed similarities and differences. When we listed our combined ideas on a chart, they included length (using their feet), width (using their hands), depth (using pile of rocks), area (using boxes), capacity (using cup measure), weight (using scale to weigh cup

> Learning goals: Students will use knowledge of nonstandard measurement to order objects, recognize that objects can be sorted in a variety of ways and think flexibly when dealing with size.
>
> Free exploration challenge: Find three objects that you can order by size.
>
> Prediction: Is there more than one way to order the three objects?
>
> Comparison: pair/share—Can you guess my measurement sorting rule?
>
> Focused exploration/application: Which measurement sorting rule would change my order? Which rule would not change the order? Where would a fourth object fit? How do you know?
>
> Recording: Record your favourite sorting rule in your math journal. What did you notice about measurement and sorting in this activity? Did you notice any patterns?
>
> Extension: How could you measure a water puddle after a rainy day?

Figure 50.6 Grade 1 lesson (four years later).

and multiplying or repeated addition), temperature (using a thermometer) and time (counting how many minutes/hours for sun to dry up puddle).

This shows what these children are able to accomplish when *their* thinking, not only the teacher's, becomes the focus of the lesson.

CONCLUSION

The lesson-study approach helped these teachers create shifts in their thinking and teaching, and develop useful knowledge of mathematics teaching. They reported deeper and more meaningful understanding of the following inquiry teaching:

- questioning techniques that guide and enrich student thinking
- open-ended, thought-provoking questions to motivate students to discuss and understand mathematics at a deeper level
- student-centered strategies for listening to students and observing their problem-solving behaviours
- strategies that allow students to assume ownership of their knowledge and knowledge construction.

With these shifts, students were more involved in mathematics as a way of problem solving, reasoning, communicating, and connecting to their world.

The teachers also valued the collaboration during the lesson study; that is, being a part of the group process of planning, observing, debriefing, and sharing specific examples from classroom work. As one teacher explained, "It was a powerful experience to hear the student communication that an observer can record, but which the teacher misses as she focuses on presenting the lesson and maintaining classroom management." In general, their collegial focus on mathematics instruction increased, and they articulated and demonstrated a renewed interest in mathematics. Most important, they learned an approach that they can use for ongoing learning and growth in their teaching. They offer their experience in the article as an example of an approach other teachers could adapt in their quest to inquiry-based teaching.

REFERENCES

Alberta Education. (1996). *Alberta Program of Studies for K–9. Western Canadian Protocol for Collaboration in Basic Education.* Edmonton, AB: Alberta Education.

Burns, M. (1988). *Mathematics with Manipulatives.* New Rochelle, NY: Cuisenaire Company of America.

Curcio, F. R. (2002). *A user's guide to Japanese lesson study: Ideas for improving mathematics teaching.* Reston, VA: National Council of Teachers of Mathematics.

Fernandez, C., & Chokshi, S. (2002). A practical guide to translating lesson study for a U.S. setting. *Phi Delta Kappan, 84*(2), 128–135.

Lewis, C., Perry, R., & Hurd, J. (2004). A deeper look at lesson study. *Educational Leadership, 61*(5), 18–22.

National Council of Teachers of Mathematics (NCTM). (1991). *Professional standards for teaching mathematics.* Reston, VA: Author.

Shimahara, N. K. (2002). Teacher professional development in Japan. In G. De-Coker (Ed.), *National standards and school reform in Japan and the United States* (pp. 107–120). New York, NY: Teachers College Press.

Stigler, J. W., & Hiebert, J. (1999). *The teaching gap: Best ideas from the world's teachers for improving education in classrooms.* New York, NY: Free Press.

COMMENTARY

MATHEMATICS EDUCATION FOR THE TWENTY-FIRST CENTURY

Olive Chapman

INTRODUCTION

Turn of the millennium; twenty-first century society; twenty-first century skills; twenty-first century learning; twenty-first century classroom; technology revolution; globalization; terrorism; world economy; environmental issues: these are some of the ideas that one might associate with the first decade of the new millennium, the 2000s, or the aughts. How does mathematics education fit into this picture? The ten articles in this decade of the aughts directly or indirectly offer particular ways to address this. Thus, the approach I take in this commentary is not to comment on each article in an isolated way, but to use them to paint a picture of mathematics education for the twenty-first century. In particular, I address the following themes that emerged for me from the papers: numeracy in the twenty-first century, cognitive science learning and teaching, mathematical processes, connecting mathematics to our world, and practice-based professional development.

Selected Writings from the Journal of the Mathematics Council of the Alberta Teachers' Association, pages 447–457

NUMERACY IN THE TWENTY-FIRST CENTURY

Numeracy, also described as quantitative literacy or mathematical literacy, simply stated refers to an individual's ability to understand and use mathematics effectively at school, at work, and in everyday life. Werner Liedtke's article draws our attention to the nature and importance of numeracy to students and society. As he points out, a highly numerate population is critical in ensuring the nation's ongoing prosperity, productivity, and workforce participation. Individuals who are numerate are better prepared to participate and engage in a world that is increasingly focused upon creativity and innovation and focuses upon knowledge creation and sharing. Thus, as Liedtke suggests, it is important for parents and teachers to actively engage children from an early age in appropriate activities to develop the foundation necessary to become numerate.

Given the direct relationship between numeracy and society, Liedtke's article opens the door to consider what constitutes numeracy in a twenty-first century context. Goos, Dole, and Geiger (2011) argued that a description of numeracy for this new period needs to better acknowledge the rapidly evolving nature of knowledge, work, and technology. They developed a model (Figure P5.1) to represent the multifaceted nature of numeracy in the twenty-first century.

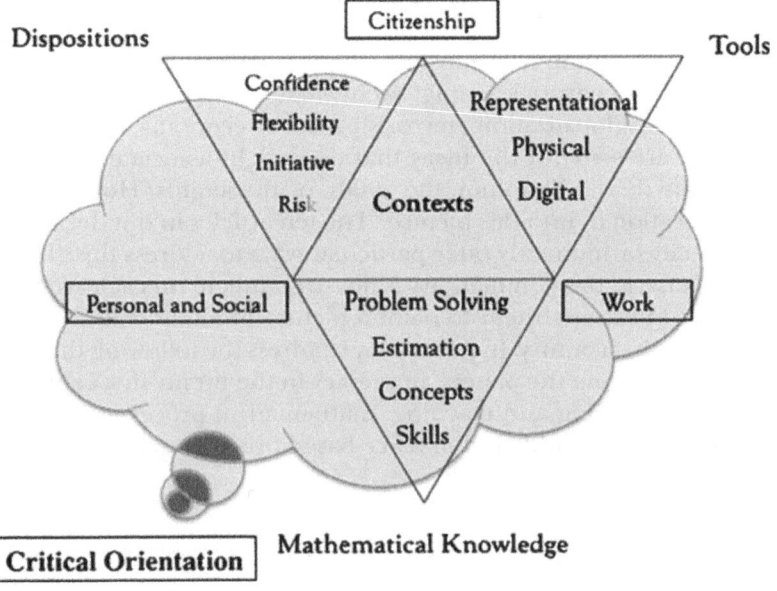

Figure P5.1 A model for numeracy in the 21st century (Goos et al., 2011, p. 33).

Based on this model and the authors' explanation of it (Goos et al., 2011, p. 34), a numerate person requires mathematical knowledge that includes not only concepts and skills, but also higher-order thinking such as problem-solving strategies and the ability to make sensible estimations. A numerate person has positive dispositions—a willingness and confidence to engage with tasks, independently and in collaboration with others, and apply their mathematical knowledge flexibly and adaptively. Being numerate involves using tools that may be representational (e.g., symbol, diagrams), physical (e.g., models, measuring instruments), and digital (e.g., computers, calculators). One needs to be numerate in a range of contexts including personal contexts, work-related contexts, citizenship-related contexts, and different curriculum contexts. The model is grounded in a critical orientation to numeracy—for example, it recognizes that a numerate person critically discusses and considers alternative solutions that give different results and have different real life consequences, evaluates the reasonableness of results, and is aware of appropriate and inappropriate uses of mathematical thinking to analyse situations and draw conclusions.

This model, then, suggests a possible scope of mathematics education for the twenty-first century. It reflects the multidimensional orientation of school mathematics that is more relevant to this new millennium. Each of the 10 articles addresses some aspect of this model which will be highlighted throughout this commentary. Thus they all contribute to our understanding of numeracy and mathematics education in this decade and beyond. They raise issues about curriculum, teaching, and learning as practised in our mathematics classrooms and offer suggestions of what we need to consider and should do to address these issues in order to transform how we engage and prepare our students for a changing world. These suggestions, as those about numeracy represented in the model (Figure P5.1), require ways of thinking about mathematics, the mathematics curriculum, and mathematics teaching and learning that are different from what are practised in traditional mathematics classrooms. This makes teachers' learning and professional development, also addressed in a couple of the articles, a critical component in realizing the change to the twenty-first century context.

The numeracy model and the articles also imply a shift or extension in the rationale for learning mathematics that includes a critical thinking component. Lynn McGarvey's article is explicitly related to this shift as she examines, "Why do we teach math to children?" She raises concern about the lack of, or inadequate depiction of, an explicit rationale in our past and present provincial mathematics curricula. As she points out, the rationale for learning mathematics when given or implied in the curriculum consists of: for the purposes of future progress of education; for its utility; for later mathematics learning; for future careers; for individual growth; and, as a way of trying to understand, interpret, and describe our world.

While these are relevant, by themselves they are inadequate in not considering, as McGarvey puts it, "How does math shape our thinking? How has it defined our culture? What has math allowed us to create? How has it constrained or perhaps disallowed possibilities for knowing, seeing, doing, and being." These are important considerations to make mathematics education more relevant to prepare numerate citizens for the twenty-first century world. They can be supported by what we now know from cognitive science and the importance of critical mathematics education and cultural relevant mathematics to a numerate society. These are addressed in the articles and the discussion in the remainder of this commentary, which highlights these and other ways in which mathematics education in this decade of the 2000s should have been evolving to meet the demands of a rapidly changing world.

COGNITIVE SCIENCE APPROACH TO LEARNING AND TEACHING

Brain-based research is offering us new ways of understanding learning. Brent Davis's article highlights some of these ways based on developments in cognitive science. Davis provides a view of mathematical intelligence that is about coming up with solutions to real problems, with answers that go beyond routine responses and that enable the person to go further than he or she could before taking on the problem. He highlights the relationship to understanding mathematical intelligence and teaching to support its enhancement. For example, he points out an important relationship between practice and intelligence; that is, the latter could be improved through the former. However, this practice must be contextualized and be meaningful and purposeful to the learner. Some of the other insightful notions related to learning in the article include: "Every lived experience entails a physical transformation of one's brain." "Every experience you have contributes to the ongoing restructuring of the brain." "For a learner to develop mathematical intelligence and robust mathematical understandings, she or he has to be aware of how mathematical concepts can be interpreted in different ways." "Intelligence is greatly enabled by a facility with contemporary tools." The article also suggests that experiences that force learners to think outside the box will enhance or extend intelligence and points out the importance of working with others and others' ideas in the production of mathematical knowledge.

In the twenty-first century classroom, all students could benefit from these ideas from a cognitive-science perspective of learning; that is, learning of mathematics can be enhanced by meaningful and purposeful practice, challenging mathematical tasks, multiple interpretations and representations of mathematical concepts, technological tools, and collaboration with others. This perspective also suggests how important teachers are to brain development. For example, if every experience impacts brain structure, then the type of experiences teachers offer students in their classrooms can enhance or limit brain development, in general, and mathematical development, in particular. Teachers who engage students in "worthwhile mathematical tasks" (National Council of Teachers of Mathematics [NCTM], 1991) and mathematical processes (discussed in the next section) and provide opportunities for creativity and critical thinking are more likely to contribute in a positive way to enhancing students' learning and mathematical intelligence and their preparation to be productive citizens in the twenty-first century context.

Many teachers continue to depend on a textbook to script their lessons. However, while textbooks can be useful references, using them as scripts can be problematic. For example, as Ann Kajander and Miroslav Lovric point out in their article, they saw evidence of secondary and postsecondary textbook formats that might have been intended to simplify the learning process for students. They note consequences of this—for example, incomplete understanding, misconceptions, inaccuracies. In general, such simplification, whether in the textbook or resulting from teachers choosing to do so to save students from being challenged, is in conflict with a cognitive science perspective of learning and denies students the experience to enhance their brain structures.

MATHEMATICAL PROCESSES

In the late 1980s the field of mathematics education began to emphasize the importance of mathematical processes in the teaching and learning of mathematics (e.g., NCTM, 1989). In this decade of the 2000s, these processes were reinforced by professional standards (e.g., NCTM, 2000) and included as integral aspects of K–12 mathematics curricula. Table P5.1 shows the processes of NCTM (2000) and the current Alberta curriculum (Western and Northern Canadian Protocol) and Ontario curriculum. These mathematical processes are not only essential to learning mathematics with understanding and enhancing mathematical intelligence from a cognitive science perspective but are also significant in preparing students for a twenty-first century society. Through them, teachers can create classrooms to support the development of students who are critical and creative thinkers, problem solvers, risk takers, and collaborators—that is, students with skills considered to be necessary for a twenty-first century global economy and society. These processes allow students to create

TABLE P5.1 Mathematical Processes in School Curricula

NCTM	Alberta	Ontario
Problem solving	Problem solving	Problem solving
Reasoning and proof	Reasoning	Reasoning and proving
Representation		Representing
Connections	Connections	Connecting
Communication	Communication	Communicating
	Mental mathematics and Estimation	
	Technology	
	Visualization	
		Reflecting
		Selecting tools and computational strategies

knowledge that connects them with their world in ways that are personally meaningful and relevant.

As with the case of numeracy, these processes require different interpretations and teaching approaches than those in a traditional classroom. Jennifer Holm's article provides an example of this in her "perception of problem solving in the elementary curriculum." She highlights the nature of problem solving, students' role, and the teacher's role as they should be interpreted. This view of problem solving as way of thinking, learning, and teaching not only allows students to develop understanding of genuine problem solving but also allows them to be autonomous learners, develop mathematical thinking, and engage in most of the other processes. Thus, problem solving, when appropriately interpreted, by itself could significantly transform mathematics classrooms into twenty-first century oriented places that promote creative thinking; exploration, discovery, and knowledge creation; risk taking; and collaboration, sharing of ideas, and opportunities to argue and defend solutions and strategies.

In general, teaching through inquiry, another key idea in this decade, is recognized as important to support students' learning of mathematics with understanding and their development of the mathematical processes and mathematical inquiry and collaborative skills. Inquiry-based learning allows students to actively engage in in-depth exploration of mathematics concepts and to be actively engaged in the construction of mathematical knowledge with deep understanding; to make connections between prior, existing, and new knowledge and experience; to work and learn collaboratively; and to take responsibility for their own learning. Students are given the opportunity to pose questions, make conjectures, and direct their own investigations and find their own answers. So there are many similarities between inquiry and genuine problem solving as ways of learning and teaching. The article by Olive Chapman, Krista Letts, and Lynda MacLellan illustrates how teachers can transform their teaching to an inquiry-oriented perspective. This article is discussed later as part of teacher learning.

This section on processes will be incomplete without a mention of technology. While technology is not synonymous with twenty-first century learning, it is an integral aspect of it, and can play a central role in transforming classrooms to support the mathematical processes and inquiry-based learning. However, while it provides a tool for exploration/inquiry, in a twenty-first century classroom it should also be about connecting students with their world and communicating with others, enabling them to learn from others and to share their own ideas. So, while it is not dealt with explicitly in the articles, as a tool it is relevant to all of them.

CONNECTING MATHEMATICS TO OUR WORLD

The importance of connecting mathematics to the world is embodied in the process of connections. As stated by NCTM (2000), instructional programs should enable students to "recognize and apply mathematics in contexts outside of mathematics" (p. 64). However, this could be interpreted in ways that are inhibiting or liberating in terms of students' learning and relationship to the world depending on what and how contexts are used. In a twenty-first century perspective, such contexts should humanize mathematics by connecting it to students' day-to-day lives, to broader society, and to issues of equity. Three of the articles offer ways to address this.

Florence Glanfield's article draws attention to the situation that across Canada there has been an attempt to humanize mathematics in courses such as applied and consumer mathematics. These courses are generally for students who are not likely to take postsecondary programs that require mathematics as a prerequisite. While such courses allow these students to develop basic numeracy, there is a perceived issue with how "pure" and "applied" are used to label school mathematics curricula. For example, such labels could suggest that a lower level of mathematics is needed for applications and that "applied" mathematics is not important for students who are in "pure" mathematics route. This could denigrate applications to practice exercises for mathematics concepts, such as traditional word problems at the end of a unit. The bottom line is that all students should have opportunities to engage in "pure" and "applied" mathematics in appropriate and meaningful ways to facilitate their learning and connect them to the world.

David Wagner's article offers a different way of thinking of such connections for all students by considering the relationship between mathematics and society in the context of teaching for peace and living in an unpredictable

world. He offers two views of peace, stating that one can be equated with focusing on product and one on process, which is analogous to two extreme
views of mathematics teaching. While some balance is necessary between the
two, giving the edge to process seems to be more relevant to a twenty-first century context. As Wagner suggests, the process view of peace involves "moulding ourselves to the world in order to find ourselves in it." A similar claim can
be made for a process view of mathematics and the connections it allows students to have with the world. Wagner notes the importance of relevant mathematics tasks to achieve such connections, that is, tasks that have potential
for building awareness and help us to see our place in the world. He offers
two ways of doing this in relation to peace: (1) through word problems that
express an interest in a peaceful world, and (2) through ethnomathematics
(includes mathematical-rich cultural artifacts from other cultures) and critical mathematics (e.g., use tasks structured to make a variety of approaches
possible and a diversity of answers correct; engage students in critical think-
ing/dialogue). He concludes that, in general, mathematical tasks/activities
that encourage thoughtful action can format the world for peace.

Wagner's suggestions regarding word problems and critical mathematics can be generalized for other concerns of society—for example, environmental, economical, and equity issues—which have implications
for world peace and a numerate society. Of particular importance in a
globalized world is how we acknowledge culture. Wagner mentions ethnomathematics, which illustrates the culturally influenced uses of mathematics through its applications. However, in general, there should be a
focus on culturally relevant mathematics and culturally relevant instruction that employs various methods of infusing culture into mathematics
instruction. Leonard and Guha (2002) explain that "culturally relevant
teaching embeds student culture into the curriculum to maintain that
culture" (p. 114). They further state that "including aspects of students'
culture into mathematics problems is one way to avoid the cultural deficit
model and help students and teachers value the culture of the community" (p. 114). For example, Aboriginal cultures are seldom recognized
for their uses of mathematical processes, yet it is through an examination
of such applications in daily living practices that our understanding of our
Aboriginal peoples can be enhanced.

The article by Gladys Sterenberg, Liz Barrett, Narcisse Blood, Florence
Glanfield, Lisa Lunney Borden, Theresa McDonnell, Cynthia Nicol, and Harley Weston draws attention to culturally relevant mathematics from a Canadian Aboriginal perspective. It provides us with meaningful examples of activities that can be used to support culturally responsive mathematics education.
It illustrates "learning mathematics from place"—that is, exploring mathematics concepts through activities at actual cultural sites and through the
creation of models in the classroom. The focus is on making real connections

by exploring actual places, such as through field trips to aboriginal sites. The article also highlights the importance of relationships within the aboriginal culture and impact on mathematics education, in particular, relationship of understanding and awareness of relationship with self, students, content, and relationship within and outside mathematics. In general, the article indicates how indigenous knowledge can offer a basis for teaching and learning mathematics by both Aboriginal and non-Aboriginal students and teachers. For example, it highlights the roles language (particularly in understanding the link between indigenous knowledge and mathematics), place, community, and culture play in Aboriginal learning of mathematics that are also relevant to non-Aboriginal students' learning. The characterization of indigenous epistemologies as being experiential, storied, relational, contextual, and holistic has implications for how we view mathematics, mathematical tools, and mathematics pedagogy and is a view consistent with twenty-first century numeracy and cognitive science perspectives of learning.

PRACTICE-BASED PROFESSIONAL DEVELOPMENT

For teachers to be able to implement and sustain a twenty-first century mathematics classroom, they should understand mathematics and learning in ways discussed for students. Thus teacher education programs and professional development approaches also need to reflect the demands of the twenty-first century and implications of cognitive science research for learning. The 2000s have seen significant changes to mathematics teacher education as proposed by the research field, such as a shift in focus to the nature of mathematics knowledge for teaching and practice-based, inquiry-oriented approaches to teacher development. I address the latter here based on two of the papers.

Practice-based models of teacher learning are based on the idea that professional development needs to be strongly linked to the professional practices of the teachers involved. The general view is that learning activities for the teachers should be purposefully connected to the curriculum they are teaching, student learning or work in their classrooms, content situated in an environment that models effective teaching, pedagogy of their classrooms, and a collaborative environment. Approaches that provide opportunities for teachers to explore important mathematical and pedagogical ideas that relate to their practice include teachers studying episodes of their practice by creating and analyzing narratives of their teaching; analyzing cases of sample teaching or problem situations; analyzing self-created videos of their teaching or researcher-created videos; creating, teaching, and analyzing experimental lessons during lesson study; and practicing "noticing" (Mason, 2002).

Mason (2002) defines noticing as "a collection of practices both for living in, and hence learning from, [classroom] experience and for informing future practice" (p. 29). This focus on classroom experience is a practice-based approach to teacher learning. Julie Long's article provides an example of this way of using noticing as a basis of understanding her practice. Her focus was on writing accounts-of and -for experience from her teaching. While this might seem to be a simple process, from a noticing perspective, as reflected in Long's article, it could be challenging. Noticing requires more than simply writing a description of a situation in one's teaching. For example, and more importantly, it requires one to learn something that he or she did not know before, or, as Long puts it, "draw a lesson out of the experience." Long provides an example of this based on her reflection of accounts from her teaching that contributed to her understanding of how discussion, challenge, and emotion were important to students' learning. In general, Mason's noticing includes teachers engaging in self-reflection as professional development to understand their practice with the goal of developing sensitivities to their students and to be awake to possibilities to transform their teaching. Chapman and Heater (2010) provide an example of a Calgary high school mathematics teacher's self-development in inquiry-based teaching in which noticing played an important role.

The article by Olive Chapman, Krista Letts, and Lynda MacLellan provides an example of a practice-based, school-based professional development approach to help teachers develop understanding of, and facilitate change of practice to, inquiry-based teaching. This approach shows how teachers can engage in lesson study to learn *though* and *for* inquiry that is relevant to a twenty-first century classroom. The approach was effective in helping the participants to create meaningful shifts in their thinking and teaching and to develop useful knowledge of mathematics teaching. One important feature of this approach was allowing teachers to have autonomy of it and their learning, thus making the professional development about them—that is, their perceived needs, their thinking, and their teaching. This allowed them to contextualize and personalize the inquiry process so they all could make sense of it from their perspectives and in a similar way that supported their learning collectively and individually. As a result, they got to develop their own ways of understanding key ideas being investigated through their own sense making. In addition, their choice of "communication" as the focus of their lesson study emerged as an ideal anchor for their inquiry of inquiry-based teaching. It allowed them to consider inquiry-based teaching in relation to the triad relationship among content, teacher, and students through communication. This relationship is of importance in the implementation of inquiry-based teaching. In general, this lesson-study approach offers teachers experiences that are in harmony with a twenty-first century perspective of learning.

CONCLUSION

The ten articles in this section of the aughts present a vision of mathematics education that is relevant to the twenty-first century context. They address learning for both students and teachers with implications for curriculum and teaching. They illustrate the relevance of *delta-K* to our provincial mathematics education community as a vehicle to inspire changes in our classrooms that will produce numerate twenty-first century citizens to ensure our province's ongoing prosperity, productivity, and workforce participation.

Dr. Olive Chapman is Professor of mathematics education and Associate Dean of Undergraduate Programs Education in the Faculty of Education, University of Calgary. She is Editor-in-Chief of the *International Journal of Mathematics Teacher Education*. Her research interests include prospective and practicing mathematics teacher thinking, learning, and change; mathematics knowledge for teaching; mathematical thinking, problem solving, problem posing, and contextual/word problems; inquiry-based mathematics pedagogy, and inquiry-based discourse to facilitate mathematical thinking. She teaches mathematics education courses at the undergraduate and graduate levels and supervises graduate students in mathematics education.

REFERENCES

Chapman, O., & Heater, B. (2010). High school mathematics teacher self-development in inquiry-based teaching. *Journal of Mathematics Teacher Education, 13*(6), 445–458.

Goos, M., Dole, S., & Geiger, V. (2011). Improving numeracy education in rural schools: A professional development approach. *Mathematics Education Research Journal, 23*(2), 129–148.

Leonard, J., & Guha, S. (2002). Creating cultural relevance in teaching and learning mathematics. *Teaching Children Mathematics, 9*(2), 114–118.

Mason, J. (2002). *Researching your own practice: The discipline of noticing.* New York, NY: RoutledgeFalmer.

National Council of Teachers of Mathematics. (1989). *Curriculum and evaluation standards for school mathematics.* Reston, VA: Author.

National Council of Teachers of Mathematics. (1991). *Professional standards for teaching mathematics.* Reston, VA: Author.

National Council of Teachers of Mathematics. (2000). *Principles and standards for school mathematics.* Reston, VA: Author.

FINAL COMMENTARY

LOOKING BACK ON OUR SELECTED WRITINGS FROM FIFTY YEARS OF *DELTA-K*

Egan J. Chernoff and Gladys Sterenberg

Looking back on it now, our approach to choosing the 50 articles was, in essence, to solve a simpler problem. In other words, we decided to first worry about whether or not we could accurately reflect and celebrate *delta-K* and the teaching and learning of mathematics in Alberta for one decade—just one. But, where to begin? We began with the seventies because there was no doubt (for either of us) that we had to include an article on the metric system. Our task had now been "reduced" from fifty articles from fifty years to nine articles (and one article on the metric system) from the seventies. While our concerns were not fully alleviated by the end of our discussion, we both agreed that reading through the ten years of *delta-K* from the seventies and putting together our own top ten lists (plus a few "extra" articles) was a manageable task. At this point we went our separate ways (for a few weeks).

Given the number of articles that were published in delta-K during the seventies, there was a good chance that the next time we convened our lists would not contain any of the same articles. Fortunately, this was not the case.

However, as we soon found out, just because we both had a number of the same articles on our lists, our task was not any easier. Case in point: While we both had a metric system article on our list, it was not the same article.

Once we had agreed on which metric system article to include and, more importantly, why we were including it, we then discussed the articles we both had in common on our lists, the articles one person had on their list and the other person had on their "extras" list, the articles that only one person had on their list and, finally, any articles that only one person had on their extras list. This discussion, and our ensuing "escalated" discussions, was, in essence, our process for choosing our fifty articles to represent 50 year of *delta-K* and the teaching and learning of mathematics in Alberta.

Of course, and as others have pointed out in this volume, we do not contend our 50 articles to be "the" or even "the most representative" articles from the past fifty years. There is no doubt, if different individuals were to conduct the same exercise, they would come out with a different set of articles. In fact, if we were to conduct the exercise over from scratch—either knowing what we know now or, for that matter, not knowing what we know now—we, too, could have come out with a different set of articles. Many times, our decision to include one article instead of another was not easy. For example, there were numerous instances where a chosen article was taken out based on a subsequent discussion. Then, in other instances, an article originally "in" and then "out," was put back in again. Looking back, we thoroughly enjoyed this part of the book project.

Given the success that we had with our top ten (plus extras) list approach, we used the same process for each of the next four decades. Once we had a better picture of the entire 50 years, that is, once we had completed our task for three of the five decades, we soon realized that our approach of treating the decades (somewhat) independently was not going to hold up. As such, once we completed our task of finding ten articles for each of the five decades, we now had to look back at the entire 50 articles, as a whole.

Once again, in the interest of full disclosure, our final 50 articles were "set" a number of times. In fact, certain articles that we had included and then excluded at the "decade level" came back into the mix at the "anniversary level." Our anniversary level discussion definitely impacted our decade level decisions. For example, over the past fifty years there have been certain individuals who have published in *delta-K* more than others. However, you will not find a representative number of their articles in this volume. This decision, along with many others, was made early on at the decade levels, and was strictly enforced once we were at the anniversary level. Finally, we had our 50 articles.

In terms of the book, choosing 50 articles was but one component of our project. Key for us and the book (and we hope you agree) was to have each decade begin with an introduction providing historical context, and conclude with a commentary providing a contemporary view of the themes emerging from each decade, each from prominent members of the Alberta mathematics education community. We were extremely fortunate (thanks to the hard work of Gladys) to have: Liedtke and Kieren, Puhlmann and Pimm, Bonifacio and Loewen, Smith and Simmt, and Mercer and Chapman provide (respectively) the introductions and commentaries for each of the decades. Consequently, in the space that remains, we wish to provide a few brief comments on *delta-K* and the teaching and learning of mathematics in Alberta over the past 50 years.

As mentioned, for us, the article we included on the metric system was unique, in that we used it as a starting point for the book. However, that article is also unique in another manner: its topic, the metric system, was found only in one decade, the seventies. For the most part, topics that appeared in *delta-K* over the past 50 years either appeared in a few decades or in all five decades. These topics are now commented on in turn.

A common thread between the diverse topics that appear in a few, but not all, of the first five decades of *delta-K* is the "backward and forward compatibility" of the topics. Topics that appear in some, but not all, of the first five decades of *delta-K* are, in a sense, "forward compatible," but not necessarily "backward compatible." Let us provide an example. There is no doubt that articles on math teacher anxiety, appearing in the pages of *delta-K* in the 1980s and 1990s, have appeared in subsequent issues of *delta-K* (in the 2000s) and, further, will continue to appear in future issues of *delta-K*. After all, math teacher anxiety is an established area of research in the field of mathematics education and this research is now in the hands of mathematics teachers in classrooms around the world. However, working in the "other" direction for a moment, having an article on mathematics teacher anxiety published in the early 1960s seems a little less plausible. Consider the following thought experiment. Imagine a smoke-filled teachers' lounge in the early 1960s where people (read: men) are pouring over recent submissions to *delta-K*. There are five slots left for articles and six articles to choose from. The topics of the six articles are: geometry, logic, deduction, proof, axioms, and mathematics teacher anxiety. It is just our conjecture, but we think we know which one of those topics did not belong—at the time. To be clear, mathematics teacher anxiety (most likely) existed, but did not appear in the pages of *delta-K* (and other journals) in the 1960s. In other words, we are not arguing that mathematics teacher anxiety did not exist in the sixties; however, we are arguing that if one had mathematics teacher anxiety it was probably not discussed and, surely, as a topic would not grace the pages of a mathematics teacher journal. The same is probably true for a number of other issues in the sixties, that is, they (most likely) existed, but were not discussed in mathematics teacher journals.

Times change and, at some point in the nineties, the math-teacher-anxiety-bell was rung. As one knows, it is hard to unring a bell. The topics that have appeared in a few, but not all of the decades of *delta-K* (e.g., gender, anxiety, Indigenous knowledges, and others mentioned above) represent important moments in the history of the teaching and learning of mathematics in Alberta.

Topics that appeared in a few of the decades of *delta-K* are diverse in nature. Articles concerning the mathematics teacher become predominant, albeit in different forms, from the seventies onward: teacher preparation in the 1970s; teacher as facilitator in the 1980s; math teacher anxiety in the 1990s; and professional development in the 2000s. Other topics, such as gender, the use of literature, and theories of learning (especially from the field of psychology) appear in articles from the eighties onwards. Topics such as constructivism, manipulatives, problem solving, anxiety, equity, and multiculturalism appear in the pages of *delta-K* from the nineties onwards. In the aughts, topics included humanism, the influence of cognitive science, Indigineous knowledges, and applying mathematics to real life.

As we continue our look back over the past 50 years, a few topics stood out. For example, as seen in the articles in this book, the University of Alberta has long been a hub for research in mathematics education in Canada (and beyond). As another example, which we attempted to capture in the "Researcher in the Classroom" chapters of this book, research, although different over the decades (e.g., methodology), has been a consistent component of the mathematics teaching and learning scene in the province of Alberta. Further, a number of themes have been a mainstay for nearly half a century, that is, certain topics have appeared in each of the past five decades of *delta-K*.

There have been two related questions, albeit in different forms, that have continued to crop up over the past 50 years: (1) Why teach mathematics, and (2) What is mathematics? Different versions of these questions and different responses are preserved in the pages of *delta-K*. Looking back to the articles from the sixties, there appears to be little doubt of the status held by geometry. Reading the articles from the sixties, logic, deduction, proof, axioms, and rigour represent the very essence of mathematics and why mathematics was being taught: to teach logic, deduction, proof, axioms, and rigour. However, as seen in the articles from the very next decade, geometry was not necessarily on the way out, but was, if you will, knocked a bit off of its perch. For example, articles from the seventies began to address teaching mathematics for different reasons. In addition, calculus was also being questioned as the holy grail of school mathematics. If calculus, then, was not at the peak of Mount School-Mathematics, what topic would take its place? As documented in the pages of *delta-K*, statistics became a new topic for the mathematics classroom. Statistics courses, being given further consideration in the eighties, were accompanied by specific courses for those who did not "like" mathematics. These courses were put "on the books" and were tried to various degrees in different schools. Due in part to the developments of previous decades, the related questions of "why teach mathematics" and "what is mathematics" resulted in "new" topics, such as estimation, mental math, problem solving, numeracy, and number sense. Said topics have become mainstream and, concurrently, lightning rods for those involved in the teaching and learning of mathematics.

Questions surrounding the nature of mathematics and mathematics teaching have been a mainstay in the pages of *delta-K* over the past fifty years. Interestingly, questions of this nature show no signs of abating. Recent articles by Andrew Hacker (in *The New York Times*), Nicholson Baker (in *Harper's* magazine), E. O. Wilson (*The Wall Street Journal*), and others are currently questioning the necessity of the teaching and learning of mathematics. Further, other individuals such as Arthur Benjamin—who advocates for statistics, not calculus, as the peak of Mount School-Mathematics—are questioning certain aspects of the teaching and learning of mathematics. Of course, those (too) close to the topic are up in arms over presentations and articles of this nature. What may appear to some as present-day arguments over the nature of mathematics and the teaching and learning of mathematics, we see, have been appearing and are preserved in the pages of *delta-K*.

This volume also demonstrates how central certain organizations and associations have been to the teaching and learning of mathematics. The articles from the sixties show not only the establishment of provincial organizations—the British Columbia Association of Mathematics Teachers (BCAMT), the Mathematics Council of the Alberta Teachers' Association (MCATA), the Saskatchewan Mathematics Teachers' Society (SMTS) and others—but also the close ties between those organizations during the early years. Looking beyond provincial borders, the pages of *delta-K* preserve the attempts to establish a Canadian Association of Mathematics Teachers in the seventies, which as discussed in this volume, may live on in the form of the Canadian Mathematics Education Forum (which will be held for the fifth time in 2014). The notion of a Canadian mathematics teacher society or association, we contend, had to have been influenced, in part, by the national mathematics teacher organization south of the border: the National Council of Teachers of Mathematics (NCTM).

As evident in the pages of *delta-K*, the MCATA became involved with the NCTM early on. In addition to the official affiliation of the MCATA with the NCTM, the influence of the NCTM on the teaching and learning of mathematics in North America has been felt in Alberta. As we move through the pages and decades of *delta-K*, we see the moves of the NCTM from the eighties (e.g., *An Agenda for Action*) to the nineties (e.g., the various Standards) to the aughts (e.g., the Processes and more recently the Common Core Standards) and resultant changes to the teaching and learning of mathematics in the province of Alberta.

It is important to note that the aforementioned provincial organizations have continued to develop and strengthen. The MCATA (and, for that matter, the BCAMT and SMTS) continues with strong membership numbers, holds annual conferences, and consistently publishes issues of its journal. Albertan members continue to play a role in regional, national, and North American conferences, associations, and organizations, which will continue to influence the teaching and learning of mathematics in the province of Alberta (and beyond). Said influence is especially true on the use of technology in the mathematics classroom.

Articles on technology have been consistently found in the pages of *delta-K* over the past 50 years. Technology has increasingly become an integral part of the mathematics classroom; however, as seen in the articles chosen for this volume, the technology under discussion changes with the decades. As we move from the sixties to the seventies to the eighties to nineties to the aughts, the focus shifts, respectively, from the beginnings of the use of the microcomputer to calculators becoming a staple of the math classroom (as they replace the slide rule and various math tables) to the microcomputer's main foray into the classroom to the domination of the graphing calculator to dynamic software. (We hope you enjoyed, as we did, the article on use of the overhead).

The articles in this book allow us to see that although the technology under discussion changes, many of the issues remains the same. For example, we see, from the articles in the sixties, that the beginnings of the use of microcomputers in the mathematics classroom had a focus on rote learning. Similarly, today, many of the "apps" dedicated to the teaching and learning of mathematics also focus on rote learning. As another example, there are similarities between the modern-day discussions of Massive Online Open Courses (MOOCs) and the previous discussion of using the videocassette recorder, better known as a VCR, during the 1980s. The impact of technology can be seen in courses such as geometry where the approach to teaching the topic, which clearly has changed since the sixties, has embraced technological advances (e.g., dynamic geometry software).

According to Moore's law, many of the digital electronic devices in the mathematics classroom will continue to improve at (approximately) an exponential rate, which means (and we have no qualms in saying it here): Technology is not a passing fad. Further to this point and for those who are still hung up on the good ol' "calculator debate," we ask that you start to embrace other forms of technology into your discussions and debates on how technology should be used in the classroom. Some such as Conrad Wolfram have made the case for "Computer-Based Mathematics," which, in essence, would eliminate calculating from the mathematics classroom; of course, this notion is a foundational source of contention with the various forms of the new math.

The "new math" is a prominent topic that has appeared in many pages of *delta-K* over the past 50 years. However, what is meant by new math has (also) changed over the past 50 years. The term "new math" being thrown around today means something entirely different than when the term was used in the 1960s. (Ah the sixties, a simpler time when the "new math" actually meant "new math.") There is no doubt that the "new math" and the math wars have had a major impact on the teaching and learning of mathematics. As seen in the pages of *delta-K*, many of the main topics under discussion (e.g., approaches to the teaching and learning of mathematics; the use of technology; whether mathematicians or mathematics educators should be writing the curriculum, assessment, and exams; the readability of textbooks and others) can be couched within a "new math" discussion.

As demonstrated, on the one hand, certain topics (e.g., anxiety and others) have graced the pages of *delta-K* for but a few of the decades since the beginnings of the MCATA. On the other hand, also as demonstrated in this volume, certain topics (e.g., technology) appear in all five decades of *delta-K*. While, to us, it has been interesting to see when particular topics made their way into *delta-K*, we are particularly fond of those topics that have appeared in all five decades.

Topics that have appeared in all five decades of *delta-K* have provided us with many opportunities to get "lost" in this book. (Worthy of note, this is not a last-ditch effort during the penultimate paragraph to try to convince the readers that this is a good book.) Consider, for example, certain discussions on technology that are found in the pages of *delta-K*. Sure, the conversation during the eighties was specific to the VCR, but when one swaps out a tape in the VCR for, say, a YouTube video on a tablet or a MOOC presentation, much of the conversation is similar. Likewise, the discussion found in the pages of *delta-K* during the seventies—when the calculator was replacing slide rules and log tables—has many components of the eventual conversation that will occur once we stop printing Z-tables in our math textbooks (those texts that are just "too wordy") and, finally, just let students use their graphing calculators (or, more accurately, their graphing calculator apps they will have on their tablets). Getting "lost" in this book also occurs for topics other than technology. Consider, as one last example, the articles that we have read through the decades about the new math and the curriculum. Certain discussions on this topic could be pulled out of an article and it would be near impossible to establish which decade the article was from: "Clearly, we need to get back to basics." Whether this statement stems from the new math and back-to-basics discussion from the sixties and the seventies or from the last few years (when the math wars made their way to Canada) is hard to determine. Yes, more context would allow one to better guess which decade the comment originated from, but that is not the point. The point is that certain topics (e.g., new math, technology, and others) transcend decades.

Jean-Baptiste Alphonse Karr once said, "plus ça change, plus c'est la même chose." This aphorism, loosely translated, states: "the more things change, the more things stay the same." As demonstrated, certain topics, especially those found in all five decades of the existence of the MCATA and *delta-K* (e.g., the nature of mathematics, why teach mathematics, mathematics teacher organizations and associations, technology, curriculum, new math, the math wars, and others) do change; yet they also, at a deeper level, stay the same. We want to be clear, although Karr's aphorism can be considered pessimistic, we do not believe pessimism applies in this particular situation. First, we use Karr's aphorism to draw attention to an important point: *delta-K's* housing of significant topics on the teaching and learning of mathematics. Second, and finally, this is, after all, a *celebration* of 50 years of *delta-K*, the Mathematics Council of the Alberta Teachers' Association, and the teaching and learning of mathematics in the province of Alberta—a celebration that we hope you have enjoyed as much as we have.

CPSIA information can be obtained at www.ICGtesting.com
Printed in the USA
LVOW04s2109170914

404629LV00003B/22/P

9 781623 967000